High-Pressure Shock Compression of Condensed Matter

Springer
New York
Berlin
Heidelberg
Hong Kong
London
Milan
Paris
Tokyo

High-Pressure Shock Compression of Condensed Matter

L.L. Altgilbers, M.D.J. Brown, I. Grishnaev, B.M. Novac, I.R. Smith, I. Tkach, and *Y. Tkach:* Magnetocumulative Generators
T. Antoun, L. Seaman, D.R. Curran, G.I. Kanel, S.V. Razorenov, and *A.V. Utkin:* Spall Fracture
J. Asay and *M. Shahinpoor* (Eds.): High-Pressure Shock Compression of Solids
S.S. Batsanov: Effects of Explosion on Materials: Modification and Synthesis Under High-Pressure Shock Compression
R. Cherét: Detonation of Condensed Explosives
L. Davison, D. Grady, and *M. Shahinpoor* (Eds.): High-Pressure Shock Compression of Solids II
L. Davison, Y. Horie, and *T. Sekine* (Eds.): High-Pressure Shock Compression of Solids V
L. Davison, Y. Horie, and *M. Shahinpoor* (Eds.): High-Pressure Shock Compression of Solids IV
L. Davison and *M. Shahinpoor* (Eds.): High-Pressure Shock Compression of Solids III
A.N. Dremin: Toward Detonation Theory
R. Graham: Solids Under High-Pressure Shock Compression
Y. Horie, L. Davison, and *N.N. Thadhani* (Eds.): High-Pressure Shock Compression of Solids VI
J.N. Johnson and *R. Cherét* (Eds.): Classic Papers in Shock Compression Science
V.F. Nesterenko: Dynamics of Heterogeneous Materials
M. Sućeska: Test Methods for Explosives
J.A. Zukas and *W.P. Walters* (Eds.): Explosive Effects and Applications

Tarabay Antoun
Lynn Seaman
Donald R. Curran
Gennady I. Kanel
Sergey V. Razorenov
Alexander V. Utkin

Spall Fracture

With 283 Illustrations

Springer

Tarabay Antoun
M.S. L-206
Lawrence Livermore National Laboratory
7000 East Avenue
Livermore, CA 94551
USA
antoun1@llnl.gov

Lynn Seaman
Poulter Laboratory
SRI International
333 Ravenswood Avenue
Menlo Park, CA 94025
USA
lynn.seaman@sri.com

Donald R. Curran
Poulter Laboratory
SRI International
333 Ravenswood Avenue
Menlo Park, CA 94025
USA
curran@unix.sri.com

Gennady I. Kanel
Institute for High Energy Densities
Russian Academy of Sciences
IVTAN
Izhorskaya, 13/19
Moscow 127412
Russia
kanel@ficp.ac.ru

Sergey V. Razorenov
Institute of Problems of Chemical Physics
Russian Academy of Sciences
Chernogolovka
Moscow Region 142432
Russia
razsv@ficp.ac.ru

Alexander V. Utkin
Institute of Problems of Chemical Physics
Russian Academy of Sciences
Chernogolovka
Moscow Region 142432
Russia
utkin@ficp.ac.ru

Editors-in-Chief:
Lee Davison
39 Cañoncito Vista Road
Tijeras, NM 87059
USA
leedavison@aol.com

Yasuyuki Horie
M.S. F699
Los Alamos National Laboratory
Los Alamos, NM 87545
USA
horie@lanl.gov

Library of Congress Cataloging-in-Publication Data
Spall fracture/Tarabay Antoun . . . [et al.].
p. cm. — (High-pressure shock compression of condensed matter)
Includes bibliographical references and index.
ISBN 0-387-95500-3 (alk. paper)
1. Fracture mechanics. 2. Shock (Mechanics). 3. Strength of materials. I. Antoun, Tarabay. II. Series.
TA409 .S715 2002
620.1′126—dc21 2002070473

ISBN 0-387-95500-3 Printed on acid-free paper.

Printed in the United States of America.

9 8 7 6 5 4 3 2 1 SPIN 10879169

www.springer-ny.com

Springer-Verlag New York Berlin Heidelberg
A member of BertelsmannSpringer Science+Business Media GmbH

Preface

The main objective of this book is to present a comprehensive and up-to-date treatment of the shock-induced dynamic fracture phenomenon known as *spall fracture*. Spall fracture is of practical importance in virtually all applications involving rapid loading by explosives, impact, or energy deposition. It is also of scientific importance in studies of the elementary strength of materials because the extremely high loading rates prevalent during spall experiments make it possible to attain stress levels that approach the theoretical strength of the material. Due to its practical and scientific importance, spall fracture has been the subject of numerous investigations conducted over the past several decades. A variety of innovative experimental techniques, measurement diagnostics, and rate-dependent constitutive models of the spall process have been developed in recent decades (see, for example, the volume of the present series *High-Pressure Shock Compression of Solids II: Dynamic Fracture and Fragmentation,* by Davison et al., 1996, and the 1987 review article by Curran et al.[1]). An extensive literature has accumulated in both English- and Russian-language publications, but much of the work conducted in the former Soviet Union has not been readily accessible to Western readers. An important objective of this book is collection, comparison, and cross-correlation of results of comparable investigations conducted in the West and in the Soviet Union. Our intent is to provide new insights and ideas for directing future work. An equally important goal is to make results obtained in the former Soviet Union readily available to Western readers and to create a reference source for fracture kinetics data, experimental techniques, measurement diagnostics, methods of interpreting experimental measurements, constitutive models, and methods and results of numerical simulations of spall phenomena. We hope this work will be useful to students seeking an understanding of spall fracture, to engineers dealing with applications involving dynamic loading and fracture of materials, and to scientists studying the physics of strength.

The subject is treated using a multifaceted approach that emphasizes the various aspects of the study of spall: experimental, analytical, and numerical. Experimentally, the techniques used to perform spall experiments are discussed

1. Curran, D.R., L. Seaman, and D.A. Shockey, "Dynamic Failure of Solids," *Physics Reports* 147(5&6), 253–388 (1987).

along with the various measurements that can be used to characterize the response of spalling materials. Also presented is an extensive compendium of experimental spall data that encompasses a wide range of materials and spans the full spectrum of ductile and brittle behaviors.

Analytically, the focus is on the development of constitutive models for spall fracture. Modeling approaches ranging from the relatively simple empirical models to more complex, microstructurally based models are discussed. The discussion includes a review of past work and a rudimentary presentation of the basic equations used to describe elastic-plastic material behavior. The emphasis of the analytical studies presented in the book is on the development of thermodynamically consistent constitutive models using the Nucleation-And-Growth (NAG) approach to describe spall damage processes in brittle and ductile materials.

Numerically, we discuss a wide range of issues relating to constitutive model implementation in computational finite-element codes. Recent state-of-the-art developments in the area of standardized interfaces that facilitate implementation of constitutive models in computational codes are presented. We also discuss previously unpublished research dealing with the implementation of damage models in Eulerian codes, with special focus on the development of appropriate advection methodologies for the damage variables.

The three aspects of the study of spall fracture discussed in the book are not considered to be mutually exclusive. Instead, experimental, analytical, and numerical studies are viewed as interdependent, with experimental results guiding model development and numerical simulations with the models used to gain deeper understanding of the complex, nonlinear, and often inelastic processes associated with spall fracture.

Because the study of spall fracture is of a multidisciplinary nature, the material presented in the book draws on several scientific and engineering disciplines, and requires some familiarity with the basic principles of continuum mechanics, thermodynamics, and fracture mechanics. Of particular importance are those aspects of continuum mechanics and thermodynamics applicable to constitutive modeling and to wave propagation in solids, and those aspects of fracture mechanics applicable to the development of criteria for crack propagation and void growth within the framework of a microstatistical approach. Basic knowledge of metallurgy and experimental mechanics is also helpful in allowing the reader to better understand some of the material discussed in the book.

The authors endeavored to make the book easy to read and—to the extent possible—self-contained. Illustrations are often used to enrich the presentation and to facilitate the discussion of the many topics covered. References are often used as sources of additional information to enhance the reader's knowledge and understanding of the topic under consideration. References are also used in cases where it is not practical to present a complete derivation of an equation used in the text, or when a topic is covered with less vigor or depth than the reader may desire, probably because it is related, but not central to the subject being discussed.

In the past, the use of advanced spall fracture models, like the NAG models, in large-scale computational studies involving dynamic failure of material was not always practical and economically feasible. Computational resources of sufficient power to perform realistic simulations of practical problems were often lacking, as were detailed spall data of sufficient quantity and quality to calibrate the models. Recent technological advances have helped remove many of the technological obstacles of the past few decades. Computer-aided imaging tools now take the place of laborious time-consuming measurements to characterize damage distributions in spalled samples. Fast, massively parallel computers, coupled with three-dimensional finite-element codes are now available at many of the institutions at the forefront of spall studies. The availability of these tools and resources has made it possible to use NAG models in technologically important applications, such as in the design of debris shields for the target chamber of the National Ignition Facility (NIF).[2] Other programs, like the U.S. Department of Energy (DOE) Accelerated Scientific Computing Initiative (ASCI) and the Defense Advanced Research Project Agency's (DARPA) Advanced Insertion of Materials (AIM) program, are poised to follow. Advances made over the past several decades that up to now have been largely confined to the laboratory are being brought to bear to solve problems of practical and technological importance. These new applications are transforming the field of spall fracture from one of a purely scientific nature to a field that places heavy emphasis on engineering applications. With this new focus come exciting possibilities for new applications and for further developments.

Acknowledgments

This book originated as a research project funded by the Defense Special Weapons Agency (DSWA)—now the Defense Threat Reduction Agency (DTRA)—and was conducted jointly in the High Energy Density Research Center (HEDRC) and in the Institute of Chemical Physics, both of the Russian Academy of Sciences, and in the Poulter Laboratory of SRI International. CDR Kenneth W. Hunter was the DSWA technical monitor.

The initiation of this joint U.S.-Russian effort would not have been possible without the active support of Charles W. Martin (then at the Ballistic Missile Defense Organization (BMDO) and now at ARES Corporation) and R. Jeffery Lawrence (then at DSWA and now at Sandia National Laboratories). Thanks are also due to Dr. Michael Frankel of DSWA for solving a number of unglamorous but important administrative problems that arose during the effort.

2. When completed, the NIF will be an experimental laboratory facility that uses 192 laser beams to induce fusion reaction in small samples. NIF experiments will produce conditions of high energy and density similar to those found at the center of the sun.

The Russian portion of the work was performed under the general supervision of Academician V.E. Fortov, Director of the High Energy Density Research Center. The SRI portion of the work was performed under the general supervision of Dr. James D. Colton, Laboratory Director, and was based largely on the contributions of two of the present authors, Donald R. Curran and Lynn Seaman, and on the contributions of T. Barbee, D. Shockey, D. Erlich, R. Crewdson, and many other past and present SRI researchers to whom the authors express their sincere gratitude. The authors also express their appreciation to Thomas Cooper for important contributions to the sections of the book dealing with numerical simulation codes, to Kitta Reeds for editing the manuscript, and to Lee Gerrans for assisting with the illustrations.

Livermore, California, USA	Tarabay Antoun
Menlo Park, California, USA	Lynn Seaman
Menlo Park, California, USA	Donald R. Curran
Moscow, Russia	Gennady I. Kanel
Moscow, Russia	Sergey V. Razorenov
Moscow, Russia	Alexander V. Utkin

Contents

Preface v

1 Introduction 1
1.1. Historical Background 1
1.2. The Material Failure Process 3
1.3. Material Characterization 5
1.4. Experimental Methods and Data Analysis 11
1.5. Constitutive Relations for the Evolution of Damage 15
1.6. Qualitative Description of Spall Processes 26
1.7. Objectives and Organization 34

2 Wave Propagation 37
2.1. Conservation Relations for Wave Propagation 37
2.2. Theory of Characteristics 39
2.3. Analysis of the Shock Wave 44
2.4. Graphical Analysis of Experimental Designs 49
2.5. Temperature in Shock and Rarefaction Waves 57

3 Experimental Techniques 59
3.1. Experimental Procedures Used to Produce Shock Waves 59
3.2. Techniques Used to Measure Shock Parameters 66
3.3. Spall Fracture Experimental Procedures 76

4 Interpretation of Experimental Pullback Spall Signals 93
4.1. Estimating Spall Stresses from Experimental Data 93
4.2. Influence of Damage Kinetics on Wave Dynamics 111
4.3. Estimating Spall Fracture Kinetics from the Free-Surface Velocity Profiles 126
4.4. Summary of the Information Obtained from the Spall Signal 133

5 Spallation in Materials of Different Classes 137
5.1. Metals and Metallic Alloys 139
5.2. Metal Single Crystals 152
5.3. Constitutive Factors and Criteria of Spall Fracture in Metals 159
5.4. Brittle Materials: Ceramics, Single Crystals, and Glasses 162
5.5. Polymers and Elastomers 169
5.6. Dynamic Strength of Liquids 173

6 Constitutive Modeling Approaches and Computer Simulation Techniques 175
6.1. General Constitutive Modeling Approaches 175
6.2. Fracture Modeling Approaches 197
6.3. Fracture Model Implementation 203

7 Nucleation-and-Growth-Based Constitutive Models for Dynamic Fracture 217
7.1. Nucleation and Growth of Voids and the Model for Ductile Fracture 218
7.2. Nucleation and Growth of Cracks and the Model for Brittle Fracture 236

8 Applications of the Nucleation-and-Growth Fracture Method 267
8.1. Ductile Fracture of Commercially Pure Aluminum 268
8.2. Brittle Fracture of Polycarbonate 272
8.3. Fracture and Fragmentation of Rock (Quartzite) 276
8.4. Fracture and Fragmentation of a Solid Rocket Propellant 281
8.5. Fracture of Beryllium Under Impact and Thermal Radiation 286
8.6. Fracture of Steel and Iron Under Impact 289
8.7. Discussion 299

9 Concluding Remarks 301
9.1. Conclusions 301
9.2. New Applications 303

Appendix Velocity Histories in Spalling Samples 305

References 379

Index 399

1
Introduction

This book contains an exposition of recent investigations of a dynamic fracture phenomenon called *spall* (or *spall fracture, spallation, scabbing*, or by other names). The exposition describes how measurements of the spall process can provide information about the basic physical processes that govern the strength of solids. We summarize past experimental data, including data obtained by researchers in the Former Soviet Union and previously not readily available to Western readers. We describe experimental techniques, experimental interpretation, mesomechanical constitutive modeling of the failure process, and provide a library of data and constitutive model parameters for several important engineering materials.

In its simplest form, spall fracture occurs when two strong plane decompression waves under uniaxial strain conditions interact to produce a region of tension in the interior of a material body. The interacting decompression waves arise, for example, when a compression pulse is reflected from a stress-free surface. The tensile stress field in the interaction zone develops at the highest rate possible for the material in question (strain rates of 10^4 to 10^6/s are typical). This tension is maintained for an interval that depends on a variety of external conditions but usually falls in the range from 10^{-6} to 10^{-8} s. The stress that can be imposed is essentially unlimited in comparison to the quasi-static strength of even the strongest materials.

The above unique experimental conditions (uniaxial strain and high strain rates) serve to make this technique a very powerful one for studying the microscopic processes that underlie and govern material strength. That fact is the foundation for this book.

1.1. Historical Background

The earliest observations of spall fracture seem to have been made by B. Hopkinson [1914] and the phenomenon was subsequently studied in some depth by Rinehart and Pearson [1964], and by Kolski [1963]. This early work has been extended by many others who sought to explore the metallurgical aspects of material failure due to application of very large stress for brief intervals and to develop criteria for its occurrence. Even the earliest data showed clearly that spallation was an evolutionary process in which complete failure of the material arose through nucleation and growth of microfractures in the sample (Smith [1963]). Nevertheless, early investigators analyzed spall as a discrete event and sought criteria for its occurrence. The first criterion proposed was simply that spall occurred at a critical tensile stress characteristic of the material. Later, criteria involving stress rate (Breed et al. [1967]) or stress gradient (Skidmore [1965]) were proposed to explain observations that are now interpreted as results of gradual evolution of damage. Tuler and Butcher [1968] proposed a cumulative damage criterion according to which spall occurred when an integral depending on the stress history at the point in question exceeded a critical value. Tobolsky and Eyring [1943] and Zhurkov and colleagues [1965] were apparently the first to introduce the concept of damage as a rate process obeying Arrhenius [1889] rate equations for bond breaking and healing. Each of these criteria incorporates parameters that characterize specific materials and are to be determined experimentally. Except for the Arrhenius rate descriptions, spall was supposed to occur instantaneously at the time and place where the criterion is first satisfied, and therefore ignored the gradual softening of the material that occurs as the level of damage evolves. Despite this, the criteria (except for the critical stress criterion) incorporated features that captured some aspects of the stress history prior to fracture.

A variety of innovative experimental techniques, measurement diagnostics, and constitutive rate models of the spall process have been developed in recent decades (see, for example, Barbee et al. [1970], Seaman et al. [1971], Shockey et al. [1973], Seaman [1980], the review by Curran, Seaman, and Shockey [1987], and the recent studies by Nigmatulin et al. [1991], by Nemat-Nasser and Horii [1993], by Meyers [1994], and by Davison et al. [1996]). An extensive literature has accumulated in both English- and Russian-language publications, but much of the work conducted in the Former Soviet Union has not been readily accessible to Western readers. An important objective of this book is collection, comparison, and cross-correlation of results of comparable investigations conducted in the West and in the Soviet Union. Our intent is to provide new insights and ideas for directing future work. An equally important goal is to make results obtained in the Former Soviet Union readily available to Western readers and to create a reference source for fracture kinetics data, experimental techniques, measurement diagnostics, methods of interpreting experimental measurements, constitutive models, and methods and results of numerical simulations of spall phenomena. We hope this work will be useful to engineers dealing with applications involving dynamic loading and fracture of materials and to scientists studying the physics of strength.

1.2. The Material Failure Process

Classical investigations of fracture of solids concerned themselves with analysis of the *static* stability of an existing macrocrack. The first successful theory of this type was presented by Griffith [1921], who postulated that a crack in an elastic body would become unstable and grow if the elastic energy released in an incremental extension of the crack exceeded the energy required to form the additional surface area produced. This fracture criterion has since been generalized in numerous ways, including accounting for the plastic work done in extending a crack in a ductile solid (see the volumes edited by Liebowitz [1968–1972]).

The spall process typically nucleates as many as a million microscopic voids or cracks per cubic centimeter in a solid sample. Attempting to describe the behavior of each void or crack and their interactions would be a formidable job, although recent advances in computing power are beginning to produce promising results. The currently most productive approach is that of mesomechanics, where the behavior of the individual voids or cracks are averaged over a "relevant volume element" (RVE), and the RVE represents a continuum point in space.

In the framework of mesomechanics, when a body is subjected to sufficiently high levels of tension and/or shear, a statistical distribution of microcracks or microvoids starts to be nucleated, and the distribution evolves as nucleation proceeds and the nucleated microcracks or voids grow and coalesce within the material. The entire process is dynamic, i.e., the microcracks and microvoids are nucleated at heterogeneities with a stress- and temperature-dependent nucleation rate, and grow gradually over the period during which the stress is maintained. The collection of defects produced is called *damage. Fragmentation* occurs when damage evolves to the degree that the body separates into distinct parts, or fragments. The damage accumulation process is controlled by the entire stress history prior to fragmentation, and depends on the material at issue, its purity, microstructure, etc. Furthermore, the stress history itself is affected by the material softening caused by the evolving damage. Details of the damage mechanisms and the morphology of the damage produced also depend on these variables. Practical interest often lies with avoidance of fracture, in which case the focus is on very low levels of damage. In other important applications, the fragmentation process and the resultant fragment size and velocity distributions are of interest. It is clear from this discussion that fracture is a rate process and cannot be characterized by a simple material property such as "fracture strength."

Close examination of material near the tip of a macrocrack of the sort described by the classical stability theories usually discloses that it contains a *process zone* in which microcracks and voids nucleate, grow, and coalesce to produce growth of the macrocrack. Thus, the microscopic and macroscopic views of fracture phenomena are connected through events occurring in the process zone.

It is interesting that energy balance principles such as underlie the Griffith criterion for stability of a static crack have recently been applied to describe fracture at the other extreme in which a body is fragmented into small particles by sudden application of very large stresses (Grady [1988], Grady and Kipp [1996]).

We do not, in general, distinguish between static and dynamic failure—it is all dynamic. Material failure evolves on a time scale ranging from nanoseconds to years and successful models must capture the responses observed over this entire range. Spall is a tensile failure that results from the nucleation, growth, and coalescence of microfractures or microvoids produced in concentrations of the order of $10^6/cm^3$ when large stresses are imposed for short times. Because of the short duration of load application, the maximum tensile stress attained during the spall process is usually greatly in excess of the stress that produces fracture under static loading. On the other hand, the tensile stress at which microfracture or microvoid *nucleation* begins is generally equal to the static value.

The maximum tensile stress attained during a spall process is often referred to as the *spall strength.* The forgoing discussion shows that this parameter is not a basic material property, but depends on the loading conditions. Nonetheless, the spall strength so defined is a very useful parameter because its value at very high imposed strain rates may approach the theoretical strength of the material (the strength that the material would have in the absence of defects).

The actual strength of a solid is significantly influenced by its crystallographic structure, by microscopic defects and texture introduced into this structure, by distinctions between laboratory samples and real components in size and form, and by differences among testing conditions. In contrast with fracture produced at low strain rates, spall is not affected by conditions at the surface of the sample because the fracture originates deep within the material.

Despite recent promising results of molecular dynamics (MD) computations, it is presently too complicated to construct a generalized theory of fracture and strength which would account for all of the influential factors. That is, MD and similar computations are proving valuable for guiding the development of mesomechanical models that average material behavior on the molecular and microstructural levels, and for aiding in interpretation of experimental data, but our computing power is not yet adequate to use MD to compute the behavior of engineering-scale structures.

Consequently, mesomechanical models are used to link failure on the microscopic and sub-microscopic levels to failure on the continuum level. The quantitative mesomechanical fracture models in use today are based primarily on empirical or semi-empirical relationships, reinforced by insights gained from MD results. Even so, the mesomechanical models themselves introduce complexities that have traditionally been resisted by the engineering community, which usually prefers to use simpler (but often inaccurate) models.

In fact, the additional complexity that arises from taking account of evolving damage and its effects on material response is more than compensated by the gain in information relative to that obtained from simpler models. It does, however, necessitate use of numerical methods to simulate experiments and solve problems arising in the application of the theories.

A promising approach to this issue is to compare MD, mesomechanical, continuum, and engineering model computations for simple geometries and loading paths to relate the input parameters for each level of model to each other. In this way, it may be possible to obtain simple engineering models that correctly reflect at least some of the underlying physics of failure.

1.3. Material Characterization

A significant portion of this book is devoted to experimental measurements of the spall process. Since an important goal of the book is to provide experimental data for future experimenters to validate, or for future computational modelers to simulate, it is obvious that the materials should be thoroughly characterized.

However, "thoroughly characterized" is an evolving concept. In the early days of shock physics, material strength at shock pressures was thought to be unimportant, as were microstructural properties. Materials studied in early shock literature are often identified simply as "aluminum" or "iron," for example.

The discussion of the previous section on the material failure process shows that we now need as much information about the material as it is possible to obtain. In the data compilation presented in this book, we have made every effort to present the material properties as thoroughly as possible. However, our own understanding of important properties has also evolved, and in some cases we wish that we had measured and recorded more properties than we did.

Thus, in the following paragraphs we discuss the properties that we currently recommend be measured before undertaking spall experiments. Hopefully, by following these recommendations, future experimenters will produce data certain to be of lasting value to the scientific community.

We organize our discussion of material characterization as follows. First, we discuss classic mechanical properties. Then, we discuss polymorphic phase transitions, thermal properties, characterization of the microstructure, and chemical composition.

1.3.1. Mechanical Properties

Although our main focus is on spall fracture, compressive waves often precede the tensile stress pulse that causes failure.[1] For this reason, it is important to account for processes that take place during the passage of the compression pulse

1. Dynamic tensile failure may also be induced by direct tensile loading without precompression such as in the case of the direct tensile Hopkinson bar test.

in order to be able to properly interpret the spall signal. The compression pulse often has important time-dependent structure and it may be influenced by many factors, the most important of which are the continuum mechanical properties.

These properties are needed in the continuum constitutive relations and in the continuum equation of state. They include the solid density, strength, porosity, phase boundaries, and thermal properties like the specific heat and thermal expansion coefficient. We discuss next strength and porosity, and then, in separate sections, discuss phase changes and thermal properties as well as microstructural and chemical properties.

Solids have finite shear strength; and under dynamic loading, the strength can be a complicated function of state variables that depend on the history of deformation. At the strain rates of interest in spall investigations (10^4 s^{-1} or higher), the strength of most materials is strain rate-dependent. Many materials also exhibit strain hardening and temperature softening characteristics. In addition, the strength of frictional materials like rock and concrete may also be pressure dependent in a Mohr-Coulomb-like fashion. Finally, the material may exhibit a Bauschinger effect characterized by differing loading and unloading paths. Though these features seem complicated, adequate models for describing them are available in the literature as discussed later, in Chapter 6. Here, we simply wish to point out that the structure of the shock wave in a material with finite strength depends on the magnitude of the peak stress (e.g., Duvall and Fowles [1963]). Below some threshold stress, the shock front will have a two-wave structure consisting of an elastic precursor with a peak stress magnitude proportional to the shear strength (which as described above may be history dependent), followed by the main shock that carries the material up to the peak stress. Above the threshold stress, the disturbance propagates into the material as a single shock front and the precursor is said to be overdriven.

Spall in porous materials should be afforded special consideration due to the added complications associated with porous compaction. Porous compaction is highly dissipative and even a small amount of porosity causes significant attenuation in the peak stress. Additionally, depending on the geometry of the voids and the properties of the matrix, compaction may cause damage in the vicinity of the compacted pores, which will have little effect during loading, but will significantly alter the development of spall fracture during unloading. Dynamic compaction has received considerable attention over the past three decades (e.g., Herrmann [1969], Carroll and Holt [1972a and 1972b]). Seaman et al. [1974] reviewed many of the widely used porous compaction models within the context of developing a comprehensive porous material equation of state for use in numerical simulations.

1.3.2. Phase Transitions

Polymorphous phase transitions[2] are an important aspect of the high-pressure behavior of crystalline solids. Under sufficiently high pressure, many such solids undergo phase transitions during which atoms of the solid rearrange themselves to form new crystallographic structures. Associated with the phase transition are changes in volume and other thermodynamic state variables. In problems involving wave propagation, phase transformations are afforded special consideration because they can lead to complicated wave structures.

Shock-induced phase transitions have been an integral part of the study of shock waves in solids ever since the discovery by Bancroft et al. [1956] of the α–ε phase transition in iron at 13 GPa. The discovery of this α–ε transition in shock wave experiments, before it had been identified from static high-pressure measurements, led to a series of studies aimed at further understanding the kinetics of the transformation. Notably, the plate impact experiments performed by Barker and Hollenbach [1974] provided highly resolved measurements that quantified the stress level at which the transition occurs, the effects of deformation rate on the transition kinetics, and the evolution of the reverse transformation.

Shock-induced polymorphous transformations have been observed in a wide range of materials. The results of DeCarli and Jamieson [1961] who explosively loaded graphite to produce diamond; Murri et al. [1975] who observed rarefaction shock waves during release from high-pressure states in calcite rock; and Ivanov and Navikov [1961], and Dally [1957] who observed rarefaction shock waves that led to the formation of smooth spall in iron are only a few examples. Comprehensive reviews that focus on various aspects of shock-induced phase transformations in solids can be found in the articles by Duvall and Graham [1977], and Ahrens et al. [1969], the latter focusing exclusively on the behavior of geologic materials.

1.3.3. Thermal Properties

Thermal effects are often present during shock wave loading. A shock wave propagating through a material causes a jump in temperature as well as in other thermodynamic variables including density, stress and particle velocity. Heating may also be accomplished by depositing energy directly into the material using a radiation source (e.g., laser). Regardless of how heating is achieved, increasing the local temperature of a material to near or above melting will cause the mate-

2. Here were are concerned primarily with so-called *first-order* transitions where the transition is accompanied by a change in volume, internal energy and other thermodynamic state variables.

rial to behave differently under load. Elastic properties of the material may be temperature dependent. Temperature may also affect the viscosity of the material, thus influencing its dispersion characteristics as well as other rate-dependent properties. Additionally, analytical equations of state, like the Mie Grüneisen EOS, which effectively describe the dependence of pressure on volumetric strain and internal energy at relatively low temperatures (i.e., near the reference curve, usually the Hugoniot) become increasingly less accurate as the temperature increases and the equilibrium state moves further and further away from the reference curve. Situations like this may arise in the vicinity of a nuclear explosion or in a material irradiated with a high-power laser, and they require the use of more complex EOS forms that account for extreme thermal effects more accurately.

Another important consequence of material heating in the study of spall is the loss of strength due to thermal softening near the melting temperature. Experimental data illustrating this effect is presented later in Chapter 5. Under certain conditions, thermal softening is also known to contribute to the formation of adiabatic shear bands (Rogers [1982]). During plastic deformation, most of the plastic work is converted to heat. As the temperature approaches the melting temperature, the material begins to thermally soften. When softening overcomes strain and strain rate hardening mechanisms, the material becomes much more sensitive to inhomogeneities and the deformation begin to localize in narrow zones (the shear bands) while the material in the surrounding regions unloads. This deformation mechanism is most often associated with compression and shear. Under tension, the material usually fails due to nucleation and growth of cracks or voids before enough local deformation accumulates to significantly raise the temperature and cause shear localization.

Thermal properties also play a significant role in determining the types of interactions that take place when a material is irradiated with a laser or a similarly intense radiation source. Such interactions, which may include melting, vaporization or ionization of the target material near the irradiated surface can be very complex and they depend not only on the target properties, but also on the characteristics of the energy source (e.g., Dingus et al. [1989]).

Under certain conditions, when the laser energy is deposited into the target faster than the target material can expand, a compressive thermoelastic stress pulse is produced. This stress pulse has a nearly exponential shape (corresponding to the nearly exponential shape of the energy deposition profile) with a maximum near the irradiated surface of the target. The stress pulse propagates toward both the front surface and the rear surface of the target. That component of the stress pulse that propagates toward the free surface is reflected back into the target as a rarefaction wave. As a result, the once compressive stress pulse gradually becomes bipolar with the tensile component trailing the compressive pulse.

1.3.4. Characterization of the Microstructure

An important aspect of the study of spall is the ability to relate the observed macroscopic behavior (e.g., spall pulse) to the evolution of damage at a microstructural level. This requires detailed characterization of the microstructure both before and after the spall experiment. A wide range of techniques are available to examine and characterize the microstructure of a material whose spall behavior is under investigation[3]. Of the many techniques available, optical microscopy is among the most widely used for characterizing damage distributions in terms of numbers and sizes of cracks and voids that form during spall. This technique, which was used in many of the nucleation and growth studies reported herein, is easy to use relative to other methods and it is capable of resolutions on the order of 1 μm. Scanning electron microscopy (SEM), which is capable of resolutions on the order of 10 to 100 nm, and transmission electron microscopy (TEM), which is capable of resolutions on the order of 1 nm, are more advanced techniques that can be used for applications that require higher resolution.

The ability to examine the pristine microstructure and ascertain whether it contains pre-existing defects that might serve as damage nucleation sites under the action of dynamic tensile loading is just as important as counting the cracks and voids after a spall experiment. This can be achieved using the same microscopy techniques used to characterize the damaged material. The focus of pre-test characterizations is on determining the nature, characteristic size and geometry of preexisting defects, their distribution density, and whether they are randomly or preferentially oriented.

Other less commonly used techniques for characterizing the undamaged microstructure of a material include X-ray diffraction, a technique that has long been used to reveal the crystal structure of crystalline solids (Reed-Hill [1964], Guy [1960]). This technique is useful for identifying crystal defects, which often contribute to the inelastic behavior of the material.

Under some circumstances, it may be possible to extract more information from post-test examination of a spalled specimen than just the size distribution of cracks or voids. Techniques like fractography may be brought to bear to examine the features of the fracture surface and attempt to relate the fracture mode to the microstructure. Among recent advances in fractography is a new technology called FRASTA (FRActure Surface Topography Analysis). This technology represents a major advance in fractography that allow a failure event to be replayed in microscopic detail (Kobayashi and Shockey [1991a, 1991b]).

3. *ASM Handbook, Volume 10: Materials Characterization*, 9th ed., American Society for Metals, Metals Park, Ohio (1986), is a comprehensive reference source for information about the various methods used to perform microstructural characterization.

1.3.5. Chemical Composition

There are several ways by which chemical composition can influence the response of materials to shock wave loading. First, altering the composition of a material can greatly affect its mechanical properties. This is evident in the behavior of metals and metal alloys. Compared to pure metals, metal alloys are generally less ductile and they tend to have a lower melting point. Furthermore, depending on the elements present in a metallic alloy, and on their proportions, the physical properties of the alloy can be varied over a wide range. Density, strength, fracture toughness, and plastic deformation are among the properties that can be altered through the use of various alloying techniques. These properties are of fundamental importance in the study of wave propagation and spall.

Second, chemical impurities in an otherwise homogeneous material can became nucleation sites for spall damage. This becomes increasingly important as the mismatch between the mechanical properties of the matrix and the second phase particles becomes more pronounced. Examples of impurities and second phase particles serving as damage nucleation sites in aluminum, steel and beryllium are shown in Figures 1.9 through 1.11.

Inclusions and second phase particles influence not only the behavior of metals, but also the behavior of nonmetallic multiphase materials like concrete and solid propellants. In concrete, a material made up of aggregates embedded in a cement matrix, interactions between the cement matrix and the aggregates are primarily responsible for the nonlinearities observed when the material is subjected to stresses above the elastic limit (e.g., Antoun [1991]). In a similar fashion, solid rocket propellants, which typically consist of grains of a high explosive, metal, and oxidizer embedded in a polymer matrix, fracture by debonding of the polymer from the grains because of excessive shear or tensile stresses.[4]

Third, the chemical composition of a material can significantly influence interactions that take place during energy deposition (e.g., laser, x-ray source). In this case, mismatch in the energy absorption characteristics of the various phases in a multiphase system causes complex interactions that may have a significant effect on the material response. To illustrate this point, let's consider the behavior of biological tissue during photoablation. Porcine reticular dermis is a biological tissue similar to the human cornea in composition and photoablation characteristics. It is composed of two primary energy-absorbing components: water, which accounts for about 70% of the total mass, and collagen fibers, which form the reticular extracellular matrix (ECM) and account for the majority of the remaining 30% of the total mass. Stress histories were measured in samples of this material which were irradiated using two lasers carefully chosen to selectively target either the water or the ECM of the tissue (Venugopalan

4. The spall behavior of solid rocket propellants and propellant simulants are discussed in more detail later in Chapters 5 and 8.

[1994]). The experimental results showed fundamental differences between the case where the extracellular matrix (ECM) is the primary chromophore of the laser radiation and the case where tissue water is the primary chromophore. These results were successfully modeled assuming different failure mechanisms for the different lasers used in the experiments (Antoun et al. [1996]).

1.4. Experimental Methods and Data Analysis

Mesomechanical investigations of spall fracture are based on the evolutionary damage model. The basis of these investigations is largely experimental. Experiments can be designed to impose stresses near the ultimate strength of the material or to produce stages of fracture ranging from absence of any apparent damage to complete fragmentation of the sample material. Spallation can be induced, measured, and characterized over length scales ranging from micrometers to centimeters and time scales of nanoseconds to microseconds, with the possibility of varying the strain rate, temperature, and load orientation (relative to internal structure that may exist in the material.) Experiments permit measurement of stress waveforms and careful recovery of samples for microscopic examination. Different classes of materials, ranging from steel to water, can be tested with this method. For these reasons, spall tests are useful tools for characterizing material failure over a wide range of conditions unattainable using conventional testing methods. Spall test results complement results of more conventional testing and provide information that can be used to improve our understanding of damage accumulation and failure in a wide range of applications.

Figure 1.1 shows several of the tests that can be used to study the mechanisms of dynamic fracture at various stresses, strains, and strain rates. The plate impact test is the most widely used of the high-rate experiments. In this test a flat flyer plate is caused to impact a flat target plate simultaneously over its surface, as illustrated in Figure 1.2. The loading conditions are then especially simple: The only nonvanishing strain component is the one normal to the plane of the wave, a state called *uniaxial strain*. At later times unloading waves originating at the specimen edges relax the uniaxial state of strain but, by that time, the reverberating stress waves in the specimen have produced the microdamage to be measured. When the tensile stress required to initiate fracture is large compared to the stress at which inelastic flow is initiated in the material, the stress field produced in the experiment is nearly isotropic.

Figure 1.2 shows an arrangement for launching a flyer plate using a gas gun, the technique most favored in the West. As discussed later, most of the analogous former Soviet Union experiments used high explosive techniques to launch the flyer plate. In plate impact tests, reverberating stress waves produce a series of tensile pulses in the target plate. The amplitudes of these pulses increase with increasing impact velocity and their duration increases with increasing sample

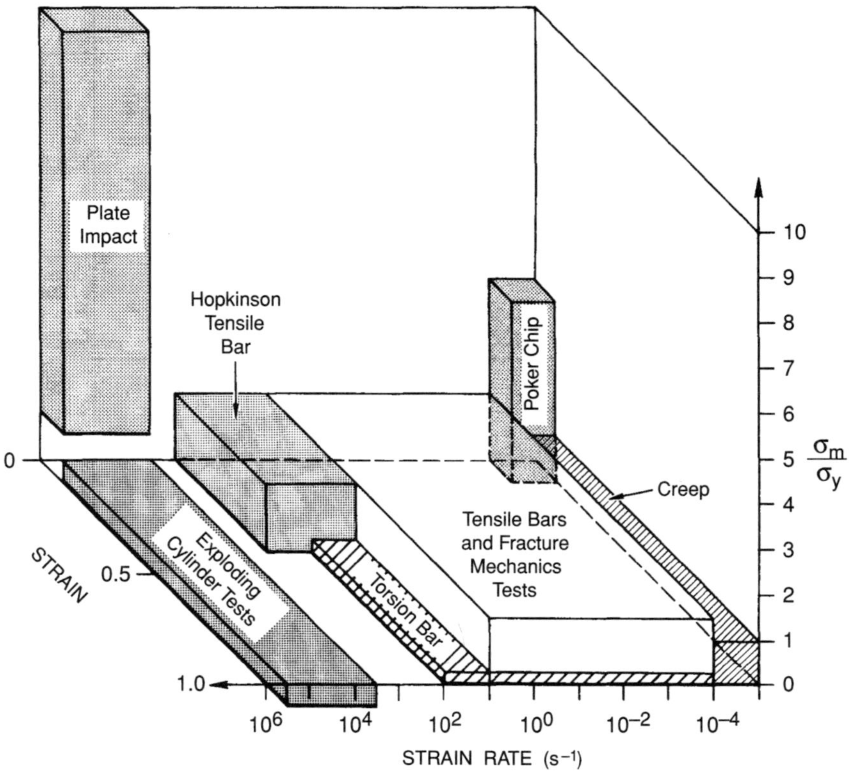

Figure 1.1. Ranges of stress, strain, and strain rate attained in various mechanical tests (from Curran et al. [1987]).

and target thicknesses. The amplitude and duration of a pulse also vary with position in the sample. If the first tensile pulse is sufficiently strong to nucleate and grow microdamage, the effect of the evolving damage field is usually to attenuate the second tensile pulse enough to prevent further damage. In some high-amplitude cases, the second and subsequent pulses can produce further damage, usually appearing on adjacent planes. The experiments discussed in this book provide insight into the kinetics of the various modes of nucleation and growth of damage. To achieve this objective, the experiments must be designed to provide well-controlled and measured stress and strain histories.

It is impossible to introduce a sensor into a sample without influencing its resistance to tensile stresses. Because of this lack of a technique for direct measurement of spall strength, several indirect methods have evolved. Each of the methods uses a different approach to determine the dynamic tensile stress. Sometimes large discrepancies are apparent among results obtained using different methods. Choosing the method of investigation that provides the most com-

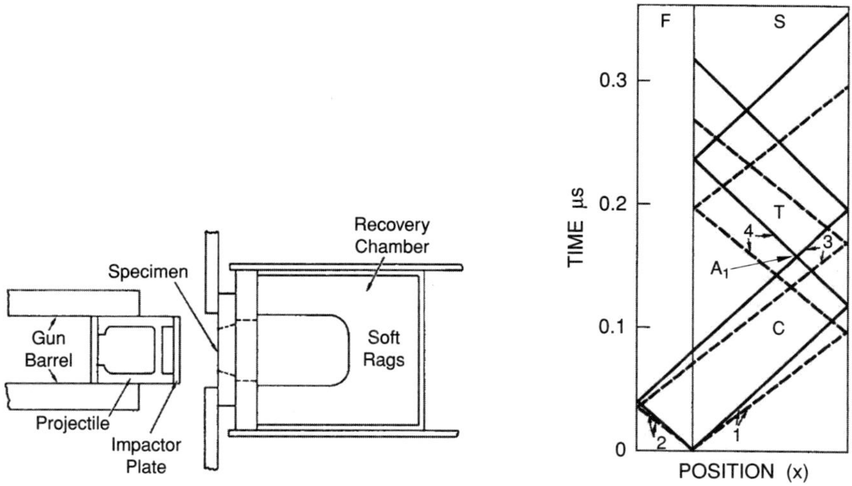

(**a**) Plate impact experiments for studies of dynamic fracture

(**b**) Distance-time plot showing wave paths and compressive (C) and tensile (T) regions in a one-dimensional impact

Figure 1.2. Schematic and wave dynamics of a plate impact test configuration used in the study of spallation (from Curran et al. [1987]).

plete and valid information is important, as is understanding of the capabilities and limitations of each method.

No target diagnostics are shown in Figure 1.2, but a laser interferometer (such as VISAR (Barker [1998]) could be used in this configuration to measure the velocity history at the stress-free surface of the sample. In such tests, the impact velocities are measured and the sample carefully recovered for post-test microscopic evaluation. The recovered samples are sectioned and polished to reveal microscopic damage, which is quantified by carefully determining the number and size distributions of the microscopic cracks, voids, or shear bands. In an iterative manner, to be described later, these distributions are then correlated with calculated stress histories.

In a variation of the plate impact test described, a material having shock impedance lower than that of the target material is placed at the rear surface of the target and particle velocity or stress history measurements are made of the wave transmitted into this buffer material. Figure 1.3 shows a schematic of the evolution of the stress pulse in this kind of experiment. Many plate impact experiments like those shown in Figures 1.2 and 1.3 have been performed. As discussed later, valuable information regarding the kinetics of microdamage evolution is contained in the shape of the transmitted wave in the case of experiments with a buffer and in the shape of the free-surface velocity profile in

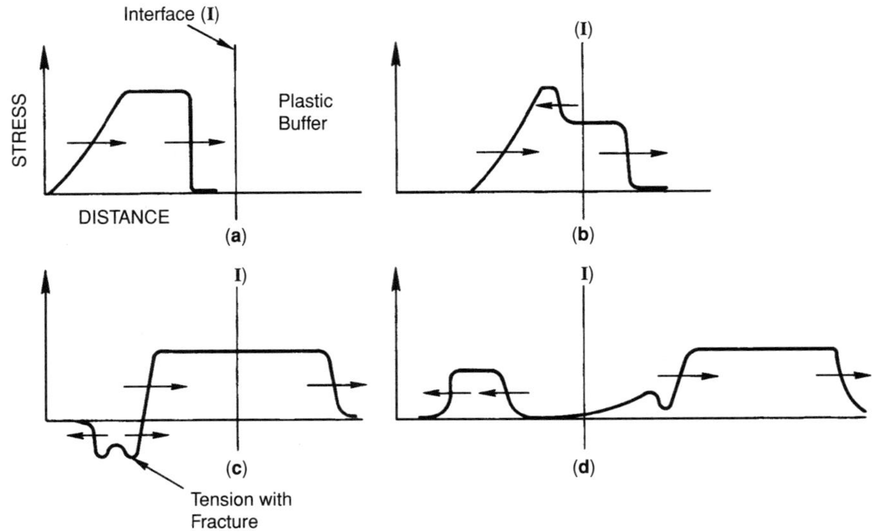

Figure 1.3. Development of a fracture signal (from Curran et al. [1987]).

the case of experiments without a buffer. However, as we will discuss later, these profiles contain more information about the early stages of damage evolution than they do about the later stages leading to fragmentation.

As discussed by Curran et al. [1987], creep tests and tensile tests using either round or notched bars are also suitable for providing useful microdamage kinetics data at the lower extreme of the strain rate range. Intermediate strain rates are attained with Hopkinson tension and torsion bar experiments.

Figure 1.4 shows results for the evolution of ductile void damage in a quasi-static tensile test performed on a round bar. This detailed picture of damage evolution is a good example of the kind of information we wish to obtain from post-test examinations of dynamically-loaded plate impact samples.

Figures 1.5 and 1.6 show an experimental technique for obtaining similar data for the microscopic damage mode of adiabatic shear banding, a type of instability in which plastic strain localizes into microscopic patches of concentrated slip.

Post-test sectioning of the specimens to reveal and permit quantification of the microscopic damage in various stages of evolution is a key diagnostic technique used in all of the experiments shown in Figure 1.1. The microstructural damage characterizations obtained can be correlated with measurements of the wave profiles recorded during the experiments to provide a powerful tool for understanding, quantifying, and modeling damage evolution and fracture.

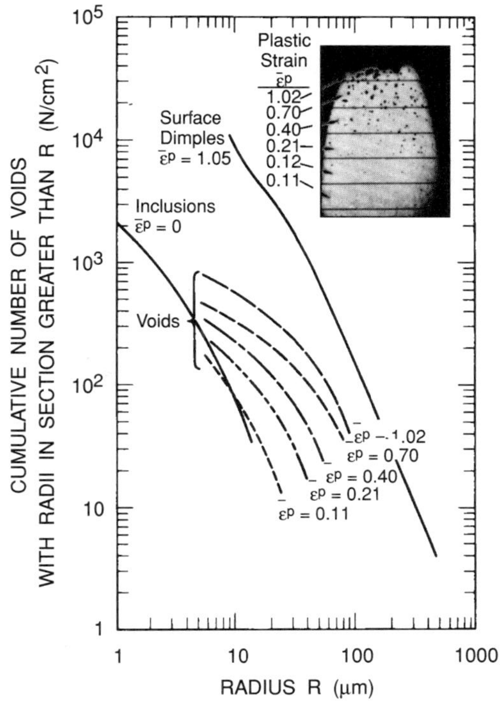

Figure 1.4. Size distributions of inclusions, voids, and surface dimples in a smooth tensile bar of A533B steel at various plastic strains (from Curran et al. [1987]).

1.5. Constitutive Relations for the Evolution of Damage

Interpretation of the experimental results, and application of the findings, are made possible by parallel development of thermomechanical theories that capture the observations and that lend themselves to numerical simulation of the experiments. Since the waves interact with the accumulating damage, they can be interpreted to yield information about the process of damage evolution. All methods of measuring the dynamic tensile stress in materials during spallation are indirect. The most effective methods involve simulation of the experiment on a computer and iteratively refining the theory underlying the simulation until it reproduces a broad range of experimental observations.

The purpose of reviewing the experiments discussed is to compile data that form the basis for development of constitutive relations. The model sought must describe the evolution of microfracture from an undamaged state to a state so extensively damaged that it has no strength. Both the formulation of mesomechanical theoretical models and the interpretation of experimental observations of material response invoke the concept of a *relevant volume element* (RVE)

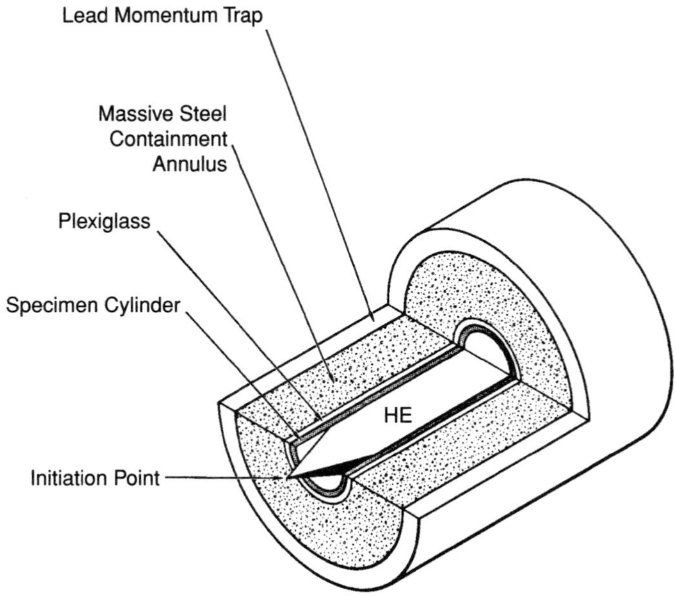

Figure 1.5. Contained fragmenting cylinder apparatus for studying shear band kinetics (from Curran et al. [1987]).

(see, for example, Nemat-Nasser and Horii [1993]). A RVE is a volume of the material in a particular object of interest that is small in comparison to the size of the object, and also small enough so that the strain is nearly constant throughout the RVE, but large enough to be representative of the material near its location. In the case of material in which damage is accumulating, this element must be large enough to contain cracks and/or voids in sufficient quantity to make the element statistically representative of the material in its neighborhood. If these conditions are not met (for example, when a thin knife blade is pushed into a material whose heterogeneities are larger than the blade thickness), then the mesomechanical approach is not appropriate.

The upper limit on RVE size arises from a desire to interpret experiments and conduct analyses on the basis of continuum-level stress and deformation fields that are constant over the element. It is understood that a finely-resolved view of each material element described by the mesomechanical theory would disclose highly inhomogeneous stress and strain fields that are represented as averages by the mesomechanical theory. The mesomechanical approach assumes that the microscopic failure processes can, on the average, be related to the stresses and strains averaged over the RVE. A key experimental challenge is to provide data regarding the anisotropic and inhomogeneous microscopic failure processes occurring within the element.

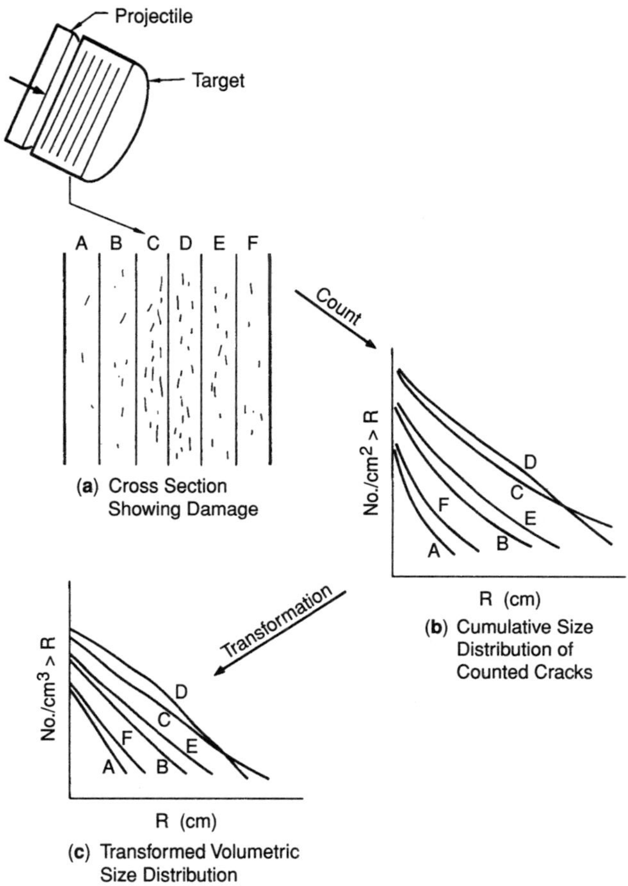

Figure 1.6. Steps on obtaining cumulative shear band distributions from contained fragmenting cylinder data (from Curran et al. [1987]).

The constitutive relations that we seek, although couched in the mathematics of continuum mechanics, are based on a description of the observed microscopic failure kinetics. Direct observation and quantification of these kinetics were the goals of the experiments to be described and summarized in this book. Our approach is to develop a mesomechanical continuum theory from knowledge of the response at the micro level, as distinguished from the alternate approach of functional forms that describe damage evolution, and then testing these forms against continuum-level measurements. We believe that the approach adopted is more efficient because basing damage evolution relations in microscopic observations strongly restricts the functional forms to be considered and adds confidence to extrapolations outside the database used to develop the relations. The experiments reviewed in this book were aimed at revealing and measuring this microscopic reality.

Table 1.1 lists the principal microscopic nucleation sites in solids or liquids and Figures 1.7 to 1.13 provide examples of the nucleation mechanisms. Once damage is nucleated at the microscopic level it can grow in three main geometric modes:

1. As ductile, roughly equiaxed voids that produce void volume by plastic flow (see Figure 1.14).
2. As brittle cleavage cracks that produce void volume by crack opening (see Figure 1.15).
3. As shear cracks or bands that produce localized slip (see Figure 1.16). The shear cracks can be of two types: brittle shear cracks or regions of localized plastic flow often called adiabatic shear bands.

The dynamic failure process has generated a large body of literature. An exposition of experimental techniques can be found in the book by Bushman et al. [1993]. A fairly recent exposition taking the point of view of materials science can be found in the compendium edited by Meyers et al. [1992].

Continuum mechanics texts usually treat fracture from the classical approach of considering the instability of a single idealized crack. The mesomechanical approach discussed earlier provides a description of evolving microscopic damage in the framework of continuum mechanics, and has been reviewed by some of the present authors (Curran, et al. [1987]) and more recently, by Nigmatulin et al. [1991], by Nemat-Nasser and Horii [1993], by Meyers [1994], and by Davison et al. [1996]. We take the mesomechanical approach in this book.

Table 1.1. Experimentally observed microscopic fracture nucleation processes.[a]

Nucleation site	Nucleation mechanism	Governing continuum load parameters	Figure reference
Preexisting flaws (voids or cracks)	Growth law	Tensile stress Plastic strain	1.7, 1.8
Inclusions and second-phase particles	Cracking of inclusion Debonding at interface Fracture of matrix near inclusion	Tensile stress Plastic strain	1.9–1.11
Grain boundaries	Vacancy clustering Grain boundary sliding Mechanical separation (solids only)	Tensile stress Plastic strain	1.12
Subgrain structure	Dislocation pileups (solids only)	Shear strain	1.13

[a]Reproduced from Curran et al. [1987]

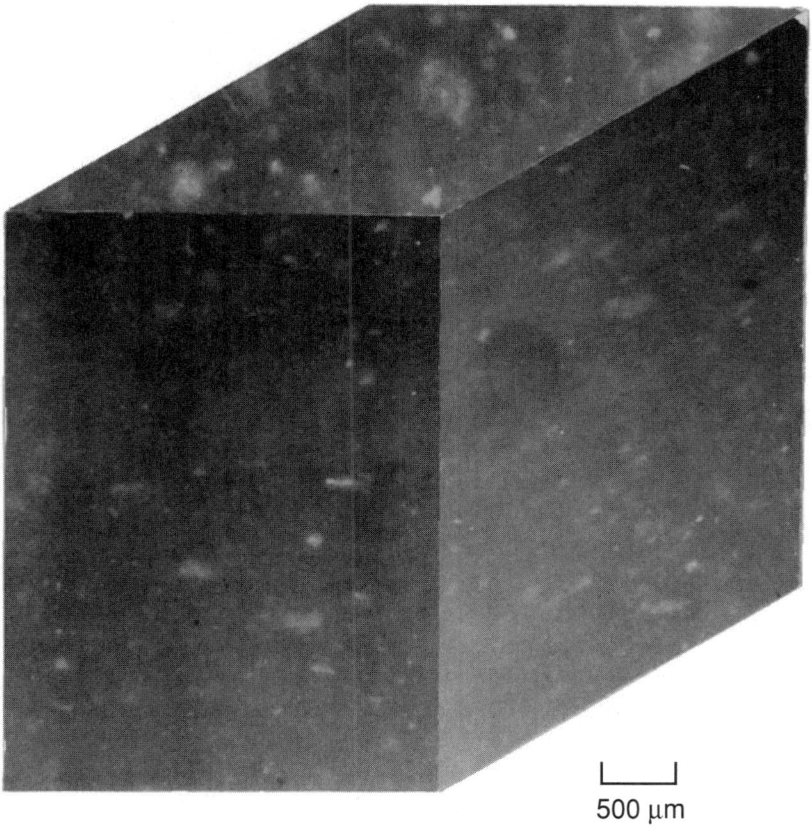

Figure 1.7. Composite micrograph of a block of Arkansas novaculite, a fine-grained quartzite rock, showing the preferred orientation of its inherent flaws (from Curran et al. [1987]).

(a)

(b)

Figure 1.8. Crack nucleation at a void. (a) Low magnification view of the fracture surface of high-purity beryllium showing fracture steps radiating from an initiation center. (b) A high magnification view of the center region of the sample, revealing the presence of a flattened void (from Curran et al. [1987]).

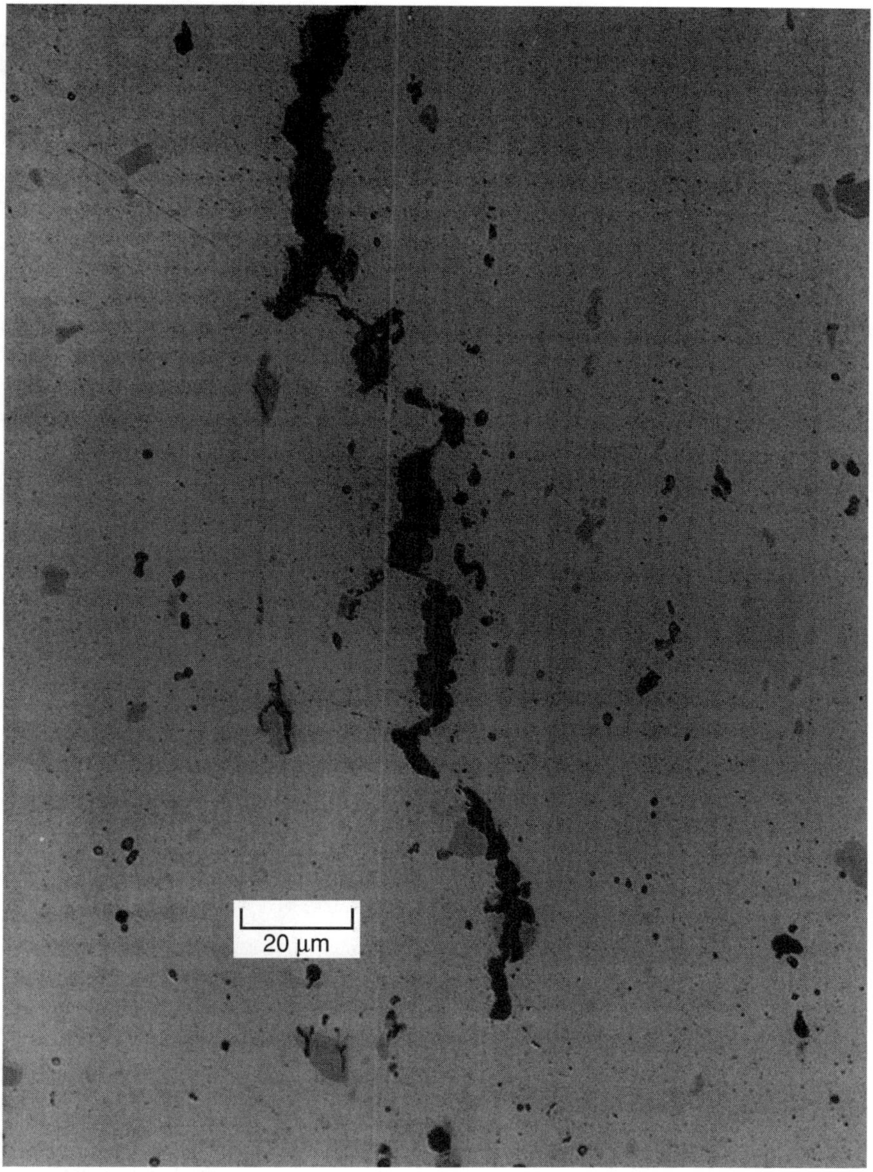

Figure 1.9. Cracking of inclusions in aluminum alloy 2024-T81 (from Curran et al. [1987]).

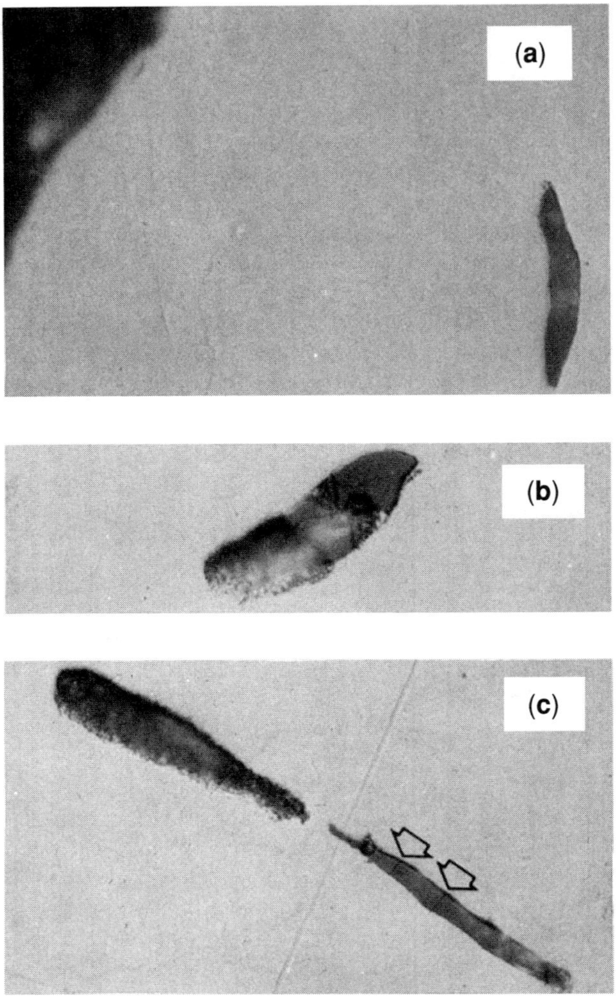

Figure 1.10. Manganese Sulfide (MnS) inclusions in a quasi-statically loaded Charpy specimen. (a) Unbroken MnS inclusions in a relatively strain-free area near the notch flank. (b) Broken MnS inclusion in highly strained region below the notch showing the void remaining where a portion of the solid inclusion has dropped out. (c) MnS inclusion in highly strained region below the notch, one of which has acquired several fractures (arrows), whereas the other has dropped out (from Curran et al. [1987]).

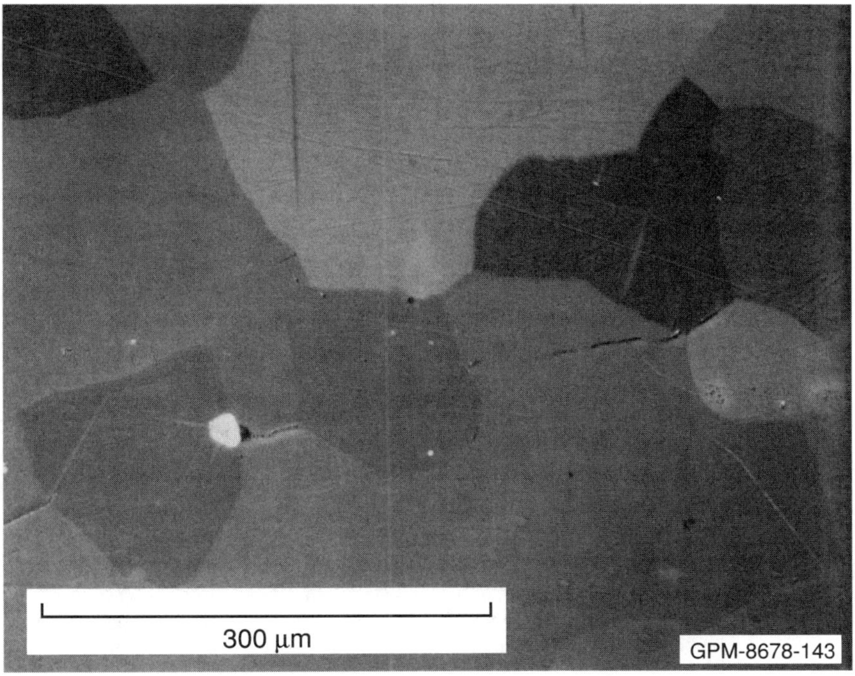

Figure 1.11. Micrograph showing nucleation of cracks and twins at oxide inclusions in high purity beryllium (from Curran et al. [1987]).

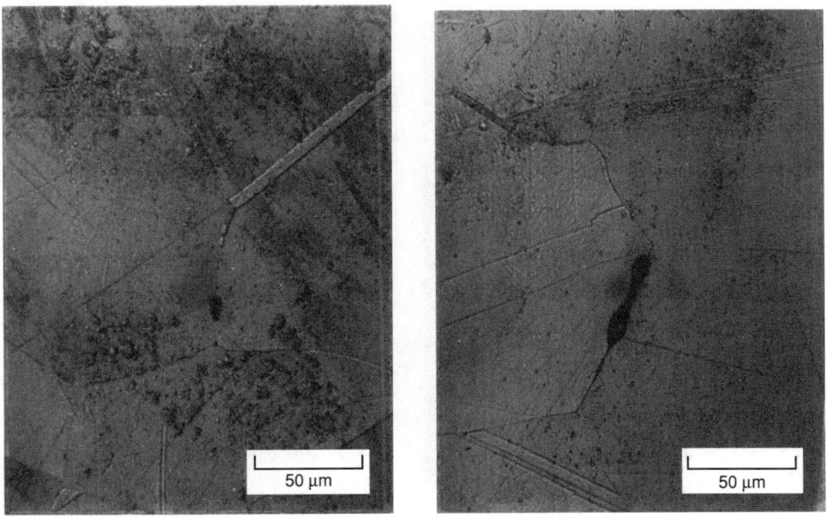

Figure 1.12. Micrographs showing nucleation of voids at grain boundaries and triple points in OFHC copper (from Curran et al. [1987]).

Figure 1.13. Micrograph showing a crack nucleation site in beryllium. A plastic flow mechanism probably operated at this site (from Curran et al. [1987]).

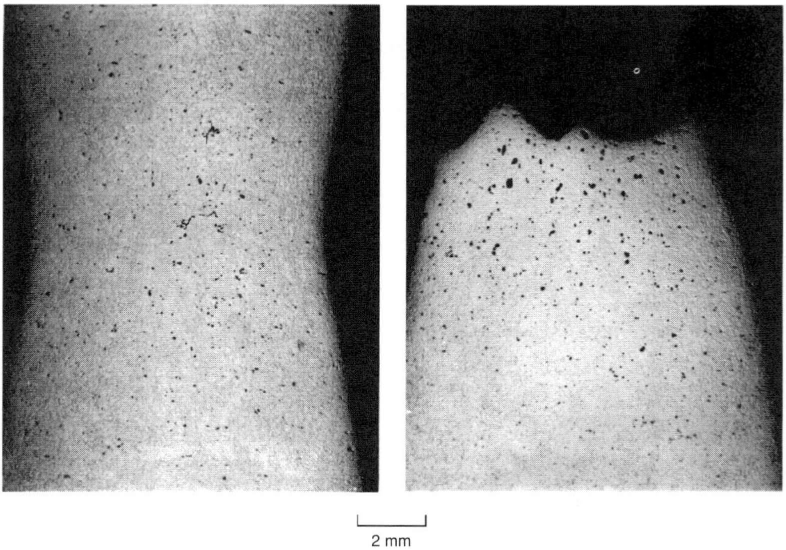

Figure 1.14. Photomicrographs of microscopic voids in sectioned specimens of A533B pressure vessel steel (from Curran et al. [1987]).

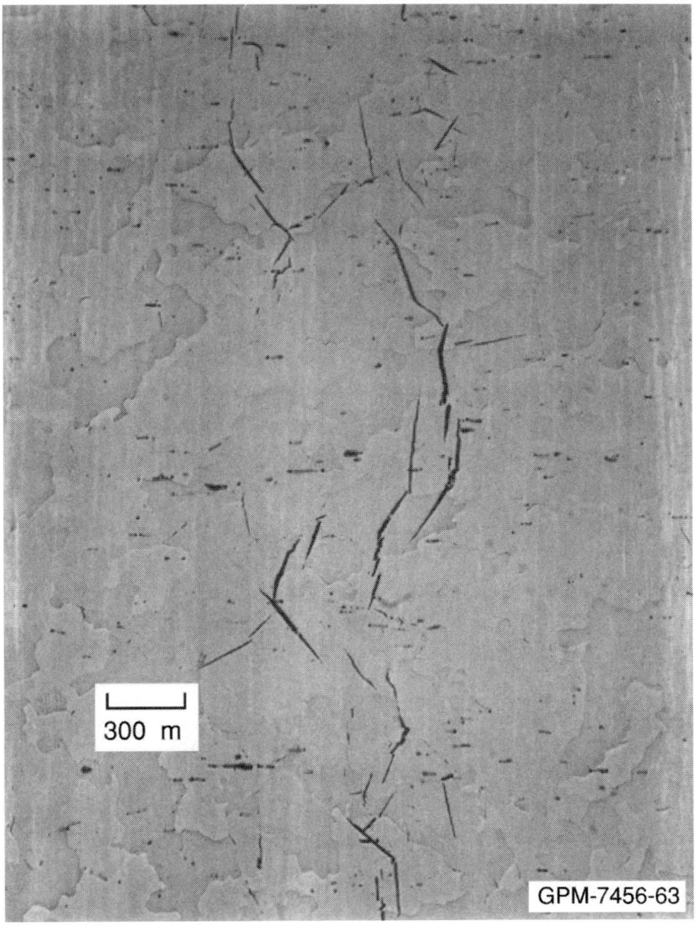

Figure 1.15. Photomicrograph of microcracks in Armco iron (from Curran et al. [1987]).

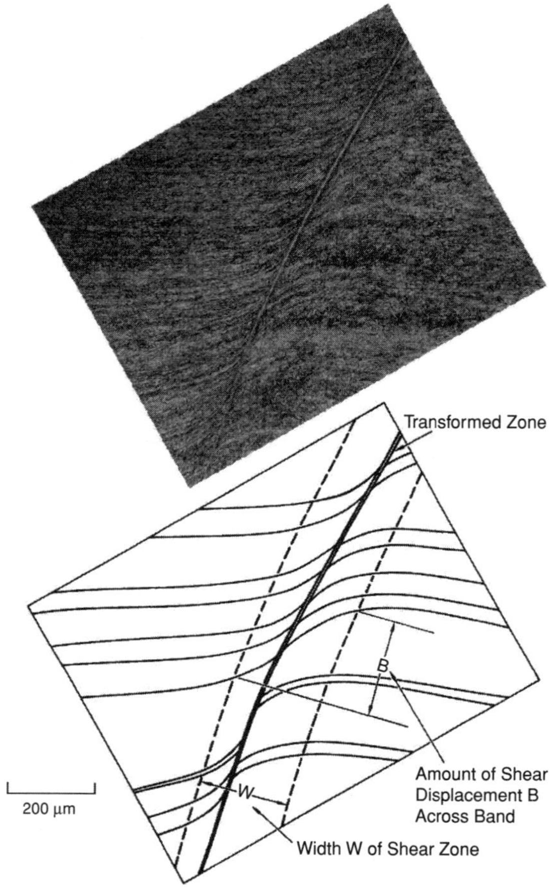

Figure 1.16. Micrograph and schematic of a shear band in a plate of rolled steel (from Curran et al. [1987]).

1.6. Qualitative Description of Spall Processes

The qualitative description of spall processes discussed in this section is based on our current view of fracture and is not necessarily applicable to all situations and all materials. Spall processes are examined in more detail in later sections, where the features associated with spall of materials of various classes are also discussed.

Spall damage occurs when rarefaction (expansion) waves within a material interact in such a manner as to produce tensile stresses in excess of the threshold required for damage initiation. Favorable conditions for spall can be produced (1) by impacts, (2) by lasers or other thermal radiation sources, and (3) by explosions. In each case, the spall-producing rarefaction waves are preceded by compression waves generated in the specimen by the initial impact, by the thermomechanical stresses associated with energy deposition, or by the detonation wave generated by the explosives. Figures 1.17 through 1.20 show examples of each of these three kinds of loadings.

Figure 1.17(a) shows a typical plate impact experimental configuration in which a 1.14-mm-thick Armco iron flyer plate is made to impact a target assembly consisting of a 3.16-mm-thick Armco iron plate and a 4.80-mm-thick PMMA buffer plate with a stress gauge sandwiched between the Armco iron and

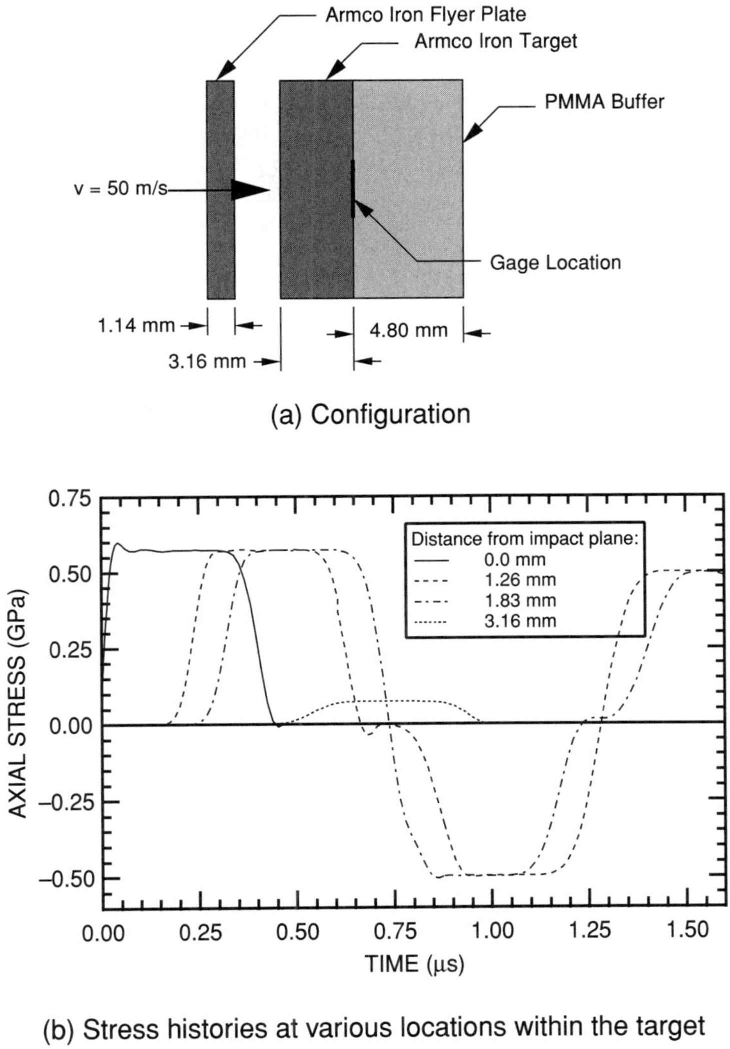

Figure 1.17. Configuration for a low-impact plate impact test in Armco iron and typical stress histories simulated using SRI PUFF.

the PMMA plates to provide diagnostic measurements during the experiment. The configuration shown in this figure was used in a series of spall experiments in Armco iron (Seaman et al. [1971]; Barbee et al. [1970]), and the data from the experiments were used to calibrate the BFRACT (Brittle FRACTure) model used in the simulations presented later in this section. The relatively low impact velocity of 50 m/s was intentionally chosen in this case to ensure elastic response throughout, so that the basic features of wave propagation in a typical spall experiment could be identified without the additional complications associated with plastic yielding or spall fracture.

The stress histories at several locations within the Armco iron target, simulated using SRI PUFF (Seaman and Curran, [1978]), are shown in Figure 1.17(b). As shown, the impact causes a square wave to propagate from the impact plane into the sample. The amplitude and duration of this stress wave can be controlled by varying the impact velocity and the thickness of the flyer plate, respectively. A wave of the same amplitude also propagates into the flyer plate. The compression waves in the flyer and target plates have uniform amplitudes of stress and particle velocity. These outward-facing compression waves are reflected from the stress-free rear surface of the impactor and from the interface between the Armco iron and PMMA plates as inward facing rarefaction waves.

The relief waves propagate toward the interior of the target plate, where they interact to produce states of tensile stress as shown in Figure 1.17(b). Tensile fracture damage occurs in the specimen if the tensile stress magnitude exceeds the threshold for spall damage.

In the foregoing example, the tensile stress wave in the target did not have a high enough amplitude to cause spall damage. In the next example, shown in Figure 1.18, the peak stress is increased to a level that causes spall damage by increasing the impact velocity from 50 m/s to 196 m/s. The stress histories shown in this figure were simulated using SRI PUFF. Spall damage was treated in the simulation by using the BFRACT fracture model, with the model parameters determined from a series of spall experiments like the one shown in Figure 1.18(b). BFRACT is described in detail in Chapter 7, but our use of the model is simply to show the effect of spall on the wave structure. As in the elastic case, stress wave interactions lead to tensile stresses in the specimen. Unlike the elastic case however, the stresses in the present example are high enough to cause spall damage under tension as well as yielding under compression, as illustrated by the kink in the stress history at a stress level of about 1 GPa. The effect of spall on the wave structure is evident in this figure. Details of the stress wave profiles in the interior of the specimen and on the gauge plane will be discussed further in later sections.

Figure 1.19 shows an example in which spall damage is induced by thermal stresses resulting from radiation deposition (e.g., lasers, x-rays) into a semi-transparent sample. Here a bipolar stress pulse develops in the sample, and there is a possibility of either front surface (left) or rear surface (right) spall depending on the parameters that affect wave interactions (i.e., Grüneisen coefficient and absorption depth of the sample material, and wavelength, pulse width, and

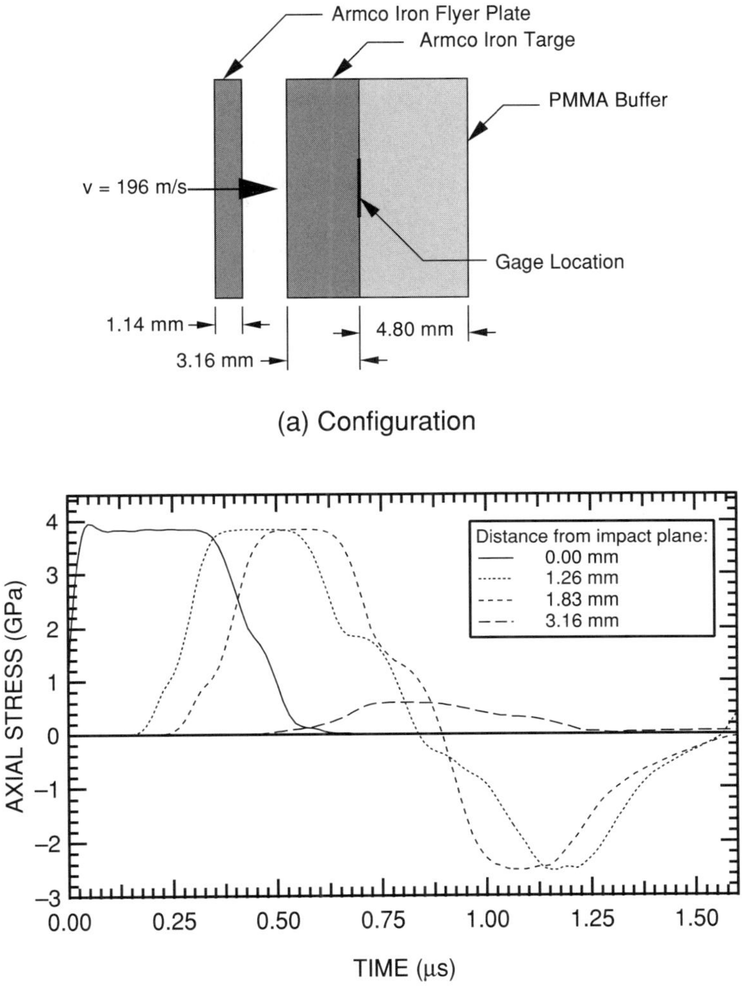

Figure 1.18. Configuration for a high-impact plate impact test in Armco iron and typical stress histories simulated using SRI PUFF and the BFRACT fracture model for Armco iron.

fluence of the laser). Front surface spall may occur when the rarefaction waves originating at the front surface of the specimen overtake the initial compression wave, attenuate it, and produce a tensile stress state of enough magnitude to cause fracture near the front surface. The triangular-shaped compression wave travels toward the rear surface of the specimen. When it reaches the stress-free back surface, the stress wave reflects back into the specimen as a rarefaction

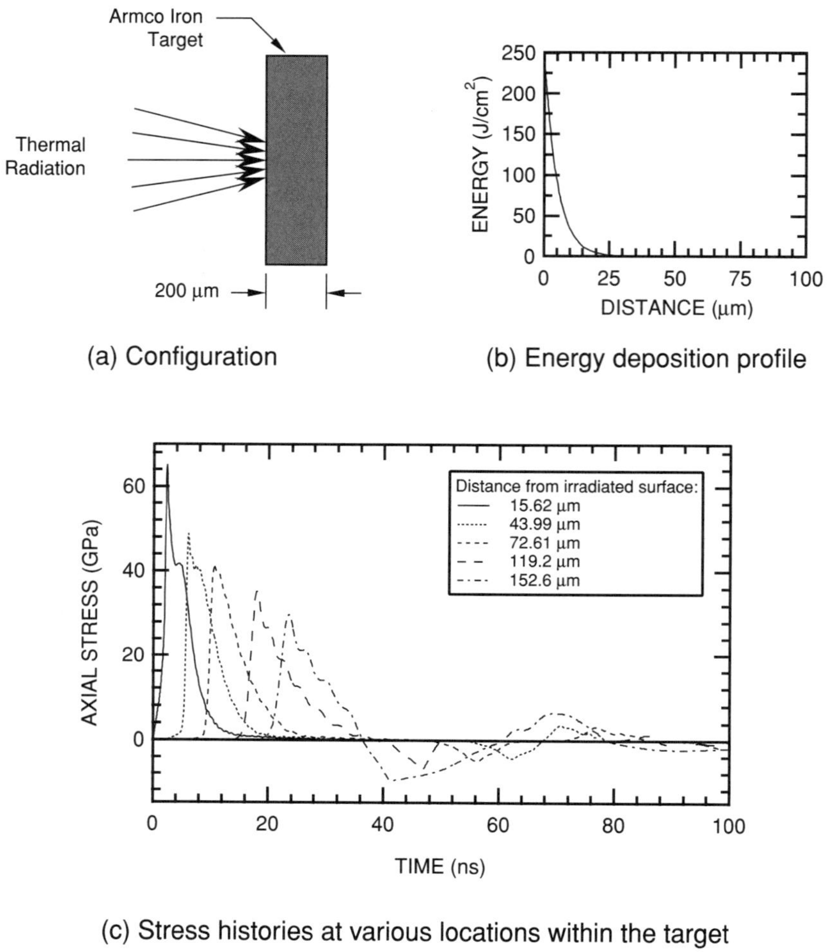

Figure 1.19. Configuration for a thermal radiation test and typical stress history in the target simulated using BFRACT.

wave. The interaction of the rarefaction wave with the compressed state could lead to tensile stresses of sufficient magnitude to cause fracture. Figure 1.19 shows a scenario in which the Armco iron plate experienced rear surface spall.

Figure 1.20 shows an explosive in contact with the sample. This case is somewhat like the impact case, except that the explosive loading provides a compressive wave with a decaying stress amplitude. The decay rate of the peak stress is rather slow; hence, the rarefaction wave reflected from the front interacts with a compressed state of essentially the same magnitude, thus providing a very small tensile stress. Therefore, the main region for spall is near the center where the rarefaction waves from the front and back surfaces intersect. The pe-

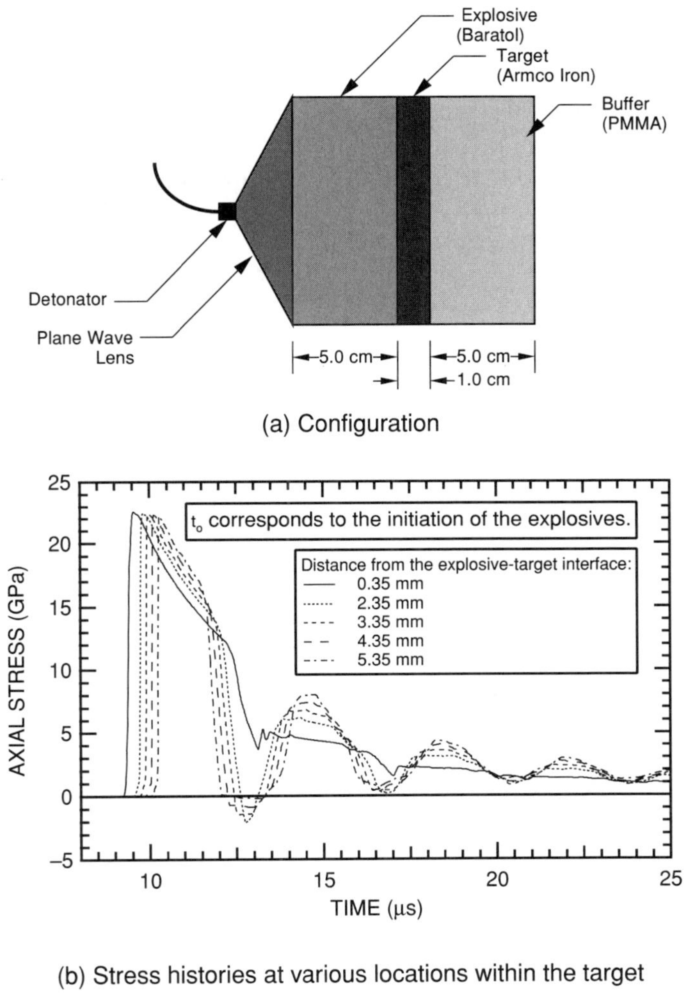

Figure 1.20. Configuration for an explosive loading test in Armco iron and typical stress histories in the target simulated using BFRACT.

riodic oscillations in the stress profiles shown in Figure 1.20 are due to wave reverberations within the sample plate.

1.6.1. Fracture Processes

Under the very rapid loading conditions that prevail during spall, fracture usually produces very many microcracks or microvoids: 10^6 or more per cubic centimeter. With so many damage sites, nucleation is a very important aspect of damage and may lead almost directly to fragmentation or shattering of the material. Here "nucleation" means the initial formation of the microcrack or microvoid by decohesion of an inclusion from the matrix material, initiation of a void or crack at a triple point where grains meet, or activation of a dormant crack or void or flaw. This aspect of dynamic fracture is different from fracture under quasi-static loading, where a single crack or a few cracks usually dominate the material response. Thus, nucleation plays a lesser role under static loading. Furthermore, in the static case surface imperfections are significant because fracture usually initiates at a surface or boundary where a small flaw already exists. In contrast, spall fracture occurs in the midst of the body; hence, it is a bulk material behavior unaffected by surface defects.

Growth of microcracks or microvoids under dynamic loading conditions is different from that in ordinary fracture mechanics in that there are myriads of damage sites, each of which grows a small amount, rather than one crack that grows from a microscopic size to the size of the structure. The crack surfaces formed under quasi-static loading and under spall conditions often look very similar in spite of these differences and the difference in the rates of loading. The surfaces are generally very rough under both fast- and slow-rate loading, because each larger crack is actually formed by the joining of many smaller cracks.

Fracture may occur under conditions of pure tension or a combination of shear and tension across the potential spall plane. Here, we focus on conditions in which tension is the primary driver for the fracture, but shear may still be present.

1.6.2. Definition of Terms

Spall fracture means fracture that occurs simultaneously over an area, not by growth of a single macrocrack, but by the nucleation and growth of many cracks, or voids at essentially the same time. Suitable conditions for such fracture occur only during wave propagation; hence, spall fracture refers to damage caused by tensile wave(s) produced when compression waves are reflected from a boundary.

In describing spall fracture, the term *damage* may have many meanings, depending on the observations and the point of view of the researcher. One definition is relative void volume or relative crack volume. The following parameter is useful for characterizing spall damage in cases where fracture occurs by nucleation, growth, and coalescence of a large number of cracks:

$$\tau = T_F \sum_{i=1}^{n} \Delta N_i R_i^3 , \tag{1.1}$$

where T_F is a dimensionless constant associated with the shape of fragments and ΔN_i are the numbers of cracks per unit volume with radii R_i. The use of ΔN_i and R_i to describe a distribution of crack sizes is described more fully in

Chapter 7. The τ factor is dimensionless, varies from 0 to 1, and controls the gradual reduction of stiffness of the material as damage increases. The term damage has often been used by other authors to describe a more qualitative factor, which describes the progress from intact to full separation, but without an explicit relationship to observed fractures.

Spall strength is a term used loosely to indicate the relative resistance of material to spallation under a specific set of conditions. The stress-strain path followed by Armco iron in Figure 1.21 illustrates several stress levels that may be associated with spall strength: the beginning of nucleation, the peak tensile stress, and the beginning of coalescence. The path certainly depends on the strain rate, stress level in the impact, and the temperature. Because the stress path does not have a square top and the peak depends on so many factors including the conditions of the test, the *spall strength*, *tensile strength*, and *fracture stress* have various interpretations. However, we will usually use the term *spall strength* to denote the peak tensile stress attained under the specific loading conditions considered.

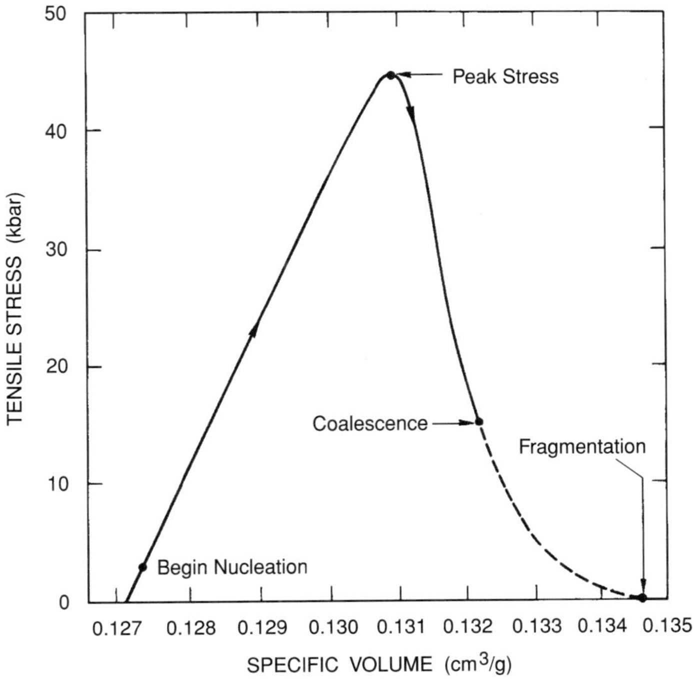

Figure 1.21. Stress-volume path for constant strain-rate loading of Armco iron to Fragmentation (Shockey et al. [1973]).

1.7. Objectives and Organization

The objectives of this book are to:

1. Describe and analyze the techniques and physical aspects of spall test methods.
2. Describe the statistical and kinetic aspects of spall fracture in materials subject to shock loading.
3. Develop mesomechanical constitutive equations for describing dynamic fracture, and elucidate the effect on damage of such factors as load duration and amplitude, orientation, and temperature.
4. Discuss common problems that arise when connecting spallation constitutive models to current types of "finite-element" computer codes, recommend constitutive model features for connecting to these codes, and discuss advection requirements for Eulerian and mixed Eulerian-Lagrangian codes.
5. Describe the experiments used to generate the reported data.
6. Provide a library of data and constitutive model parameters for several important engineering materials.

The remainder of this book is organized as follows. Chapter 2 presents the theoretical background for analyzing wave propagation. We discuss the conservation laws for continuous media, the theory of characteristics, and temperature effects in shock and rarefaction waves. Also included in this chapter is a brief analysis of the shock front and some of the thermodynamic paths, along the equation of state surface, that are important in shock wave physics. The treatment is not comprehensive, but the equations of one-dimensional motion are described in enough detail to facilitate the discussion of dynamic experiments presented in later chapters.

Chapter 3 describes the experiments that generated the data discussed in the book, and also describes the kinds of measurements that were made, and the main experimental techniques used.

Chapter 4 discusses the interpretation of the experimental data, that is, how the parameters of the evolving damage must be indirectly deduced from the data. This chapter also discusses some of the uncertainties associated with the various interpretation methodologies.

Chapter 5 discusses the observed similarities and differences in spallation behavior of materials of different classes. We describe spallation in metals and alloys, both single crystal and polycrystalline, ceramics, glasses, polymers and elastomers, and liquids.

Chapter 6 presents an overview of constitutive modeling approaches and computer simulation techniques. We include a survey of current types of "finite element" computer codes, and discuss common problems that arise when connecting spallation constitutive models to such codes. We also include a discussion of recommended constitutive model features for connecting to these codes, and discuss advection requirements for Eulerian and mixed Eulerian-Lagrangian codes.

Chapter 7 reviews some specific constitutive spall models, with emphasis on the nucleation and growth (NAG) models developed by some of the present authors for spall from ductile microscopic voids or from brittle microscopic cracks.

Chapter 8 reviews prior applications of the NAG models to a variety of solids, including aluminum, steel, beryllium, quartzite, polycarbonate, and rocket propellant. We draw conclusions about the applicability of this modeling approach to the new data presented in this book.

Chapter 9 contains concluding remarks and suggestions for promising future research directions.

An Appendix contains a collection of Former Soviet Union (FSU) data that have previously not been readily available to western readers. The data and experimental details are complete enough to allow interested researchers to either repeat the experiments or to computationally simulate them.

2
Wave Propagation

This chapter presents the laws of one-dimensional motion of compressible continuous media to the extent necessary for the discussion of dynamic fracture throughout the remainder of the book. A comprehensive account of the fundamentals of shock physics and wave propagation in continuous media can be found, for example, in the texts of Courant and Friedrichs [1948] and Zel'dovich and Raizer [1967]. The volume edited by Chou and Hopkins [1972] provides yet another useful reference for wave propagation theory and computations. In that volume, contributions from several authors cover a wide range of experimental and analytical topics including shock waves, the method of characteristics, constitutive relations, and finite-difference computational procedures. Other wave propagation references that emphasize material properties and constitutive relations are those of Rice et al. [1958] and of Duval and Fowles [1963].

2.1. Conservation Relations for Wave Propagation

The theoretical analysis of stress wave and shock wave propagation begins with the conservation equations, which underlie all solutions. Here, we consider these conservation relations only in one-dimensional planar form. This simplified description is sufficient for the study of the wave propagation problems that are of interest here. The plate impact experiments, explosive loading experiments and energy deposition experiments that are considered here are all planned to be primarily uniaxial strain tests. Under these conditions, the one-dimensional planar form of the conservation equations can be used, and in Lagrangian coordinates, these equations can be written in the following differential form:

Conservation of mass
$$\left(\frac{\partial x}{\partial h}\right)_t = \frac{\rho_0}{\rho}, \tag{2.1}$$

Conservation of momentum
$$\left(\frac{\partial u}{\partial t}\right)_h = -\frac{1}{\rho_0}\left(\frac{\partial \sigma}{\partial h}\right)_t, \tag{2.2}$$

Conservation of energy
$$\left(\frac{\partial E}{\partial t}\right)_h = -\sigma\left(\frac{\partial (1/\rho)}{\partial t}\right)_h, \tag{2.3}$$

where x is the Eulerian position (i.e., x changes with particle motion), h is the Lagrangian position (i.e., h retains its initial value and travels with the particle) expressed in terms of x as

$$h = \frac{1}{\rho_0} \int_0^x \rho dx,$$

t is time, ρ and ρ_0 are the current and initial values of the density, u is particle velocity, σ is the mechanical stress in the direction of wave propagation, and E is the internal energy density (i.e., per unit mass). The sign convention used in Eqs. (2.1) through (2.3), and unless otherwise stated throughout the remainder of this chapter, is that stress is negative in tension and positive in compression.

Stress and particle velocity measurements in shock wave experiments are normally performed in Lagrangian coordinates (i.e., the position of the sensor relative to the material particles is fixed). Therefore, it is more convenient to perform the analysis of wave dynamics in Lagrangian coordinates than in Eulerian coordinates. For this reason, Eqs. (2.1) through (2.3) were given in Lagrangian form.

The equations of motion, or conservation equations, are universal equations that apply to all materials that satisfy the underlying assumptions of continuum mechanics. However, by themselves, these equations do not provide solutions to wave propagation problems. The number of unknown independent variables in the problem (u, ρ, σ, and E) exceeds the number of available equations by one variable, and an additional equation is required to render the mathematical problem well posed. This additional equation is the constitutive relation for the material, which relates the stress to kinematical and thermodynamic variables.

When strength effects are negligible, the constitutive relation is simply the equation of state (EOS) of the material. An EOS is usually a relation in which one thermodynamic quantity is given in terms of two other quantities, for example,

$$E = \hat{E}(\rho, s),$$

where s is the entropy per unit mass. Since this is a complete EOS, the remaining two quantities (in this case pressure and temperature) are obtained as derivatives. Often in shock wave studies, the response is assumed to be adiabatic and the EOS is reduced to a relationship between the pressure, density, and internal energy. Only when temperature is required is a more complete equation of state specified.

The equation of state is unique in the sense that the state is independent of the path taken to reach it. Hence, we expect to reach the same state by an impact followed by heating or by heating followed by an impact.

When strength effects are significant, the equation of state is supplemented with a constitutive relation for the deviatoric stresses (the equation of state provides only the mean stress or pressure). A constitutive relation is like an equation of state, but has a looser definition. A stress-strain path for a material undergoing yielding is a simple example of a constitutive relation. It is not necessarily unique because the state may depend on the path, and only a small subset of the state variables may actually be specified. The behavior of elastic-plastic material is described in more detail in Chapter 6.

2.2. Theory of Characteristics

Two standard solution procedures are available for solving problems in wave propagation: the method of characteristics and the finite difference or finite element method. Here, we give a brief discussion of the method of characteristics because its mathematical procedure is closely related to the wave motion and aids in understanding the interaction of waves. The finite element and finite difference computational procedures are discussed in Chapter 6.

To discuss the method of characteristics, we consider the flow field developed during one-dimensional wave motion as illustrated in Figures 2.1 through 2.5 for a variety of one-dimensional flow situations ranging in complexity from the simple wave shown in Figure 2.1 to the plate impact experiment with spall damage shown in Figure 2.5. The coordinates are time and Lagrangian distance; because the Lagrangian position is fixed with the material, the boundaries and interfaces do not shift with time.

In this flow field, there are so-called "characteristic" lines, special paths in the time-distance space along which the usual set of coupled partial differential equations reduces to a much simpler ordinary differential equation. For wave propagation, characteristics are especially interesting because wave motion occurs along these directions. In one-dimensional flow, waves generally propagate forward and backward in space, giving rise to C_+ and C_- characteristics, as shown in these figures. An additional characteristic line follows the particle path. This latter type of characteristic line becomes more important in case the material particle is modified during the flow, thereby becoming different from its neighbors. Work-hardening, fracture, and phase changes are modifications that cause the material to differ from point to point (see Figure 2.5 for such a particle path).

Waves move in the characteristic directions with the Lagrangian sound velocity a, which is related to the velocity c in the laboratory coordinate system, by the formula

$$a = \frac{\rho}{\rho_0} c = \frac{\rho}{\rho_0} \sqrt{\left(\frac{\partial \sigma}{\partial \rho} \right)_S}, \tag{2.4}$$

where σ is the stress in the direction of wave propagation, S is the entropy, and the notation $(\ )_s$ indicates that the derivative is taken along the isentrope (i.e., path of constant entropy). Thus, a is the wave velocity with respect to the moving material. For isentropic flow, the two sets of characteristics, C_+ and C_-, show trajectories of perturbations in positive and negative directions. The slopes of the characteristics are given by

$$\frac{\partial h}{\partial t} = a \text{ (on } C_+) \qquad \text{and} \qquad \frac{\partial h}{\partial t} = -a \text{ (on } C_-). \tag{2.5}$$

The variation of the material state along these characteristics in the time-distance plane is described by the following set of ordinary differential equations:

$$\frac{du}{ds} + \frac{1}{\rho_0 a}\frac{d\sigma}{ds} = 0 \qquad \text{along } C_+, \tag{2.6a}$$

and

$$\frac{du}{ds} - \frac{1}{\rho_0 a}\frac{d\sigma}{ds} = 0 \qquad \text{along } C_-. \tag{2.6b}$$

The integrals of these equations are the Riemann integrals:

$$u = u_0 - \int_{\sigma_0}^{\sigma} \frac{d\sigma}{\rho_0 a} \qquad \text{along } C_+, \tag{2.7a}$$

and

$$u = u_0 + \int_{\sigma_0}^{\sigma} \frac{d\sigma}{\rho_0 a} \qquad \text{along } C_-, \tag{2.7b}$$

where u_0 and σ_0 are integration constants.

A simple or progressive wave is a flow field in which all disturbances propagate in the same direction. In this simple wave, the states along the characteristic pointing in the direction of wave propagation remain constant, while all states along any other path in the $x-t$ plane are described by the function $u(\sigma)$ or $\sigma(u)$ corresponding to the Riemann integral along the characteristic pointing in the opposite direction.

As an idealized illustration, a simple wave moving into an elastic medium is shown schematically in Figure 2.1. The stress is rising in steps and the C_+ characteristics each correspond to one of the steps. Another example of a simple wave is the rarefaction in a uniformly compressed nonlinear medium shown in Figure 2.2. Here, the material has been compressed to some stress state, and now the left boundary begins to move left at time zero. The C_+ characteristics each have constant wave velocities, but successive characteristics have lower velocities because the stress is decreasing and the velocity is stress dependent.

We will refer to the trajectory that describes states in a simple wave in $\sigma - u$ coordinates as the rarefaction or Riemann isentrope. This rarefaction path is generally called an isentrope, although some nonisentropic processes (yielding,

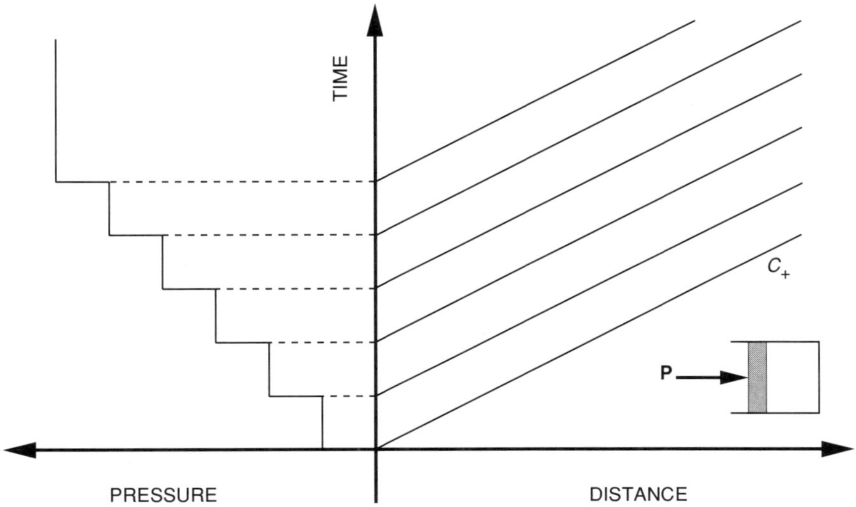

Figure 2.1. Simple wave in a linear elastic material.

for example) may be occurring. When all characteristics originate at a single point in the $x-t$ plane, the wave is referred to as a centered simple wave. The slope of the Riemann isentrope,

$$\frac{d\sigma}{du} = \pm \rho_0 a, \tag{2.8}$$

is the dynamic impedance of the material.

In normal media, the sound velocity increases with pressure. Therefore, rarefaction waves are diverged during propagation, as illustrated in Figure 2.2. Compression waves, however, become steeper and steeper during propagation and gradually evolve into discontinuities (shock waves) as indicated in Figure 2.4.

The pressure history in Figure 2.3 is typical of that generated by the detonation of an explosive charge: a shock front followed by an attenuating pressure. The pressure attenuation is due to unloading waves described by right-moving C_+ characteristics. The intersection of the C_+ characteristics with the shock front gives rise to reflected waves that travel back toward the explosive-target interface along left-moving C_- characteristics. Each intersection of characteristic lines gives rise to new characteristic lines. The wave velocities (and slopes of the lines) generally depend on the stress levels. Therefore, the new characteristic lines generated at an intersection will have essentially the same slope as the intersecting characteristics. The shock trajectory, however, represents an attenuating stress level so its wave velocity decreases gradually, causing an upward curvature in the shock trajectory.

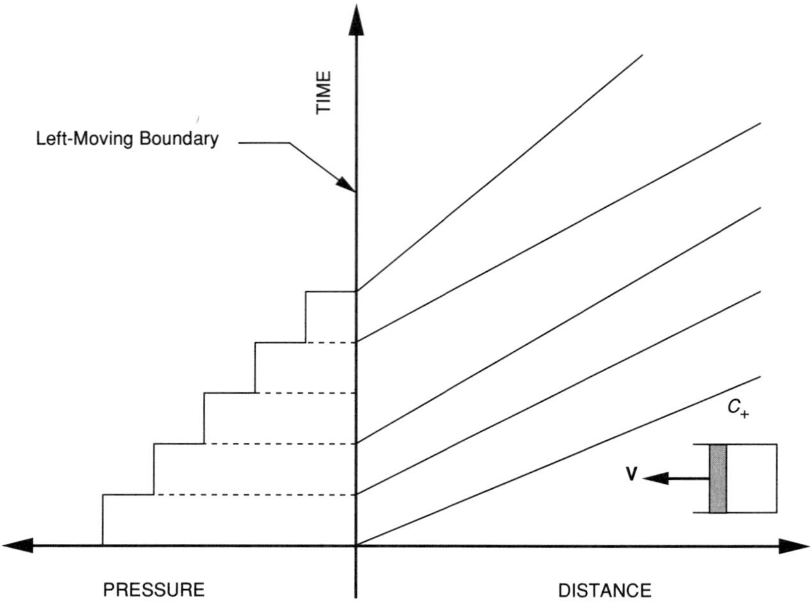

Figure 2.2. Rarefaction fan caused by a piston moving away from compressed material.

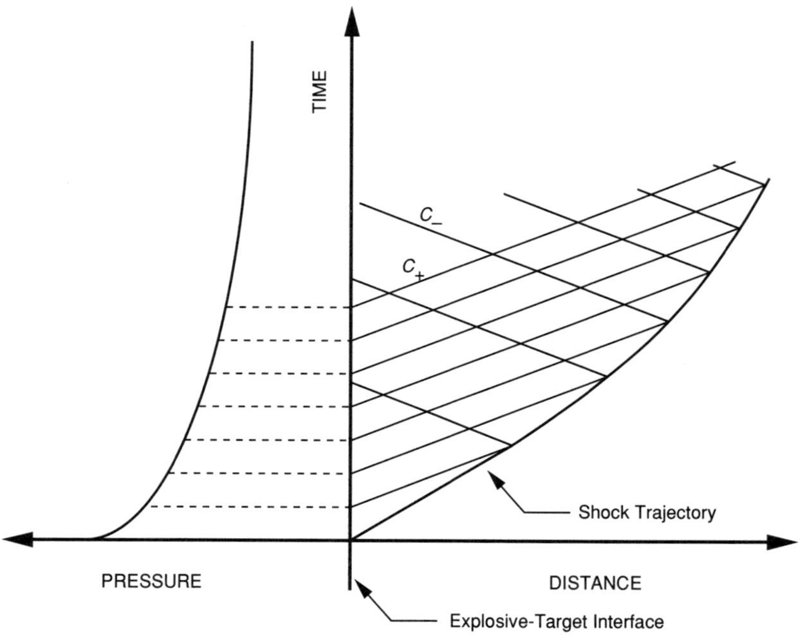

Figure 2.3. Rarefaction behind a shock, causing attenuation.

Figure 2.4 illustrates the waves caused by a gradually rising stress wave in a material that stiffens with compression (the usual case). As the pressure rises, the corresponding characteristic lines increase in velocity so the C_+ characteristics are not parallel. When the C_+ characteristics intersect, they produce a rearward-facing disturbance described by C_- characteristic and a forward-moving shock with a velocity that is intermediate to the velocities of the interacting characteristics. The shock front thus formed is concave downward, showing a gradually increasing velocity as it travels.

Figure 2.5 shows the characteristics for a plate impact configuration that produces fracture damage. Here, one shock wave is initially moving into each material after impact. When these initial C_+ and C_- characteristics are reflected at the outer free boundaries, rarefaction fans (centered simple waves) are generated. When these rarefaction fans intersect, they cause tensile stresses to arise. Here, we indicate that damage also occurs at some intersections. Since damage changes the condition of the material, the post-damage characteristics calculation must account for intersections of each C_+, C_-, and particle path. Clearly, the computation as well as the diagram becomes very complex.

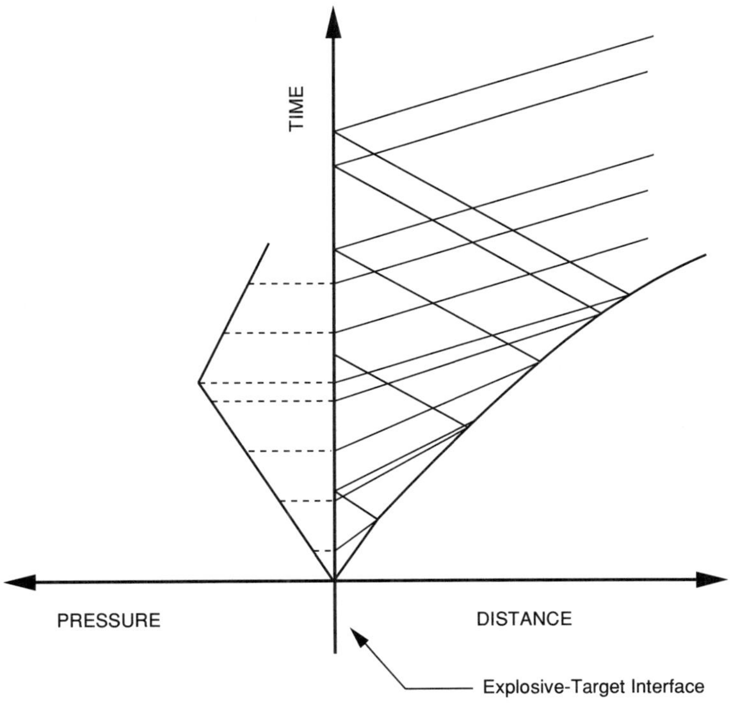

Figure 2.4. Shock formation for a gradually rising wave.

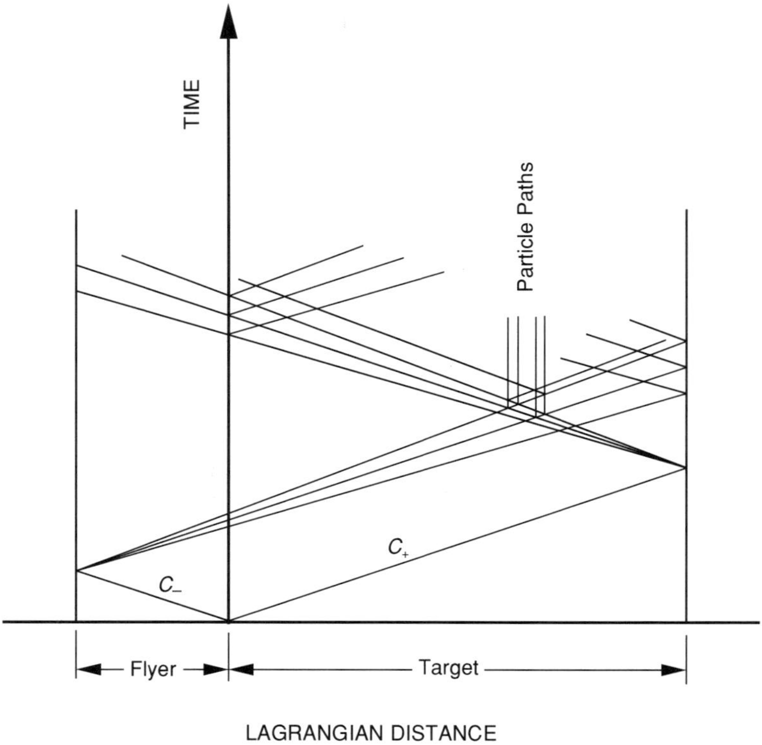

Figure 2.5. Impact of a flyer plate onto a target plate with resultant fracture damage.

2.3. Analysis of the Shock Wave

A shock wave occurs as a rapid change in stress, density, and particle velocity in the flow. The usual conservation equations for mass, momentum, and energy still govern the flow field, but the discontinuous nature of the shock wave leads to the following special forms for these equations:

Conservation of mass $$\rho_0 U = \rho(U - u), \tag{2.9}$$

Conservation of momentum $$\sigma = \sigma_0 - \rho_0 U u, \tag{2.10}$$

Conservation of energy $$\rho_0 \left(E - E_0\right) = \frac{1}{2}\left(\sigma + \sigma_0\right)\left(1 - \frac{\rho_0}{\rho}\right), \tag{2.11}$$

where ρ_0 and ρ are the initial and current values of the density, U is the shock velocity, u is particle velocity, σ_0 and σ are the total mechanical stress in the direction of propagation before and after the shock, and E_0 and E are the internal energy before and after the shock. The sign convention used here is that stress is positive in tension. Collectively, Eqs. (2.9), (2.10), and (2.11) are known as the Rankine–Hugoniot jump conditions; they are named after the two men who independently derived them.

2.3.1. The Hugoniot, Isentrope, and Isotherm

Because of their importance in shock wave work, we discuss here three important paths taken by state points across the equation-of-state surface: the Hugoniot, the isentrope, and the isotherm as shown in Figure 2.6. The Rankine-Hugoniot curve, also known as the *Hugoniot* or *shock adiabat*, is the locus of end states achieved through shock wave transitions. Such a curve is usually shown in stress-volume or stress-particle velocity space. Usually, the initial state of the material is at rest and with zero stress, but Hugoniot curves can also be defined for other initial conditions. When the peak shock stress is much higher than the yield strength, the normal stress is usually assumed equal to the pressure (mean stress) as in Figure 2.6.

Under moderate compression (less than 1 Mbar or 10^5 GPa) and for a wide range of materials, the Hugoniot of a condensed medium can usually be de-

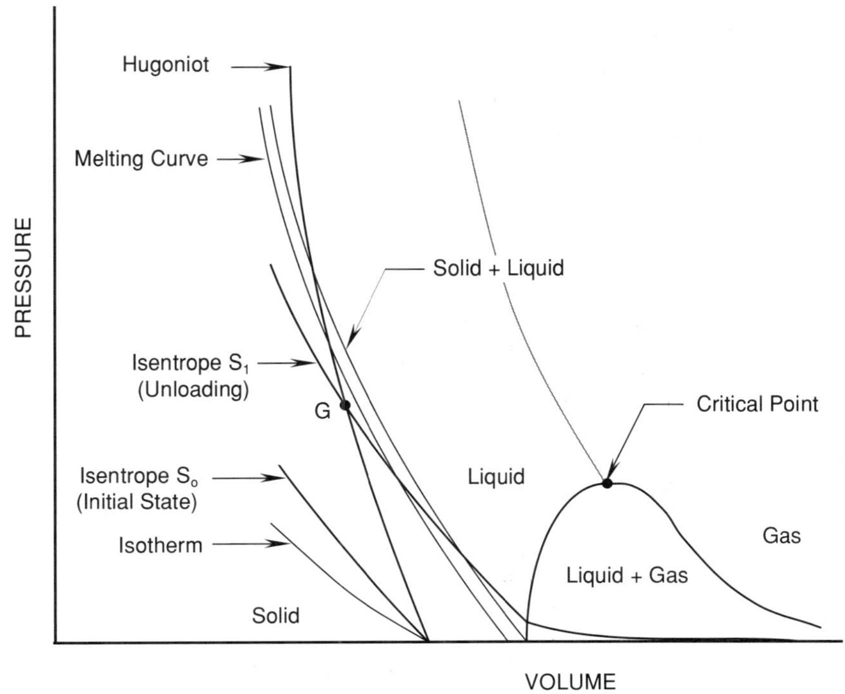

Figure 2.6. Schematic phase diagram of matter in pressure-volume space. The diagram also shows the relative positions of the Hugoniot, isotherm, isentrope, and melting curve.

scribed by a linear relationship between the particle velocity and the shock velocity,

$$U = c_0 + su, \tag{2.12}$$

where the material constant c_0 is the sound velocity corresponding to the initial equilibrium bulk compressibility of the medium and s is a dimensionless material constant. The empirical relationship expressed in Eq. (2.12) is based on a large collection of experimental shock wave data and has been in use since the 1950s (e.g., Rice et al. [1958], and Marsh [1980]). Rice et al. [1958] used least-square fitting of experimental data to determine the values of c_o and s for many solids. A summary of their results is presented in Table 2.1. As indicated, the slope of this shock velocity *versus* particle velocity relationship varies between 1 and 2 with values on the order of 1.5 being most common.

An isentrope is a series of states (of stress, energy, density, temperature, and entropy) along which the entropy is constant. Such a path is useful for reference because it includes no exchange of heat with the surroundings, no dissipation, and is reversible. Loading paths such as a Hugoniot may approximate an isentrope for weak shocks (low stress amplitude), but not for strong shocks where shock-wave compression is accompanied by an increase in entropy and irreversible heating of the material.

Figure 2.6 shows two isentropes: one corresponding to the initial or reference state (S_0), and another (S_1) that is a possible unloading path from the shock-loaded state represented by Point G in Figure 2.6. Note that during unloading the material may melt and may even reach the mixed liquid-vapor state. If the material was shock loaded to a higher stress state, the isentrope could lie above the critical point, in which case the material will vaporize during unloading.

Table 2.1. Linear least-square fits of the shock velocity versus particle velocity shock wave data for several metals (Rice et al. [1958]).

Metal c_0	**c_0 (km/s)**	s	**Metal**	**c_0 (km/s)**	s
Ag (fcc)	3.215	1.643	Ti (hex)	4.786	1.066
Au (fcc)	3.059	1.608	Th (fcc)	2.079	1.381
Be (hex)	7.975	1.091	Zn (hex)	3.042	1.576
Cd (hex)	2.408	1.718	In (tet)	2.370	1.608
Co (hex)	4.652	1.506	Nb (bcc)	4.447	1.212
Cr (bcc)	5.176	1.537	Pd (fcc)	3.793	1.922
Cu (fcc)	3.972	1.478	Pt (fcc)	3.671	1.405
Mg (hex)	4.493	1.266	Rh (fcc)	4.680	1.645
Mo (bcc)	5.173	1.204	Ta (bcc)	3.374	1.155
Ni (fcc)	4.667	1.410	Tl (hex)	1.821	1.566
Pb (bcc)	2.066	1.517	Zr (hex)	3.771	0.933
Sn (tet)	2.668	1.428			

An isotherm is a series of states along which the temperature remains constant. Such a loading path as the isotherm is common in quasi-static loading, where the material may remain at constant temperature throughout the whole loading regime. A reference isotherm is shown in Figure 2.6.

2.3.2. Estimating the Lagrangian Sound Velocity from Hugoniot Data

Sound velocities at high pressures are often necessary for planning shockwave experiments and for interpreting experimental results. An estimate of the sound velocity in a shock-loaded material can be derived using the Hugoniot. Measurements have shown that, in the pressure–particle velocity plane, the release isentrope of many materials deviates from the Hugoniot by not more than 3% for pressures up to at least 50 GPa.[1] If we assume that the Hugoniot and the rarefaction isentrope coincide on the *p*-*u* plane, we find that

$$\left(\frac{dp}{du}\right)_S = \pm\rho_0 a = \pm\left(\frac{dp}{du}\right)_H = \pm\rho_0\left(c_0 + 2su\right), \tag{2.13}$$

where the derivatives $(dp/du)_S$ and $(dp/du)_H$ are evaluated along the isentrope and Hugoniot, respectively. The Lagrangian sound velocity in shock-compressed matter can now be estimated using the results in Eq. (2.13) in combination with the Hugoniot equation, Eq. (2.12), and the momentum conservation law, Eq. (2.10),

$$a = c_0 + 2su = \sqrt{c_0^2 + \frac{4sp}{\rho_0}}. \tag{2.14}$$

When the Hugoniot deviates from the isentrope by only a small amount, the quasi-acoustic approach for treating the shock wave velocity (Landau and Lifshitz, [1959]) is satisfactory. According to this approach, the velocity of the shock wave is the average between the sound velocities ahead of the shock discontinuity, c_0, and behind it, a:

$$U \approx \frac{1}{2}\left(c_0 + a\right) \approx \frac{1}{2}\left(c_0 + c_0 + 2su\right) = c_0 + su. \tag{2.15}$$

Hence, we have a procedure for estimating the Lagrangian sound velocity that is consistent with the quasi-acoustic approach to shock wave computations.

1. For simplicity, the pressure, *p*, is used in the foregoing analysis instead of the stress, σ, thus implicitly assuming that the yield stress is much lower than the peak stress, as is typically the case in intense shock waves.

2.3.3. Rate Processes at the Shock Front[2]

The thickness of plastic shock waves or the so-called "shock front rise time" is controlled by the stress relaxation time at high strain rate within the shock wave—a quantity closely related to material viscosity. Swegle and Grady [1985] and Dunn and Grady [1987] have examined the wave fronts measured by laser interferometer and deduced the rate-dependent stress–strain relations from the thickness and shape of the wave fronts. Swegle and Grady observed that the strain rate in the plastic wave (above the elastic precursor) and the stress jump in the plastic wave are related as follows:

$$\frac{d\varepsilon}{dt} = A(\Delta\sigma)^4, \tag{2.16}$$

where ε is the compressive strain beyond the elastic precursor, A is a material constant (about 6000 $\text{GPa}^{-4}\cdot\text{s}^{-1}$ for aluminum), and $\Delta\sigma$ is the stress jump from the precursor level to the peak stress of the steady wave. In both papers, Grady et al. assumed that the rate dependence is associated only with the deviator stresses. Using this assumption Dunn and Grady deduced that the viscous stress, S_v, defined as the difference between the effective stress, $\bar{\sigma}$, and the yield strength, Y, is the product of a function $h(\varepsilon)$ of the strain only and of a power of the strain rate:

$$\bar{\sigma} - Y = sign\left(\frac{d\bar{\varepsilon}}{dt}\right) h(\varepsilon) \left|\frac{d\bar{\varepsilon}}{dt}\right|^m. \tag{2.17}$$

Hence, the plastic strain rate is

$$\frac{d\bar{\varepsilon}^P}{dt} = \left(\frac{\bar{\sigma} - Y}{h(\varepsilon)}\right)^{1/m}, \tag{2.18}$$

and we can write the differential equation for the stress as

$$\frac{d\bar{\sigma}}{dt} = 3G\left(\frac{d\bar{\varepsilon}}{dt} - \frac{d\bar{\varepsilon}^P}{dt}\right) = 3G\left[\frac{d\bar{\varepsilon}}{dt} - \left(\frac{\bar{\sigma} - Y}{h(\varepsilon)}\right)^{1/m}\right]. \tag{2.19}$$

In the above equations, $\bar{\varepsilon}_p$ is the effective plastic strain, G is the shear modulus, and m is a material constant.

Values of h are given in Table 2.2 based on the analysis of Swegle and Grady [1985] where it is assumed that $h(\varepsilon)$ is a material constant and $m = 1/2$. From this analysis, we have a general rate-dependent equation for the deviator stresses occurring in a shock front plus values of h for several metals. Furthermore, since in this analysis the maximum shear stress in the steady shock wave

2. Readers who are not familiar with elastic-plastic material behavior might find it useful to consult Section 6.1 for background information on the topic.

Table 2.2. Shock front parameter h for several materials (Swegle and Grady [1985]).

Metal	h dyne - $\sqrt{s}$ /cm^2	**Metal**	h dyne - $\sqrt{s}$ /cm^2
Aluminum	2.04E+06	Iron	5.88E+06
Beryllium	7.06E+06	MgO	8.06E+06
Bismuth	1.65E+06	Fused Silica	4.40E+07
Copper	1.17E+06	Uranium	1.24E+07

is proportional to $(\Delta\sigma)^2$, we may also deduce, through the use of Eq. (2.16), that the maximum plastic strain rate is proportional to the square of the maximum shear stress.

The analysis of Grady et al. can be applied to derive an expression for the apparent viscosity in the shock front. To do this, let us combine the definition of the plastic strain rate in Eq. (2.18) with the usual relation between the viscous stress, S_v, and the plastic strain rate:

$$S = 3\eta \frac{d\bar{\varepsilon}^P}{dt}, \tag{2.20}$$

where η is the material viscosity. Combining Eqs. (2.18) and (2.20), and solving for the viscosity we obtain,

$$\eta = \frac{1}{3}\frac{h}{\sqrt{d\bar{\varepsilon}^P / dt}}. \tag{2.21}$$

This result emphasizes that the viscosity is not constant, but decreases as the strain rate increases. Using the Aluminum parameters of Swegle and Grady, we find that the apparent viscosity is 700 Pa-s at a strain rate of 10^4 and 70 Pa-s at 10^6.

2.4. Graphical Analysis of Experimental Designs

A particularly useful approach for analyzing shock wave propagation problems relies on graphical techniques. In place of the conservation and constitutive equations normally used in analytical solutions, this approach uses a distance-time diagram (or $x - t$ diagram), and a stress–particle velocity diagram (or $\sigma - u$ diagram) to provide graphical solution to stress wave propagation problems. The distance-time diagram is used to display the relative positions, in space and time, of the materials involved in a given problem. This $x - t$ diagram is also used to keep track of the type and relative position of the stress waves, as well as their interactions. The stress-particle velocity diagram is used to find new equilibrium states, in the $\sigma - u$ space, after the impact of two materials or after the interaction of waves with one another or with material boundaries. In locating new

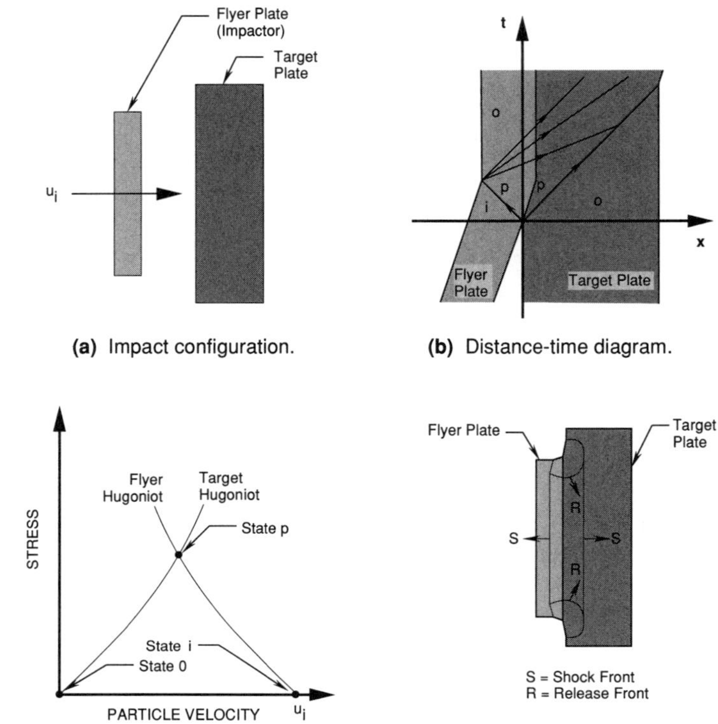

(a) Impact configuration. **(b)** Distance-time diagram. **(c)** Stress-particle velocity diagram. **(d)** Edge effects.

Figure 2.7. Generation of a compression pulse by the impact of a flyer plate.

equilibrium states, we use the fact that, at any point in a solid material, both the stress and the particle velocity must remain continuous at all times.[3] We also invoke the definition of the Hugoniot curve as the locus of all possible end states in a material attainable behind a shock front.

Let us now apply this graphical analysis technique to study wave dynamics under plate impact loading conditions. The configuration of a typical plate impact experiment is shown in Figure 2.7(a). A flyer plate approaches a target plate from the left with velocity u_i. Figure 2.7(b) shows the distance-time diagram of the wave process after impact of the flyer plate upon the plane target. The origin of the $x - t$ diagram corresponds to the moment in time the impact event is initiated. This impact causes shock waves to propagate in the impactor and target plates away from the impact face as shown in Figure 2.7(b).

3. If spall occurs, this continuity condition no longer holds, and appropriate boundary conditions should be applied at the newly formed free surfaces.

To determine the equilibrium states in the flyer and target plates after impact, we use the $\sigma - u$ diagram shown in Figure 2.7(c). By definition, all attainable equilibrium states in the shock-compressed target must lie on the target Hugoniot. The state of the target before impact defines the point at which the target Hugoniot is centered. In our example, the point (0, 0) in $\sigma - u$ space defines the initial state of the target; therefore, the target Hugoniot passes through the point (0, 0). The direction of wave propagation in the target determines the sign of the shock front velocity, U, and that of the jump of particle velocity, Δu, which in turn, determines whether the Hugoniot faces to the right or to the left. In our example, the shock wave in the target travels from left to right. Therefore, the target Hugoniot shown in Figure 2.7(c) faces to the right. Using the same reasoning, we find that the Hugoniot of the flyer must pass through the point $(0,\ u_i)$ and must be facing to the left.

Continuity of stress and particle velocity at the interface between the flyer and target plates requires that they both reach the same post-impact stress and particle velocity state. This common state is represented by the intersection of the two Hugoniots representing the target and flyer plates and is labeled state 'p' in Figures 2.7(b) and (c). When the target and flyer are made of the same material, then due to symmetry, the particle velocity of the shock compressed matter is exactly one-half of the initial impactor velocity. This, in fact, is the situation shown in Figure 2.7.

When the shock wave reaches the back surface of the impactor, it is reflected as a centered rarefaction wave, which propagates toward the target. The material behind this rarefaction fan is in a stress-free state and, if the impactor and target are made of the same material, at rest. Because the rarefaction front, which propagates with the sound velocity in shock-compressed matter, is faster than the initial shock front, the rarefaction front eventually overtakes the shock front and causes a decay in peak stress. If the impactor and target are made of the same material, the distance x where the rarefaction front overtakes the shock front can be calculated by using the following equation derived using the quasi-acoustic approach:

$$x = \delta\left(1 + \frac{2U}{su}\right) = \delta\left(1 + \frac{4U}{su_i}\right) \approx \delta\left(1 + \frac{4c_0}{su_i}\right), \tag{2.22}$$

where δ is the impactor thickness, and c_0 and s are the coefficients defining the intercept and slope of the linear relationship between the shock front velocity U and the particle velocity u [Eq. (2.12)]. Thus, stronger shock waves begin to decay earlier than weaker shock waves. In the case of elastic–plastic materials with Poisson's ratio independent of pressure, the distance x can be estimated as

$$x = \delta\frac{U(c_l + c_0) + c_l su}{U(c_l - c_0) + c_l su} \approx \delta\frac{c_l + c_0}{c_l - c_0 + \dfrac{c_l}{2c_0}su_i}, \tag{2.23}$$

where c_l is the longitudinal wave velocity and all other parameters are as previously defined. The value of the distance x computed from Eq. (2.23) is actually an upper bound estimate, and it decreases if we take into account the elastic precursor of the shock wave in the impactor.

An important factor in the design of experimental configurations for shock wave propagation experiments is the ratio of the axial to transverse dimensions of the impactor and target plates. The transverse dimensions must be large enough to ensure one-dimensional motion throughout the time period required for the measurements. The dimensions of the experiment should be chosen such that the release waves emanating from the edges of the impactor and target plates [see Figure 2.7(d)] do not interfere with the one-dimensional flow in the central region of the specimen during the experiment. Thus, longer recording times can be achieved by using impactor and target plates with large diameter-to-thickness ratios.

Figure 2.8 illustrates the wave dynamics for the impact of a flyer plate with high dynamic impedance upon a softer target. The impact configuration shown in Figure 2.8(a) is essentially the same as in the previous case. As before, the impact causes compressive shock waves to propagate in the impactor and target plates away from the impact face. These waves are shown in Figure 2.8(b), where several later wave reverberations are also shown. The $\sigma - u$ diagram for this impact configuration is shown in Figure 2.8(c). As before, this diagram is used to determine the end stress and particle velocity states in the impactor and target. Those end states are labeled with the letters A through E, while the initial states in the impactor and target before impact are designated with 'i' and '0', respectively.

The stress-particle velocity state in both the target and projectile just after impact is designated by the letter A in Figure 2.8. As the initial shock wave reaches the back surface of the impactor, it reflects as a release fan, which unloads the impactor to a state of zero stress. This state is designated by the letter B in Figures 2.8(b) and (c). The interaction of this release fan with the impactor-target interface produces a new equilibrium state in both the impactor and the target. This state is indicated by the letter C in Figure 2.8. In the target, the new equilibrium state is reached through an unloading wave that transforms the material from state A to state C. In the impactor, the new equilibrium state is reached through a compression wave that transforms the material from state B to state C. Several such wave reverberations occur between the impact surface and the back surface of the impactor. Each reverberation produces a release wave that propagates into the target material. Thus unloading of the target occurs in several successive steps, each lower in magnitude than the previous one. The resulting wave structure after two reverberations is shown in Figure 2.8(c).

The progressive unloading of the target noted in the case just discussed cannot be achieved if the impactor material is softer than the target material. The wave dynamics for this case are illustrated in Figure 2.9. On impact, the stress and particle velocity in the impactor change from state 'i' to state 'A' whereas in

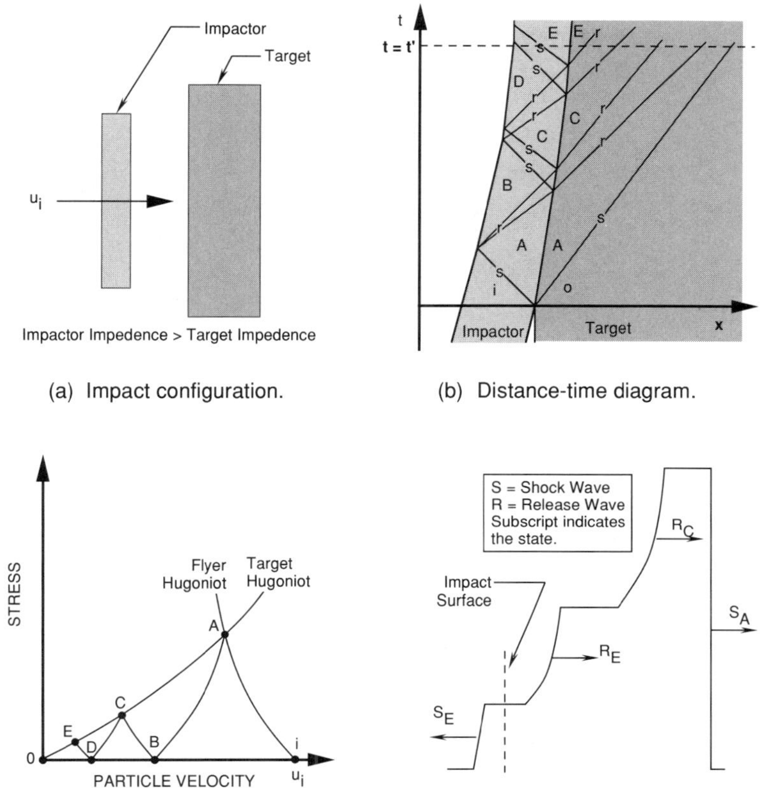

(a) Impact configuration.

(b) Distance-time diagram.

(c) Stress-particle velocity.

(d) Wave structure at time t = t'.

Figure 2.8. Wave interactions for the impact of a relatively rigid flyer plate upon a softer target.

the target material the stress and particle velocity change from state '0' to state 'A'. Interaction of the shock wave with the rear surface of the impactor causes a release fan to emerge. Behind this release fan, the impactor material is in a state of zero stress and negative particle velocity. This state is indicated by the letter B in Figure 2.9.

Wave interactions resulting from the interaction of the release fan with the impactor-target interface lead to a new equilibrium state. The continuity condition requires that both the impactor and target reach the same new equilibrium state indicated by point C in Figure 2.9. However, this is physically impossible because the interface cannot support tensile stresses. For this reason, the target separates from the projectile at the interface, and the target material unloads to a stress-free state.

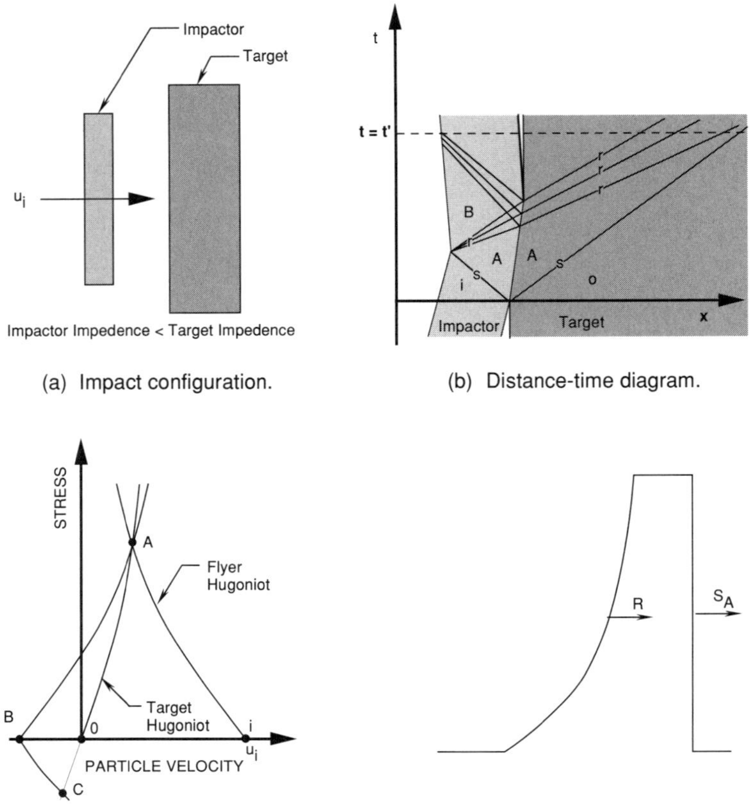

(a) Impact configuration.

(b) Distance-time diagram.

(c) Stress-particle velocity diagram.

(d) Wave structure at time t = t'.

Figure 2.9. Wave interactions for the impact of a relatively soft flyer plate upon a hard target.

Next, we examine a case involving generation of a stress pulse in a material using a laser beam, particle beam, or other radiation source. In this case, the energy is deposited in a thin layer of material near the front surface. The deposition profile is nearly exponential for lasers and X-ray sources.[4] The deposition depth depends on the light absorption characteristics of the target material and on the characteristics of the light source. The nearly instantaneous deposition of energy in a thin layer of material causes local heating at constant volume, which in turns causes the stress to increase. The highest stress magnitude occurs near the irradiated surface.

Since a free surface cannot sustain any normal stresses, a rarefaction wave

4. Stress profiles caused by radiation from a particle beam are more complex.

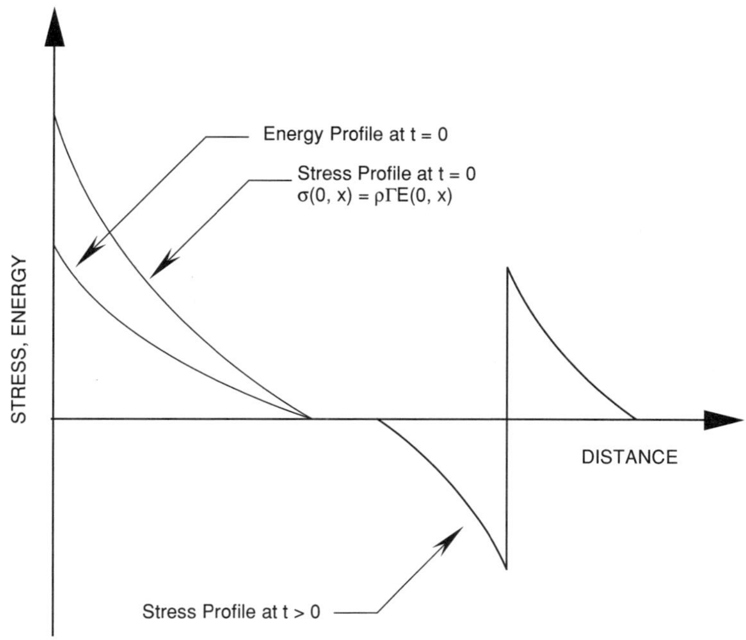

Figure 2.10. Generation of a stress pulse by instantaneous energy deposition and the evolution of a bipolar stress wave.

forms at the irradiated surface and propagates toward the interior of the sample. Meanwhile, the compression wave that forms during energy deposition also propagates into the cold interior of the sample, ahead of the rarefaction wave. Thus, a bipolar stress pulse forms as shown in Figure 2.10. The stress profile in the irradiated material at the instant of deposition is directly related to the deposited energy profile through the Grüneisen coefficient, Γ. For this reason, the thermomechanical stress that develops as a result of energy deposition is also known as Grüneisen stress.

Figure 2.11 illustrates the wave dynamics for instantaneous bulk energy deposition. The analysis assumes that the deposited energy profile has its maximum near the surface, as shown in Figure 2.10. If the deposited energy is not great enough to cause vaporization, the process can be analyzed, at least qualitatively, by using the acoustic approach. Immediately after the instantaneous irradiation of the target, the particle velocity is identically zero throughout the deposition region as well as in the rest of the target. States of particles on the $\sigma - u$ diagram (see Figure 2.11(a)) are described by points along the stress axis.

Information about changes in state of each point is propagated by sound perturbations both into the body and toward its irradiated surface. For each point x at time t, the stress and particle velocity in the new state are located on inter-

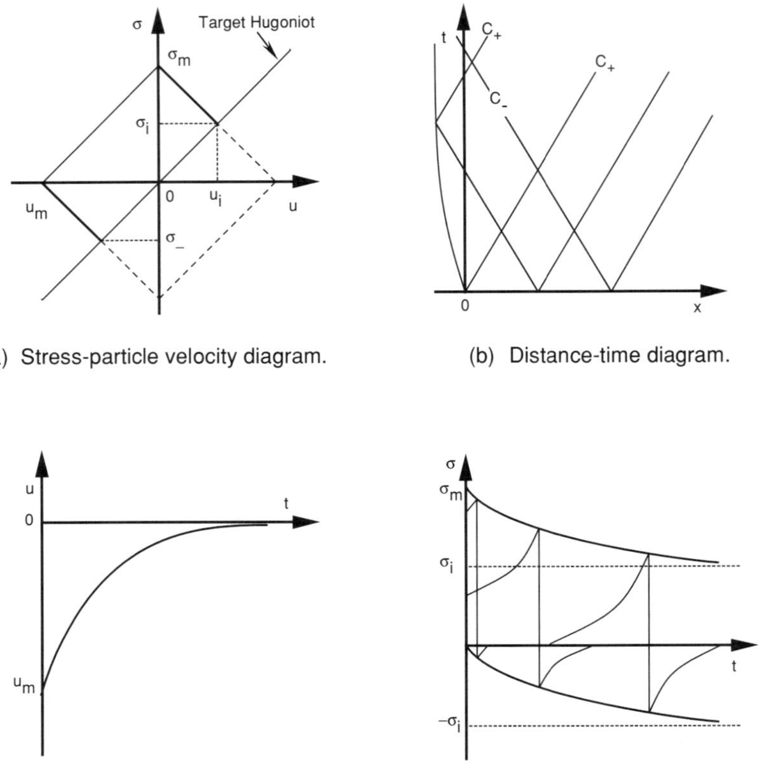

(a) Stress-particle velocity diagram.

(b) Distance-time diagram.

(c) Free-surface velocity profile on irradiated side.

(d) Evolution of the bipolar stress pulse.

Figure 2.11. Wave dynamics for problems involving instantaneous bulk energy deposition.

sections of Riemann's isentropes describing the changing states along C_+ and C_- characteristics, which pass through the point x, as shown in Figure 2.11(b). Thus maximum pressure and particle velocity magnitudes at points far from the irradiated region (i.e., where deposited energy is equal to zero) correspond on the $\sigma - u$ diagram to intersections of the lines

$$\sigma = \rho c u \tag{2.24}$$

describing states along a C_- characteristic that originates from a undisturbed area, and the lines

$$\sigma = \sigma_m - \rho c u \tag{2.25}$$

describing states along a C_+ characteristic that originates from the point of maximum stress, σ_m. Thus,

$$u = \frac{\sigma_m}{2\rho c} \quad \text{and} \quad \sigma = \frac{\sigma_m}{2}. \tag{2.26}$$

The values of u and σ obtained from Eq. (2.26) are indicated by u_i and σ_i in Figure 2.11(a). The maximum free-surface velocity toward the radiation source is

$$u_s = u_m = -\frac{\sigma_m}{\rho c}, \tag{2.27}$$

where u_m is the maximum particle velocity (with a negative sign). Thus, the maximum particle velocity is reached at the free surface of the irradiated side of the target. The free-surface velocity begins to decay almost instantaneously with the arrival of perturbations from internal layers of the targets. The resulting free-surface velocity profile is shown in Figure 2.11(c).

Expansion of the target material is accompanied by the appearance of negative stress (i.e., tension) inside the target. The tensile stress magnitudes can be obtained from the intersection of Riemann's isentropes for perturbations coming from deep layers of the target toward the irradiated surface with Riemann's isentropes for perturbations reflected by the surface. On the time-distance diagram of Figure 2.11(b), the negative pressure area is situated above the C_+ characteristic emanating from the origin.

The tensile stress in the target increases gradually until the ultimate maximum value is reached during propagation of the reflected wave into the cold region of the target. The magnitude of the maximum tensile stress is given by the relation

$$\sigma_- = \frac{\rho c u_m}{2} = -\frac{\sigma_m}{2}. \tag{2.28}$$

This maximum tensile stress is reached on cross-sections in the cold region of the specimen, where initially the deposited energy and stress are both zero.

The evolution of the stress pulse is shown in Figure 2.11(d), which show stress histories at several successive locations in the target. As shown, a bipolar stress pulse develops in the target. Initially, the compressive component of the pulse dominates the stress history. However, as noted in Figure 2.11(d), the pulse baseline continually shifts downward while the magnitude of the difference between the maximum tensile stress and the maximum compressive stress remains constant. This trend continues until the peak compressive stress is equal in magnitude to the peak tensile stress.

2.5. Temperature in Shock and Rarefaction Waves

Processes that occur in shock or rarefaction waves (e.g., plastic flow, phase changes, chemical reactions, evolution of damage) are in general rate processes that depend on the temperature of the material. In many cases such temperature dependence can be neglected, but in others (notably chemical reactions), computational simulations must calculate the temperature. A specific and important example is the case in which burning occurs on microscopic crack surfaces in the propellant in a rocket motor, thereby leading to unstable burning rates and potential detonation. Although the initial nucleation and growth of the fractures may be to a first approximation temperature-independent, the burning rate depends strongly on the temperature. Furthermore, in fracturing material the fracture mode and kinetics are also typically temperature-dependent, so if the material is being heated by radiation, plastic flow, or exothermic reactions, the fracturing process may change significantly with time.

Unfortunately, standard computer "hydrocodes" generally use a caloric equation of state that contains only pressure, specific volume, and internal energy. Knowledge of the specific heats as functions of stress, deformation, and internal energy is needed to calculate the temperature. Assuming such knowledge, various numerical procedures for calculating temperature are used in working hydrocodes, but they tend to be *ad hoc* and nonrigorous.

A thermodynamically rigorous approach to calculate temperature in materials that are undergoing evolutionary processes was reported in 1967 in an important paper by Coleman and Gurtin [1967]. This approach is described in more detail in Chapter 6 along with three other less rigorous approaches for calculating temperature in deforming and/or fracturing materials.

3
Experimental Techniques

Investigating the strength of condensed matter under shock wave loading requires the ability to create plane shock pulses in laboratory samples and to measure the evolution of these pulses inside the samples. This chapter discusses methods of producing and recording intense load pulses in condensed media. Although shock wave techniques are well documented in technical papers, monographs (e.g., Caldirola and Knoepfel [1971]; Graham [1993]), and reviews (e.g., Al'tshuler, [1965]; Graham and Asay [1978]; Chhabildas and Graham [1987]), a summary of methods will be useful here for a better understanding of the experimental results that will be presented in later chapters.

3.1. Experimental Procedures Used to Produce Shock Waves

Plane shock waves for spall strength measurements are usually generated by impacting the sample of interest with a flyer plate or by detonating an explosive plane wave generator in contact with the sample. These shock wave generation schemes produce loading pulses with durations on the order of a microsecond. Radiation energy from a laser or particle beam can be used to produce stress pulses with much shorter durations. To properly design and correctly interpret the results of shock wave experiments, we need to understand the details of the loading history in the specimen for each of the wave generation schemes used in the experiments.

3.1.1. Explosive Devices

The simplest method of producing a shock wave with a peak pressure of a few tens of gigapascals is to detonate a chemical explosive charge on the surface of the sample. Various explosive lenses have been designed to create plane shock and detonation waves with lateral dimensions up to few tens of centimeters.

Detonation of an explosive in contact with the sample creates a triangular stress history because, in detonation waves, the pressure begins to fall immediately after the shock as a result of expansion of the detonation products. Often, well-controlled stress wave propagation experiments require a square stress pulse (i.e., a stress pulse with constant amplitude) rather than the triangular pulse produced using in-contact explosives. Such a stress pulse is usually generated by using the flyer plate impact configuration in which a flyer plate, or impactor, is made to collide with the target in a planar fashion and at a well-controlled impact velocity. Then the peak stress in the target is controlled by the impact velocity and by the dynamic impedances of the impactor and target materials. The duration of the stress plateau behind the shock front is controlled by the thickness of the impactor. Experimentally, plane impactors are projected using explosive detonation facilities or ballistic devices known as "guns."

Figure 3.1 shows a typical arrangement of an explosive launching device. Such a device can accelerate metal or plastic impactors, 1 to 10 mm thick, to velocities of 1 to 6 km/s. The central region of the impactor remains flat even though the radial expansion of the detonation products leads to a pressure gradient that causes the pressure in the explosive gases to decrease with distance away from the center of the explosive charge. The guard ring shown in Figure 3.1 is placed around the impactor to compensate for the effect of this pressure gradient. The reflection of the detonation wave from the guard ring causes a momentary increase in pressure around the periphery of the impactor, which in turn produces additional inflow of the detonation products into the gap above the impactor. This gap also serves to "soften" the impact and prevent fracture in the flyer plate.

It is difficult to attain impactor velocities below 1 km/s using the launching scheme shown in Figure 3.1. An alternative explosive launching technique that produces low impact velocities is shown in Figure 3.2. With this technique, an intermediate or attenuator plate with high dynamic impedance is placed between the explosive charge and the flyer plate. Detonation of the explosive charge

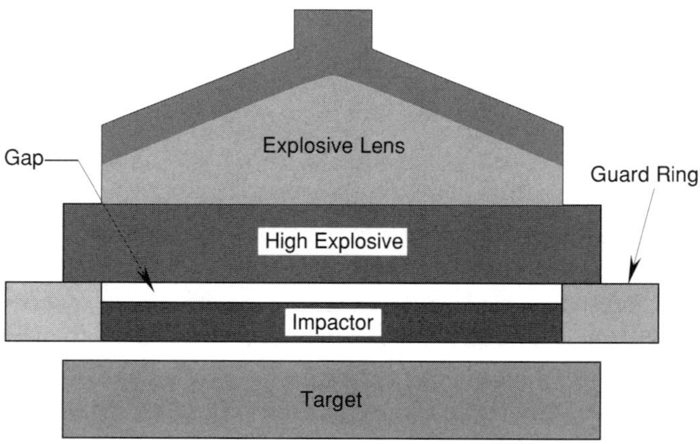

Figure 3.1. Experimental configuration for using explosives to launch a flyer plate at high velocity.

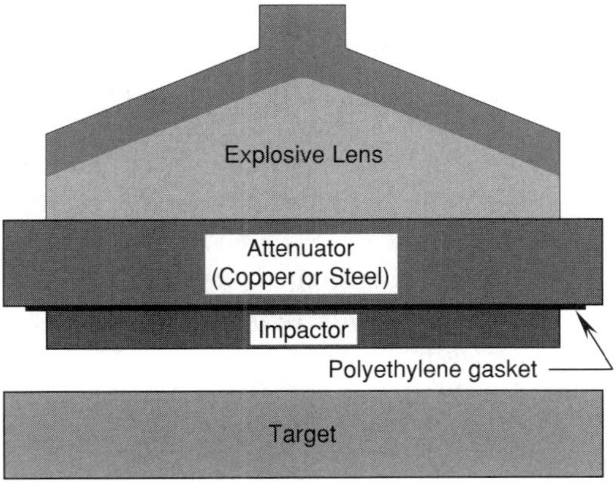

Figure 3.2. Experimental configuration for using explosives to launch a flyer plate at low velocity.

produces a plane shock wave in the attenuator. The flyer plate, which has a dynamic impedance lower than that of the attenuator, is accelerated by the shock wave and it acquires a velocity higher than that of the attenuator. This velocity difference causes the flyer plate to separate from the intermediate plate. A soft polyethylene gasket is inserted between the attenuator and impactor to prevent damage to the impactor as a result of rarefaction wave reflection from the rigid attenuator. This launching technique is also suitable for accelerating very thin impactors, such as foils or films, which are normally used to produce very short shock pulses.

Explosive launching techniques have been used to perform shock wave experiments since World War II, primarily because explosive facilities are compact and inexpensive. The impactor velocity can be easily varied over a wide range by varying the composition and density of the high explosives and the material and thickness of the flyer plate. However, using explosive materials is destructive and highly hazardous, thus requiring the use of safety measures. The experiments must be contained in specially designed containment chambers or performed at remote test areas. The explosives must be stored in specially designed bunkers where accidental detonations can be harmlessly contained. Furthermore, experiments with explosives require the availability of the technology to manufacture suitably shaped high-grade explosive charges. These constraints make it impractical in many cases to use explosive launch facilities.

3.1.2. Gas and Powder Guns

A popular alternative to the use of explosives for performing shock wave experiments is the use of smooth-bore ballistic installations such as gas guns or powder guns. With these smooth-bored guns, it is possible to vary the impactor velocity over a wide range in a reproducible and controllable fashion. Figure 3.3 shows a schematic of a typical gas gun (Fowles et al. [1970]). This gas gun barrel is 14 m long and 101.6 mm in diameter. These dimensions are usually chosen to optimize the performance of the gas gun in terms of attainable projectile velocity, which is controlled by the length of the barrel as well as the volume and pressure of the gas; and recording time, which is controlled by the diameter of the gun bore (i.e., the lateral dimensions of the impactor plate and the specimen). Gas gun dimensions vary greatly from one facility to another, but generally the bore diameter varies from 20 to 150 mm and the length of the barrel varies from 3 to 30 m (e.g., Fowles et al. [1970]). With a barrel up to 14-m long and initial compressed gas (nitrogen or helium) pressure of up to 15 MPa, a propulsion velocity of 100 to 1500 m/s can be produced.

At low impact velocities (below 100 m/s), frictional effects in the gun barrel become nonreproducible. For this reason, standard full-size gas guns are often not reliable in terms of generating reproducible low impact velocities. To overcome this inherent deficiency, the 101.6-mm-diameter gas gun at SRI International, for example, has been equipped with a "monkey's fist" that grips the projectile and holds it close to the target, thereby decreasing the travel distance of the projectile, minimizing nonreproducible friction effects, and enabling reproducible experiments at impact velocities below 100 m/s.

The flyer plate in a gas gun experiment is usually attached to a hollow projectile, which holds the plate normal to the axis of the gun barrel. To ensure pla-

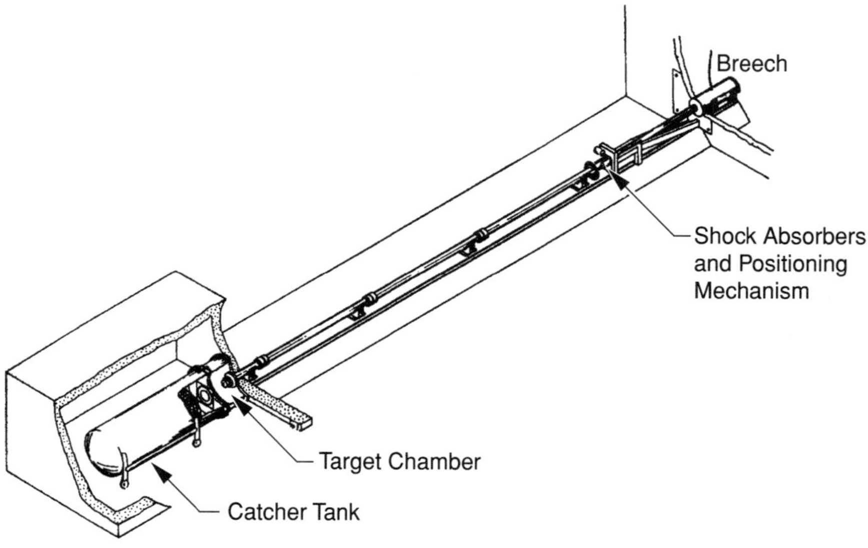

Figure 3.3. Overall view of the gas gun facility at Washington State University (Fowles et al. [1970]).

nar impact and thereby minimize impactor tilt with respect to the target plate, impact is often arranged with the projectile still partly in the barrel, as shown schematically in Figure 3.4. The gun barrel and target chamber are usually tightly sealed and evacuated before each experiment to minimize the effect of the air cushion that would otherwise develop as the projectile travels down the gun barrel and compresses the air column in its path.

The target in Figure 3.4 is configured for soft recovery. The tapered edges of the specimen allow it to easily separate from the remainder of the target plate after the impact event. The specimen is then softly recovered in the rag-filled catcher box for post-test microstructural examination.

3.1.3. Electro-Explosive Devices (Electric Guns)

The desire to extend the range of parameters attainable in wave propagation experiments has led to the development of novel shock wave generators. Promising sources of high dynamic pressure include electro-explosive devices (electric guns) and high-power pulsed laser and particle beams.

In the electric gun, the explosion of an electrically heated metal foil and the accompanying magnetic forces drive a thin flyer plate up a short barrel (Osher et al. [1989]). Such a device is diagrammed in Figure 3.5. The gun uses the energy initially stored in a fast-rise-time capacitor bank to ohmically heat and explode a bridge foil. The dense plasma produced by the electrical explosion of the foil pushes a cover polymer film, which can then be used as an impactor. In a later stage, the magnetic field of the expanding current-carrying circuit contributes to the acceleration of both the partially expanded plasma and the flyer.

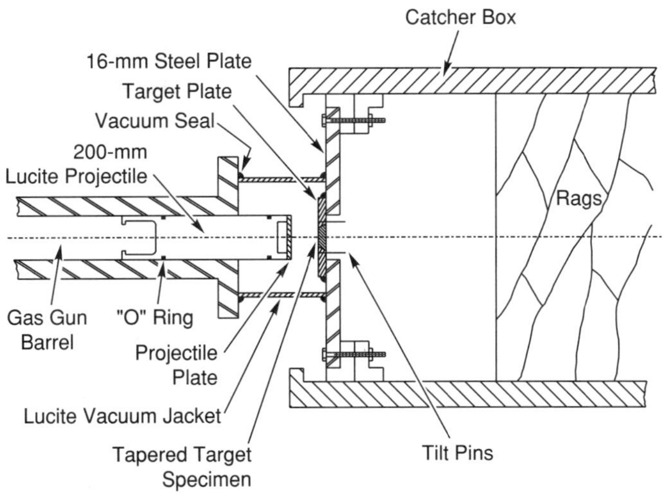

Figure 3.4. Schematic of the target area in a typical plate impact experiment (Barbee et al. [1970]).

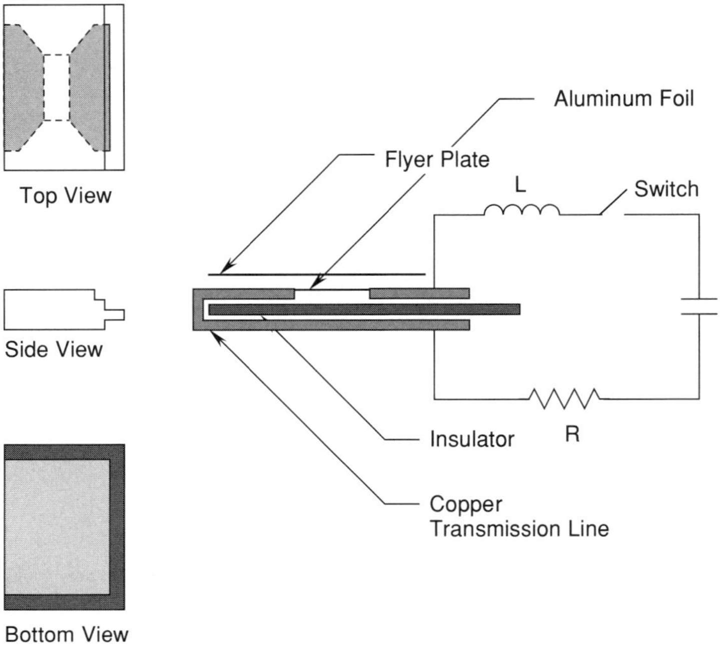

Figure 3.5. Schematic of the electric gun.

The energy density of the electro-explosive plasma may exceed the energy density achieved with chemical explosives by one or two orders of magnitude. The flyer velocity can thus be varied from ~100 m/s to 10 km/s or higher. Lateral dimensions of the accelerated film can be varied from ~1 mm to ~10 cm. Thus electric gun is an effective tool for studying the dynamic strength of materials for short duration loads.

3.1.4. Radiation Devices

High energy concentration in a sample can be achieved by focusing a powerful laser beam on a small area of the sample. Since the early 1960s, lasers have been used to generate shock waves in condensed matter by directing a short (~10^{-9} to 10^{-8} s) high-power laser pulse onto the open surface of a material. The surface layer is vaporized, and the resulting pressure in the ablation plasma produces a shock wave in the target. Only a small fraction of the energy is coupled into the target in this case. In another configuration, the ablative pressure is used to launch a thin (~1 to 10 μm) flyer plate. The maximum free-foil velocities can be modeled adequately by a rocket propulsion model, which predicts that the velocity is inversely proportional to the foil thickness.

Sheffield and Fisk [1984] transmitted laser pulses through a transparent substrate optically coupled to a launched foil, as shown in Figure 3.6(a). Their results show that water-confined foils attained peak velocities about three times higher than free foils due to tamping of the laser-induced plasma.

Figure 3.6(b) shows an advanced scheme developed by Paisley et al. [1992] to perform miniature plate impact experiments for material property studies. A plate to be launched, 0.2 to 20 μm thick, is placed on the output end of an optical fiber. Fiber diameters are typically 0.4 to 2 mm, and the flyer diameter is that of the optical fiber. A laser pulse is transmitted through the fiber and vaporizes a small amount of the flyer plate at the interface between the output end of the fiber and the flyer plate. The optical fiber provides a spatially uniform energy profile through the cross section. The laser-pulse temporal profile, the optical properties of materials, and the power density determine the optical coupling efficiency of the laser energy to the kinetic energy of the launched plate. This

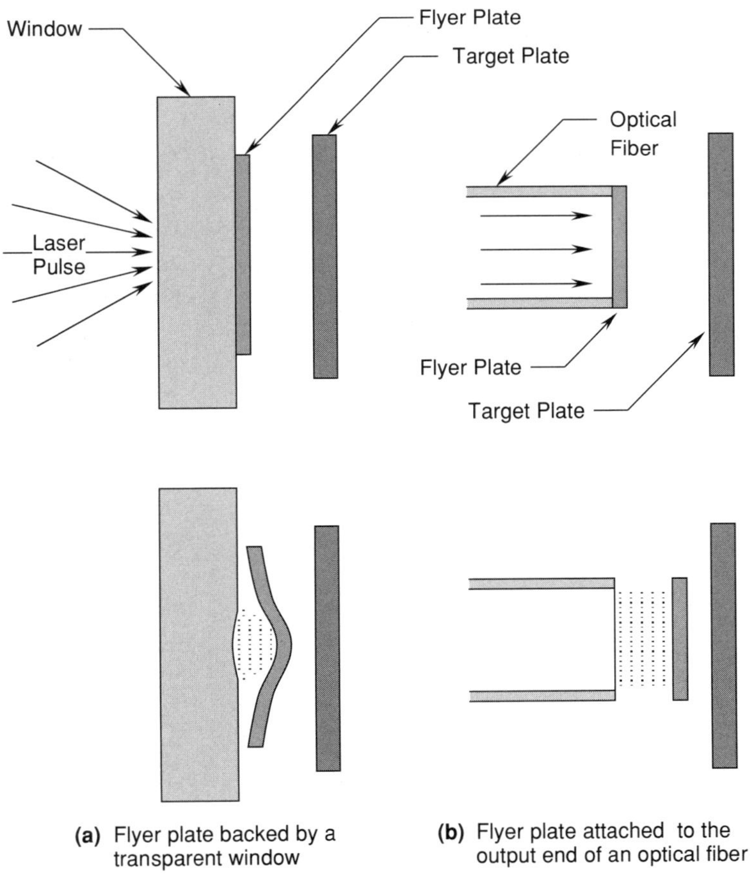

Figure 3.6. The acceleration of foils by laser-induced plasma.

miniature plate-launch technique gives any laboratory with an Nd:YAG laser and subnanosecond shock wave diagnostics the ability to study mechanical properties of materials for nanosecond load durations.

The powerful pulsed sources of electron and ion beams, developed for controlled thermonuclear fusion and other applied physics problems, are now being used as shock wave generators. Pulse accelerators with power from a gigawatt to several terrawatts or more are operated in laboratories around the world to drive intense particle beams. In shock wave applications, the particle beam extracted by the high voltage pulse from a diode is focused on a target spot with a diameter of a few millimeters. The high-energy particles are absorbed in a thin surface layer of the target, and the kinetic energy of the particles is transformed into heat (see Figure 3.7). The depth of the energy deposition zone depends on the energy and kind of particles and on the target properties. The rapid heating of the finite material layer produces a compression wave inside the target. If the beam energy is high enough to vaporize the target matter in the deposition zone, the ablation pressure from the particle beam source can be used to launch thin foil flyer plates by the same mechanisms as those discussed earlier in connection with laser beams.

3.2. Techniques Used to Measure Shock Parameters

Continuous measurements of the wave evolution inside a sample are needed to quantitatively characterize the mechanical properties of matter under shock wave loading conditions. Several techniques, using various physical principles, were developed during the early 1960s to provide direct time-resolved measurements of particle velocity or stress. This survey of these measurement techniques is not

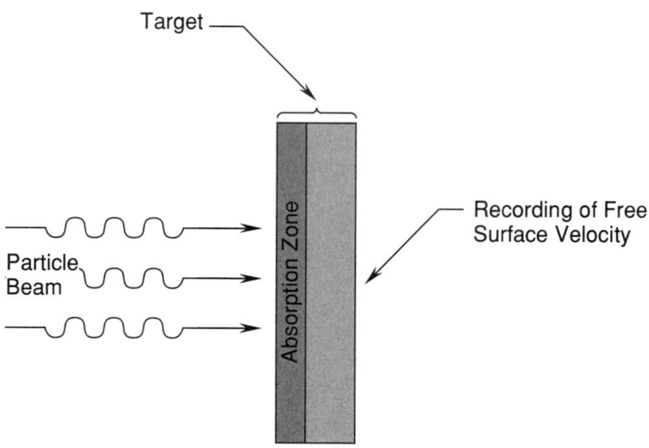

Figure 3.7. The acceleration of thin foils by electron or ion beams.

exhaustive, but it is comprehensive enough to provide a general view of the overall characteristics and the advantages and disadvantages of the methods currently used for shock wave diagnostics. In this survey, emphasis is placed on methods of measuring stress and particle velocity histories in shock-loaded specimens.

3.2.1. Methods for Measuring Particle Velocity Histories

Methods for measuring particle velocity histories in shock wave experiments are based on fundamental physical laws. For this reason, these measurements have the advantage of not relying on any sensor calibrations. Modern methods of continuous time-resolved measurements of particle velocity include the capacitor gauge, the electromagnetic gauge, and laser Doppler techniques.

3.2.1.1. Capacitor Gauge

The capacitor gauge is used to record the motion of electrically conducting surfaces. This method of measuring free-surface velocity is illustrated in Figure 3.8. The measuring capacitor C_m consists of two parallel surfaces: the sample surface and a flat electrode, with a distance x_0 between them. An external voltage is applied to the capacitor via the resistor R_i, whose resistance is low enough to ensure that the time constant R_iC_m is much less than the characteristic time of measurement. The guard ring ensures that the electric field is uniform over the region of the measuring electrode.

Motion of the free surface of the sample causes the capacitance of the gauge

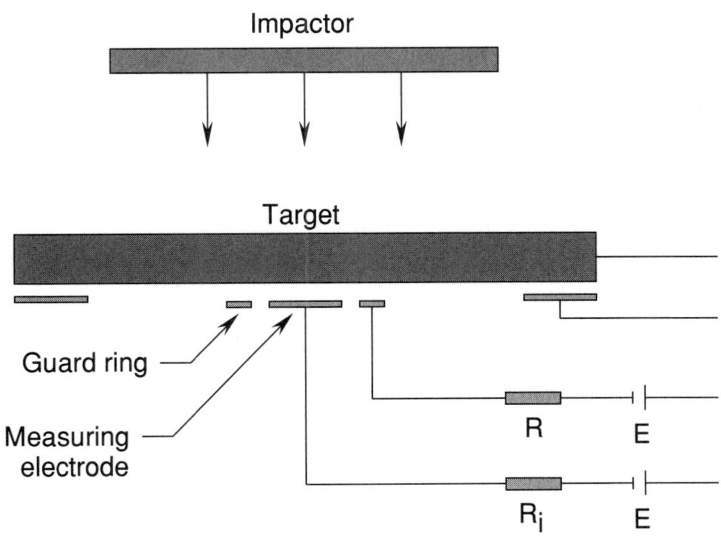

Figure 3.8. Capacitor gauge for measuring free-surface velocity histories. The signal is recorded as a current in the resistor R.

to vary, and an electric current begins to flow through the gauge circuit. This current is proportional to the rate of change of the capacitance, and ultimately, to the velocity of the free surface of the specimen, u_{fs}:

$$i(t) = U\frac{dC_m}{dt} = \frac{\varepsilon AU}{4\pi x^2(t)}\frac{dx}{dt} = \frac{\varepsilon AU}{4\pi x^2(t)}u_{fs}, \tag{3.1}$$

where U is the applied external voltage, ε is the dielectric constant, A is the area of the measuring electrode, and x, the distance between the plane electrodes at time t, is determined by integrating the current oscillogram $i(t)$. An example of a current oscillogram measured using a capacitor gauge and the resulting particle velocity history are shown in Figure 3.9.

The capacitor gauge method provides a noncontact measurement so that, in principle, its time resolution is limited only by the tilt of the shock wave with respect to the sample surface in the sensor-monitored region. Depending on the required resolution and the duration of the event, the gauge diameter and its initial distance from the sample surface, x_0, can be varied within 5 to 25 mm and 1 to 6 mm, respectively. The actual time resolution of a capacitor gauge with a 5-mm electrode diameter is ~10 to 20 ns.

With a supply voltage of 3 kV, the signal typically is 1 to 100 mV. Because of this relatively low output, the capacitor gauge is susceptible to electrical

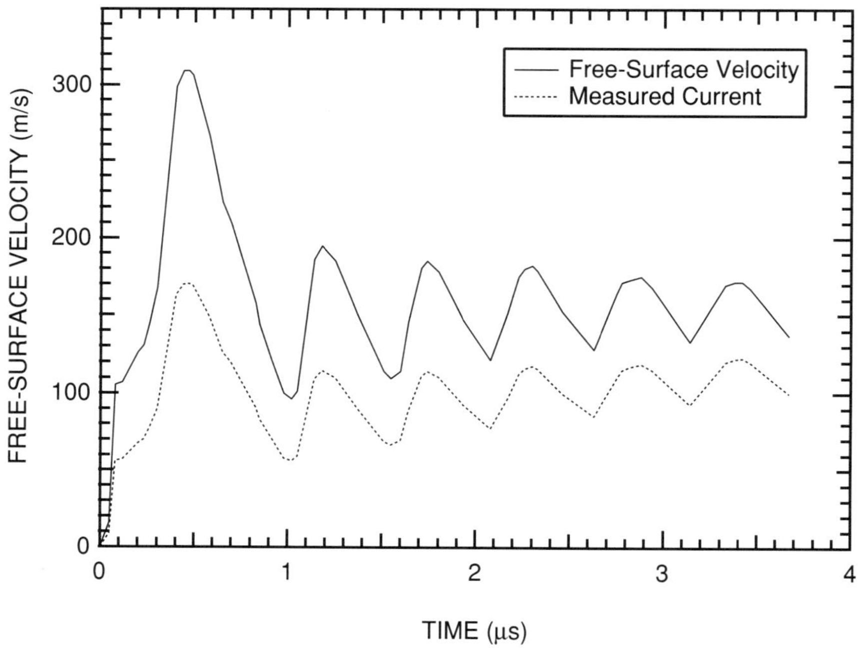

Figure 3.9. An example application of the capacitor gauge.

noise, which restricts its applications. Another limiting factor in the use of capacitor gauges is that the nonlinearity of the registration causes the accuracy of the measurement to decrease at a large shift of the sample surface in the capacitor gap.

3.2.1.2. Electromagnetic Gauge

The electromagnetic gauge is used to record particle velocity profiles in dielectric materials. The technique is based on Faraday's law of induction, which asserts that the motion of a conductor of length $\vec{l}$, when placed in a magnetic field of intensity $\vec{B}$, generates an EMF, E, that is proportional to the velocity of the conductor $\vec{u}$, as given by the relation

$$E = \vec{l} \cdot \left(\vec{u} \times \vec{B}\right). \tag{3.2}$$

The gate-shaped electromagnetic gauge made of thin aluminum or copper foil is embedded in the interior of the sample. The whole experimental assembly is placed in a constant uniform magnetic field, such that the sensitive element of the gauge is perpendicular to the magnetic lines and parallel to the shock wave front, as shown in Figure 3.10. Since the gauge is embedded within the specimen, the velocity of the sensing element of the gauge is equal to the particle velocity in the sample at the location of the gauge. This velocity is simply given by

$$u(t) = \frac{E(t)}{lB}. \tag{3.3}$$

3.2.1.3. Laser Velocimeter

The spatial resolution of the two velocity measurement techniques described above is limited by the size of the sensing element of the gauge. At best, this amounts to a few millimeters in the plane of the wave front. Since some tilt be-

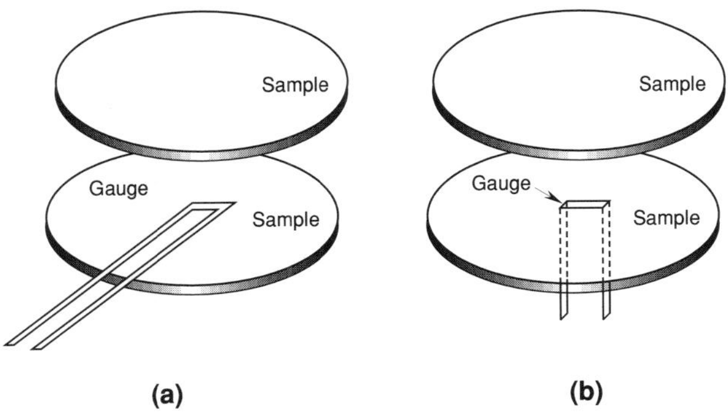

Figure 3.10. Typical electromagnetic particle velocity gauge configurations.

tween the shock front and the gauge plane almost always exists, the finite dimensions of the gauge sensor also limit the time resolution of measurements. Laser methods of recording the motion of free and contact surfaces offer much higher resolution in space and in time.

Laser velocimeters use Doppler-shifted light reflected from the target surface. Since the Doppler shift is very small for velocities of ~1000 m/s (the wavelength shift is ~10^{-2} Å), it must be recorded using two-beam or multiple-beam interferometry. The measurements thus become differential, and this provides a significant increase in their accuracy. Interferometers have become standard devices used by shock wave physicists to measure velocity histories. The laser techniques have high space resolution because the laser beam is focused down to a spot ~0.1 mm in diameter on the target surface.

Figure 3.11 illustrates the two-beam laser Doppler velocimeter VISAR (Barker and Hollenbach [1972]; Asay and Barker [1974]). In this system, the reflected beam is split equally into two beams to form the two legs of a wide-angle Michelson interferometer. In the interferometer, one leg is delayed in time by a period, Δt with respect to the other. The operation relies on the periodic variation in time (fringes) of the radiation intensity due to interference between two light beams of slightly different wavelengths. In the velocimeter, interference fringes result from the interaction between light beams reflected from a moving surface at different instants of time. If the velocity of the reflecting surface varies with time, the Doppler shift for the two beams will be different because of the time difference. The frequency of the fringes recorded by photodetectors is proportional to the acceleration of the reflecting surface and the delay time Δt.

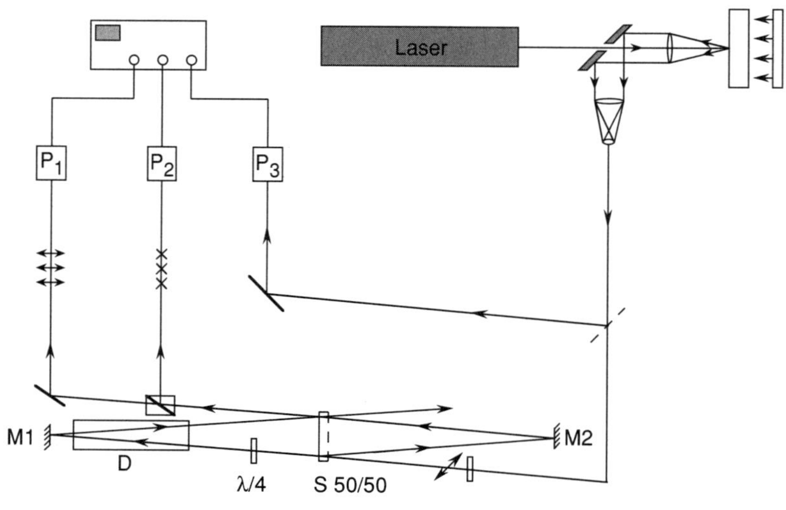

Figure 3.11. Schematic of a two-beam laser Doppler velocimeter (VISAR).

Glass etalons or a lens system can be used to introduce a temporal delay in the delay leg of the interferometer. The apparent optical path length of the two legs is maintained the same, whereas the geometric paths are different. In the case of a solid etalon, the geometrical difference is given by

$$\Delta l = l_d(1 - 1/n), \tag{3.4}$$

where l_d and n are the length and refractive index of the delay line. The delay time is then given by

$$\Delta t = \frac{2l_d}{c}(n - 1/n), \tag{3.5}$$

where c is the velocity of light under vacuum.

When the lens combination is used for delay, the delay time Δt is given by the following relation:

$$\Delta t = 2l_d / c. \tag{3.6}$$

Because of the apparent optical symmetry of the interferometer, a coincidence of wave fronts of superimposed beams is reached, and as a result, the technique can operate with both specular and diffuse reflecting surfaces.

When two beams are superimposed, fringes, $F(t)$, are produced in the interferometer and are related to the change in velocity of the reflecting surface, $u(t)$, by the following relation (Barker and Hollenbach, [1970]; Barker and Schuler, [1974]):

$$u(t - \Delta t/2) = \frac{\lambda}{2\Delta t}\frac{F(t)}{(1+\delta)(1+\Delta v/v)}, \tag{3.7}$$

where λ is the wavelength of the light used, and δ is a correction term that accounts for the dependence of the refractive index of the etalon material on wavelength, given by

$$\delta = \begin{cases} \dfrac{n}{n^2-1}\lambda\dfrac{dn}{d\lambda} & \text{for the etalon,} \\ 0 & \text{for the lens combination.} \end{cases} \tag{3.8}$$

The optical correction term $\Delta v/v$ is incorporated in Eq. (3.7) for measurement at an interface between the target and transparent window. The correction results from the change in refractive index of the window material with shock stress (Barker and Hollenbach [1970]).

In the VISAR, quadrature coding has been included to distinguish between acceleration and deceleration and to improve fringe resolution. This coding is accomplished by adding a quarter-wave retardation plate and a polarization beam splitter to provide a 90° out-of-phase shift between the two fringe signals. Two independent detectors are used to record the fringes in the two polarization components. Any change in the sign of acceleration will thus be recorded by at least one of the photodetectors as a "turn point" of oscillations in the interferogram. Fringes in the interferograms are related to the velocity of the reflecting surface by a simple sine expression. The instantaneous velocity therefore can be found from experimental interferograms, either discretely (by counting the number of fringes) or by measuring within individual fringes. The complete analysis of VISAR data for many time points is sophisticated and usually requires a computer. The accuracy of the velocity measurements with VISAR is ~1% to 2% or less; the time resolution can reach ~2 ns.

Limitations on the time resolution of VISAR measurements are associated mainly with a limited bandwidth of the oscilloscope, photodetectors, and other recording equipment. The optically recording velocity interferometer system (ORVIS) uses a high-speed electronic streak camera to record interference fringes, which improves the time resolution of the measurement to ~200 ps (Bloomquist and Sheffield [1983]). Compared with the VISAR, the ORVIS system is adjusted so that the recombining beams are at a small angle φ to each other, and the resulting pattern has a fringe separation $d = \lambda / sin\ \varphi$ When the reflecting surface is at rest, the phase difference of the two beams is constant and hence the fringes are at rest. As the surface moves, the Doppler shift causes the phase difference to change and thus the fringes to shift. A streak record of the fringe pattern that is changing position in time directly yields the time history of the surface velocity. The shift is proportional to the velocity so that shift value d corresponds, as before, to the velocity increment,

$$u_0 = \frac{\lambda}{2\Delta t(1+\delta)(1+\Delta v / v)}.$$

Compared with VISAR, ORVIS provides a higher temporal resolution, but a slightly less accurate velocity measurement.

Standard multibeam Fabry-Perot interferometers are also used as an element of the laser Doppler velocimeter (Johnson and Burgess [1968]; Durand et al. [1977]). A fringe pattern in this case is also recorded by the streak camera. As the frequency of the light from the moving target surface changes, the fringe diameter changes from its incident static value, D_1, to a new value, D_1'. The velocity of the moving surface is then calculated using the relation

$$u(t) = \frac{c\lambda}{4L}\left(\frac{D_1'^2 - D_1^2}{D_2^2 - D_1^2} + m\right), \tag{3.9}$$

where L is the distance between the plates of the Fabry-Perot interferometer, D_2 is the static diameter of the next fringe, and m is an integer. The precision of the system can be varied over the range of 0.1% to 2% and is determined by the spacing between the Fabry-Perot plates, the number of fringe jumps inserted, and the lens system. The time resolution of such velocimeters is determined by the photon fill time of the Fabry-Perot plates and is typically lower than that of VISAR and ORVIS.

In 1986, Gidon and Behar used a Fabry-Perot interferometer to measure velocity over an entire surface. In this modification, the velocity at many points for a single time is measured instead of the velocity history at a single point. Mathews et al. [1992] developed the experimental and analytical methods to make this full-field Fabry-Perot interferometer a practical diagnostic tool. Using a framing camera provides a time history of a velocity over a moving surface. A line-imaging VISAR was constructed by Hemsing et al. [1992] to measure many velocity histories simultaneously along the line on the target surface. Both versions (Mathews et al. [1992], and Hemsing et al. [1992]) use a dye amplifier that provides 600-W single-frequency power starting from a standard argon-ion laser. Baumung et al. [1994] modified the optical scheme of the VISAR/ORVIS velocimeter to allow for illumination of a line on the target surface and for measurement of the velocity history along this line with a standard argon ion laser and streak camera.

3.2.2. Methods for Measuring Stress Histories

Sensors used to measure stress histories in shock-loaded specimens include manganin, ytterbium, and carbon piezoresistance gauges; dielectric gauges; and quartz and PVDF ferroelectric gauges. Unlike particle velocity gauges, which do not require sensor calibration, all stress gauges require calibration so that their output can be related to stress in the specimen. The subject of stress sensor calibration for shock wave studies has received significant attention over the past three decades. Here, we limit our discussion to manganin gauges, the most widely used gauges for performing in-material stress measurements in planar shock wave studies.

The use of manganin gauges in uniaxial strain shock wave experiments is illustrated in Figure 3.12. The gauge consists of a 10- to 30-μm-thick grid arranged in a zigzagging pattern. The gauge is embedded in the specimen such that the active gauge element is normal to the direction of wave propagation. The gauge is electrically isolated from the specimen by a thin layer of Kapton, Mylar, Teflon, or mica. A constant electrical current is passed through the gauge. When a shock pulse passes through the gauge plane, the recorded voltage increases with pressure applied to the gauge. To increase the precision of the pressure measurements, a resistance bridge is used to eliminate the d.c. component of the signal, defined by the initial resistance of the sensor.

Manganin was first used by Bridgman [1911, 1940] as a pressure sensor under static loading conditions. Bridgman found that the resistivity of the manganin alloy increases with increasing pressure and is relatively insensitive to changes in ambient temperature. Fuller and Price [1964] and Bernstein and Keough [1964] used manganin gauges for pressure measurements in plane shock wave experiments. Since then, several investigators have contributed to the understanding and calibration of the piezoresistance response of manganin under

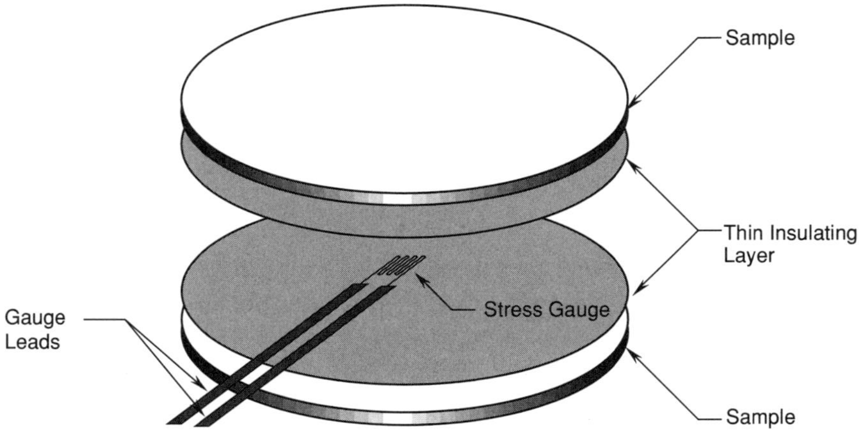

Figure 3.12. Typical manganin stress gauge configuration.

shock wave loading conditions, including Chen et al. [1984], DeCarli [1976], Gupta and Gupta [1987], Kanel et al. [1978], Lee [1973], and Postnov [1980].

The intended purpose of a manganin stress gauge is to measure the stress normal to the direction of shock wave propagation. In reality, piezoresistive materials like manganin are also sensitive to straining. Thus, if 1-D strain conditions are not maintained during the measurement, the sensor responds to both stress *and* strain along the gauge plane. In this case, it is important to be able to separate the stress component of the measured change of the gauge resistance from the strain component to obtain accurate stress measurements. For this reason, independent strain measurements are necessary when dimensional changes in the gauge are not negligible (e.g., Dremin et al. [1972]; Kanel and Molodets [1976]), such as might be expected in divergent flow situations. Simultaneous stress and strain measurements are routinely used in the flatpack series of armored stress gauges used for measuring stresses in large-scale dynamic experiments in geologic materials (e.g., Keough et al. [1993]).

Spall tests are carried out predominantly under 1-D strain conditions where dimensional changes in the plane of the gauge are negligible. Under these conditions, experimental procedures are normally used to calibrate the gauge, thus allowing the stress normal to the gauge to be determined uniquely and accurately. The procedure for calibrating the manganin gauge involves measuring the fractional change in resistance of the active gauge element, $\Delta R/R_0$, in a well-controlled uniaxial strain shock wave experiment and correlating the measured resistance change to the stress in the material at the location of the gauge, determined through some other means. Repeating this procedure at several stress levels provides the necessary data for characterizing the relationship between the fractional change in resistance of the gauge and the stress component normal to the gauge. For manganin, this relationship is shown in Figure 3.13.

Special measurements (Kanel et al. [1978]) have shown that, at pressures above 7 to 10 GPa, the change in the resistivity of manganin is reversible and does not depend on whether dynamic compression occurs by single or multiple shocks (i.e., quasi-isentropic behavior). Chen et al. [1984] also found the resis-

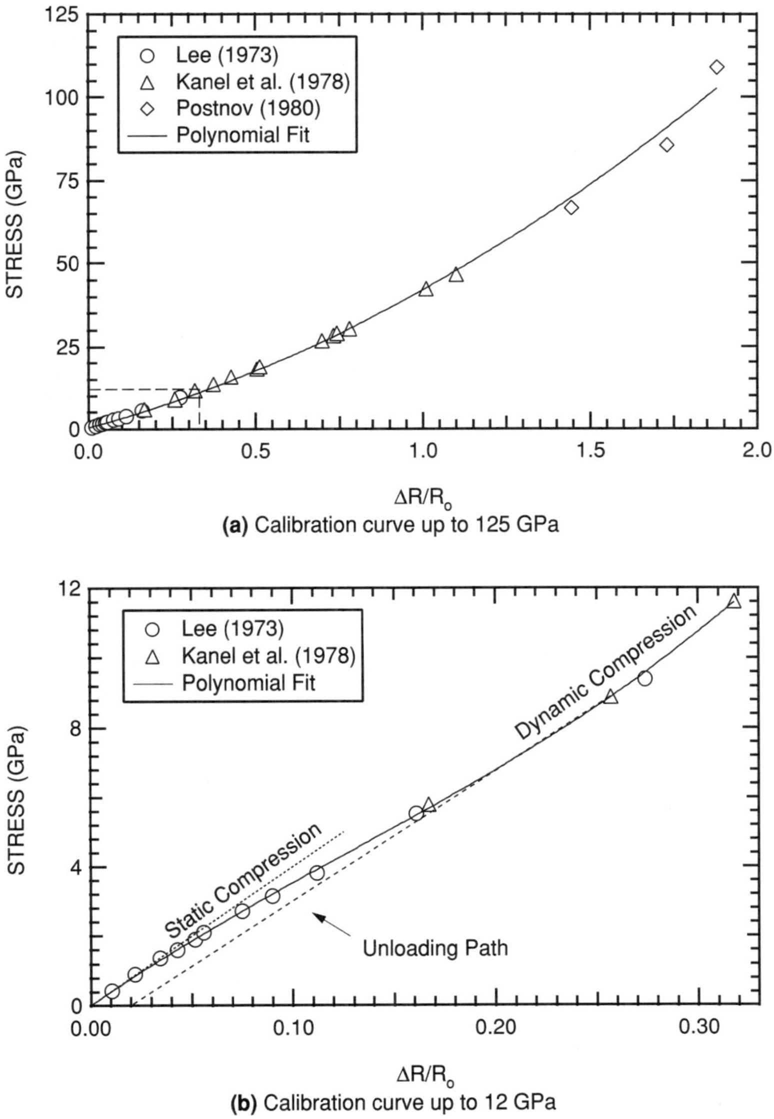

Figure 3.13. Calibration curves for the manganin stress gauge.

tivity of manganin to be history-independent. These findings are important because they imply that the resistivity of manganin can be uniquely related to stress at any instant during shock deformation, regardless of the history of deformation. Therefore, the manganin gauge can be used to measure stress in experiments involving multiple wave structures such as those encountered during plastic flow, phase transition, or fracture.

The release to zero pressure from a shock compressed state produces slight hysteresis in the gauge resistance. This irreversible component is attributed to strain hardening effects, which result in increasing concentration of defects in the gauge material during shock compression. For manganin, the residual resistance is small, usually 2% to 2.5%.[1] Below 7 GPa, the residual increment of the resistance is nearly proportional to the peak pressure and can easily be taken into account during interpretation of low-pressure measurements.

3.3. Spall Fracture Experimental Procedures

The focus of our investigation is spall fracture under one-dimensional uniaxial strain conditions. Under these conditions, the material undergoes relatively large volumetric strain and comparatively little shearing strain, a situation that is very different from the more usual one in structural analysis, where there may be large shear strains but little volumetric strain. Also under these conditions the stresses are not limited to the static yield strength, and stresses many times the yield strength are often reached.

In the shock wave tests the strain rates usually exceed 10^4 per second under tension and are even higher under the preceding compression. The durations of loading and therefore the time during which fracture occurs in samples with dimensions of about 1 centimeter (laboratory-scale tests) are about 1 μs, and often the tests are arranged so that the loading duration is only a few nanoseconds.

Both active and passive measurements can be used to assist in quantifying the damage that occurs during the spall process and in determining the fracture rate processes. Active measurements are dynamic time-dependent measurements of stress or particle velocity histories that occur at some points in the test specimen. Passive measurements include post-test examinations of the recovered sample. Measurements of both types provide valuable information that can be used to aid in the understanding of spall processes. However, neither active nor passive measurements provide a direct means for determining either the stress history at the spall plane or the rates at which the damage has occurred. Nevertheless, instrumented measurements of the wave profiles provide information

1. Ytterbium gauges, which are normally used for performing measurements at relatively low stress, exhibit a more complex hysteresis response than do manganin gauges. Gauge calibration in this case may require a more complex model than the one described here for manganin (e.g., Gupta [1983]).

that can be used to determine the tensile stress immediately before fracture whereas post-test microscopic examination of the sample provides a means of determining the fracture mechanisms and estimating the fracture kinetics.

In the remainder of this section, we first describe an experimental arrangement well suited for obtaining both active instrumented measurements and passive post-test examination, then typical results of the metallographic examination are given, and finally some stress gauge records are examined.

3.3.1. Experimental Techniques

Experimental investigations of the spall phenomena include measurements aimed at determining the fracture stress and the fracture mechanisms and kinetics. Instrumented measurements of the fracture stress at spalling are based on recording the waveform at the back free surface of the sample or at the interface between the sample and a soft buffer material. Theoretical background of the measurements will be discussed in the next chapter. Here, we note that all methods of measuring the dynamic tensile stress are indirect, because it is not possible to introduce a sensor into a sample without influencing its resistance to tensile stresses.

While different techniques have been developed to study the spall fracture phenomena over a wide range of shock load parameters, many spallation experiments are performed using a plate impact configuration (see Figure 3.4) with the flyer plate of the same material as the target. This symmetry helps ensure an accurate stress calculation. To achieve tensile pulses on the order of one microsecond, projectile plates are about 0.5 to 5 mm thick and target plates are about twice as thick.[2] The flyer and target thicknesses may be varied to provide a range of stress durations in the targets.

Two typical target designs used in spall experiments performed at SRI International are shown in Figure 3.14. In both designs, a plug with a taper (8 degrees is appropriate) is fitted into the rest of the target plate as shown. Following the initial compressive pulse during the impact, this plug separates from the rest of the target and is caught in a soft material to avoid further damage. The target plugs are then sectioned along a diameter, and the cut surface is polished and etched for metallographic examination. Stress measurements are made using manganin or ytterbium piezoresistive stress gauges mounted in a buffer material such as epoxy at the back surface of the target samples.

Even more widely used is a scheme of spall tests based on recording the free surface velocity histories. In the tests, various techniques are used to generate a shock loading pulse in the sample, including impact by a flyer plate, detonation

2. This ratio of the flyer plate thickness to the target thickness is appropriate for the post-test examination of the sample but, as will be shown in the next chapter, it is not optimal for the accurate determination of the fracture stress.

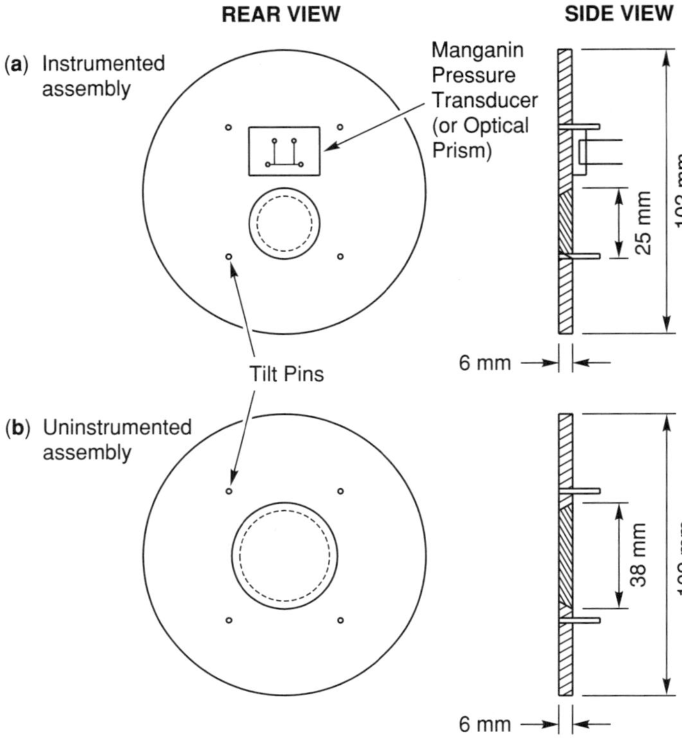

Figure 3.14. Target plate assembly showing tapered specimen (Barbee et al. [1970]).

of high explosive, and an intense radiation pulse. A wide range of load parameters (the load duration of from 1 nanosecond to 10 microseconds and the peak shock stresses from 10 MPa to 100 GPa) is covered by means of using different shock-wave generators. Capacitor gauges or laser Doppler velocimeters are used to monitor the rear free surface velocity of the sample as a function of time. The free surface velocity histories provide more precise determination of stresses at spalling because these measurements are independent on any calibrations and are not sensitive to the accuracy of the EOS of the sample and buffer materials. As before, the sample may be recovered for post-test examination.

3.3.2. Metallographic Observations of Shocked Specimens

Metallographic observations are made on the polished and etched axial cross sections of the target samples. A collection of photomicrographs of such cross sections is shown in Figure 3.15 for samples of 1145 aluminum tested with the same plate thicknesses—only the impact velocity was varied. The

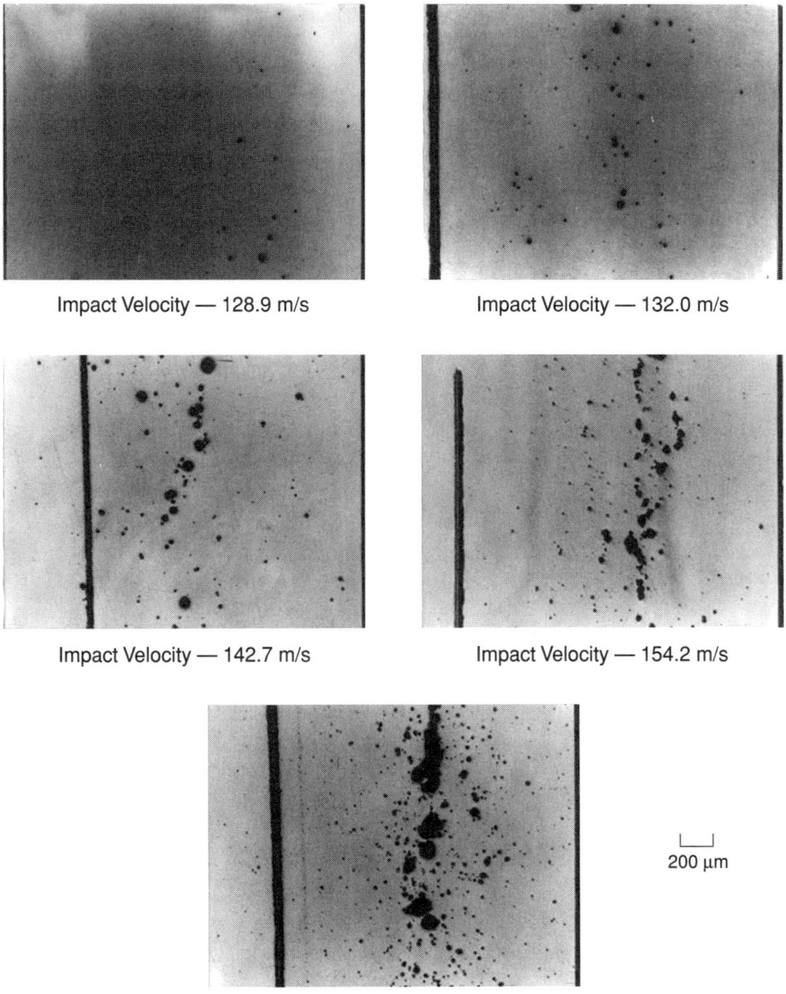

Figure 3.15. Damage observed in 1145 aluminum for a constant shot geometry (i.e., time at stress) for increasing impact velocities (i.e., stress) (Barbee et al. [1970]).

photomicrographs are arranged in order of increasing impact velocity (and therefore, increasing tensile stress) and also evidently in order of increasing damage. Damage is in the form of individually nucleated spherical voids that grow and coalesce to induce failure.

Four characteristics are apparent from these photomicrographs. First, the observed microdamage features (voids) have a circular cross section in the plane view. These cross sections are, in fact, sections through spherical voids. That the voids were spherical was verified by sectioning the samples normal to the direction of shock propagation. Circular cross sections were observed on these normal planes also. Second, the voids are distributed over some central region of the plate: there is no narrowly defined spall plane. Rather there is a narrow vertical region of maximum damage; then the numbers and sizes of voids decrease with distance away from this region on either side. From simulations of these experiments with a simple elastic-plastic model, we determined that the expected location of the spall plane (location for first occurrence of tensile stress) falls in the region of maximum damage. Third, there is a range of sizes of voids within regions with the same shock history. Fourth, at higher damage levels the interaction of the growing voids leads to the formation of large crack-like defects and finally to full separation (as seen in Figure 3.16).

Observations of full-spall samples have supplied further insights into the failure of these ductile materials. The opening of a crack resulting from void coalescence in an aluminum sample is shown in Figure 3.16(a). The impact was

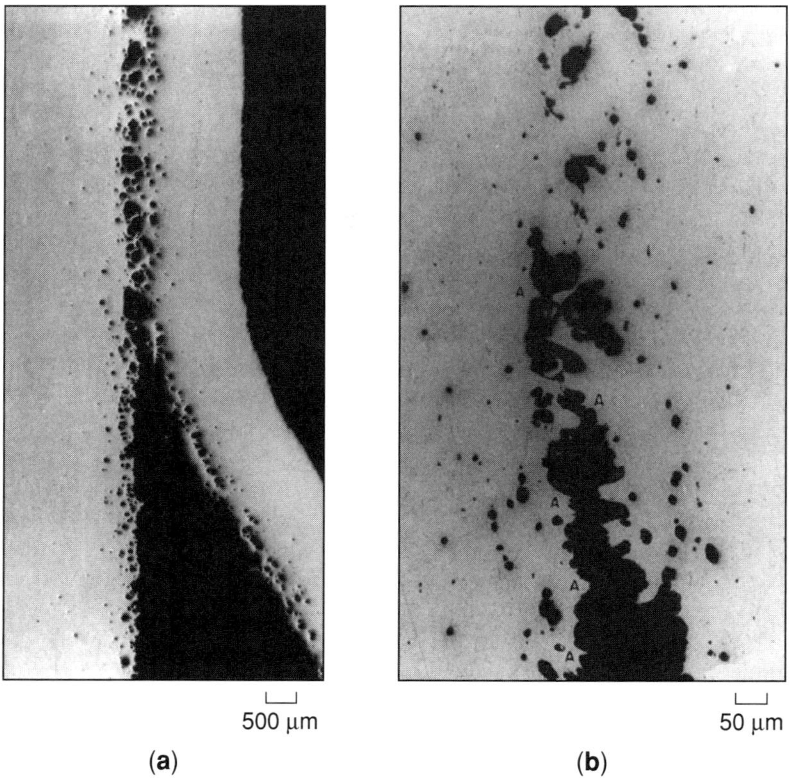

Figure 3.16. Ductile cracks. (a) Ductile crack propagation by void coalescence. (b) Tip of ductile crack shown in (a) at higher magnification (Barbee et al. [1970]).

from the left at 251 m/s and an epoxy buffer plate was on the right. The tip domain (Figure 3.16(b)) shows the region of the material corresponding to full failure or approaching full failure. This photomicrograph was made near a cylindrical edge of the target disk (down and out of the photo) where the flow is not uniaxial for the entire period of damage. Near this edge of the target plate, full separation has occurred, whereas the center is heavily damaged but not separated. The macroscopic appearance is that of a running crack with a very rough surface, but in fact the damage occurred mostly simultaneously along the damage plane and the running crack represents only the completion of separation for a portion of the distance. Near points labeled A in Figure 3.16(b), separation has occurred by elongation of voids and necking of the regions separating them. The necked regions have failed by fully ductile, knife-edge fracture under essentially uniaxial strain conditions. The macroscopic crack initiated near the edge of the plate and was most likely influenced by edge effects, which perturb the uniaxial strain state that existed in the plate during the earlier stages of damage development.

Figures 3.17, 3.18, and 3.19 show cross sections of disk-shaped target plates each of which having been impacted by another flyer plate. The sections were made along a diameter of the disk, and most photomicrographs were made of

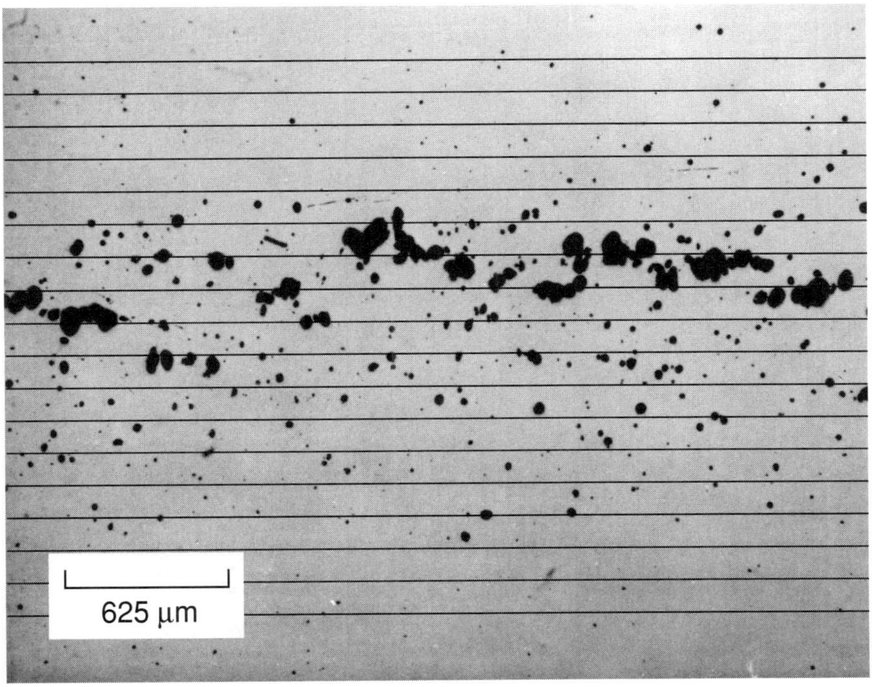

Figure 3.17. Impingement of voids and cracks in impact-loaded specimens of 1145 aluminum.

regions near the center of the disk where the material was under a state of uniaxial strain during most of the period of damage. The targets in Figure 3.17 were impacted from the top with enough velocity to produce an intermediate level of damage.

The target plate in Figure 3.17 was 6.313-mm-thick commercially pure aluminum, impacted by a 2.27-mm-thick flyer plate traveling at 145 m/s, and it was heated to 400°C before the impact. The lines drawn on the photo were used for a quantitative analysis of the statistical distribution of voids. The appearance of the fracture is that of nearly spherical voids in regions of low damage. The odd-shaped voids in the heavier damage areas in a central plane in the target were probably formed by coalescence of many smaller voids. About 40% of the plate thickness is shown in the figure, so we see that the fracture is spread over the central 20% of the plate.

Figure 3.18 shows a cross section through a target plate of Armco iron. In this case, microcracks cut through the iron grains (the grain boundaries are not

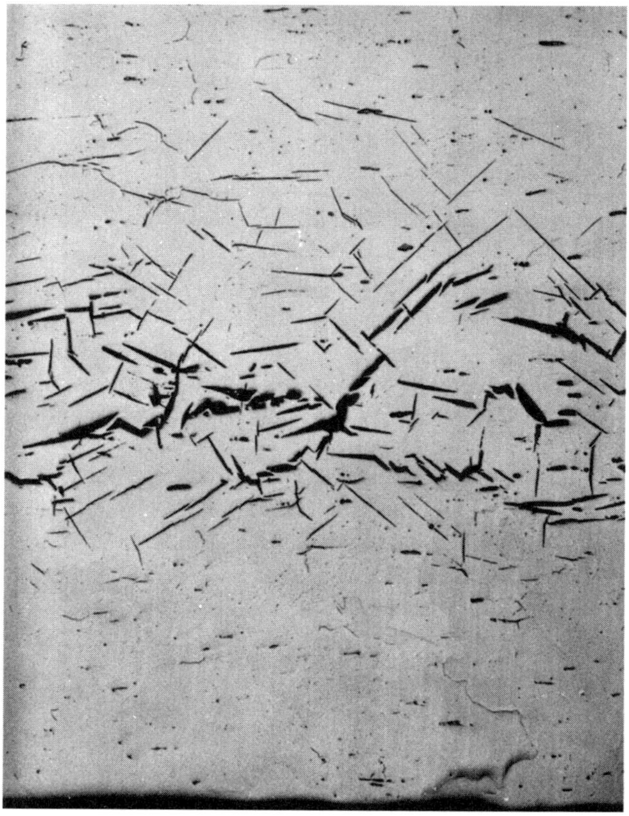

Figure 3.18. Impingement of voids and cracks in impact-loaded specimens of Armco iron (Curran et al. [1987]).

visible in the photo). The zigzag-nature of the cracks occurred because the cracks follow preferred directions in each grain and then change direction as they cross grain boundaries. A great many microcracks have formed, and they have interacted strongly so that they are almost to the point of producing separate fragments. No fracture plane has formed, but there is a central region along which there is a maximum of damage; this region would have become the fracture plane if the impact velocity had been higher.

Another Armco iron target with somewhat higher damage is shown in Figure 3.19. The 6.35-mm-thick target was struck at 149 m/s by a 2.39-mm-thick flyer plate. Again, we see a broad region of damage nearly a millimeter wide, and coalescence of the microcracks has proceeded to the point of roughly defining a fracture plane. This figure is typical in illustrating that no actual spall "plane" occurs. Rather a surface of separation wanders through a field of partially frag-

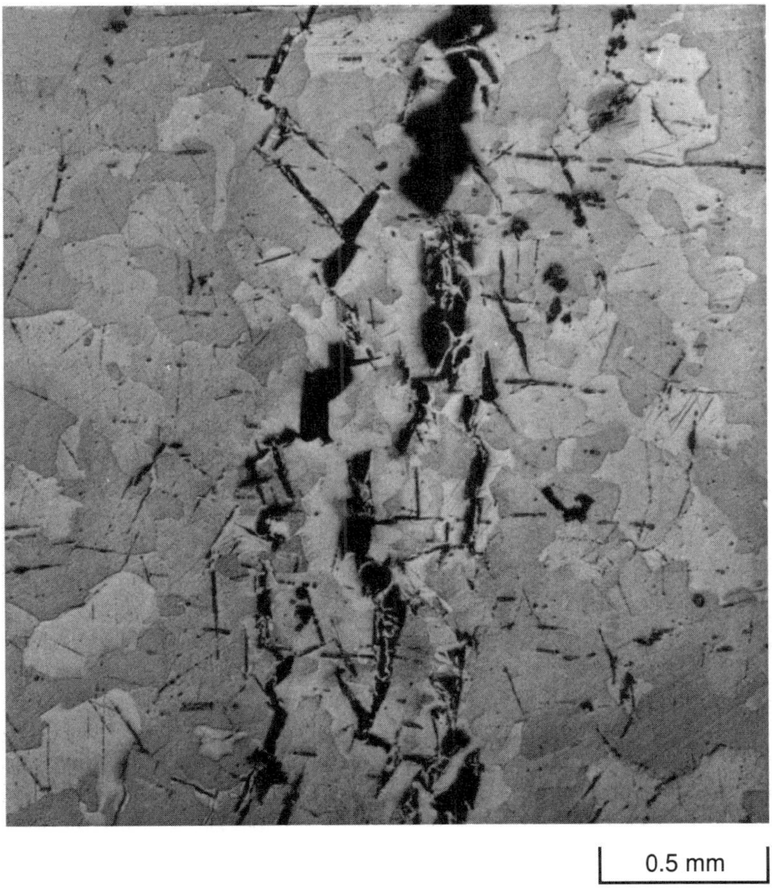

Figure 3.19. Coalesced microcracks in Armco iron (Curran et al. 1987]).

mented material. An etchant has been used on the target so we can faintly see some of the grain boundaries.

The information contained in the photomicrographs of recovered fracture samples discussed above can be used to locate and quantify the cracks, voids, shear bands, and other evidence of fracture. For these observations, we section, polish, and etch the sample, and then we examine the cross section under a microscope. Figure 3.17 above showed a cross section of an aluminum target that was polished to reveal the microvoids. The lines were drawn on the photograph to separate the sample into zones for counting the voids.

A void count made on another aluminum sample that was radiated with a laser is shown in Figure 3.20. Following counting in five successive zones, the numbers were summed to provide a cumulative size distribution. We see that the

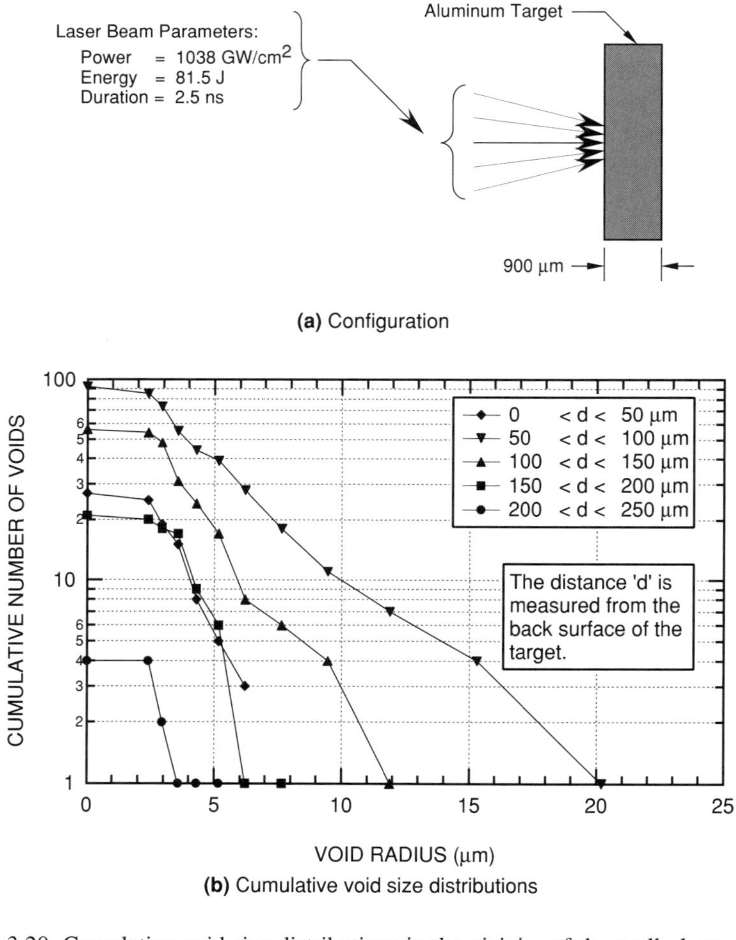

Figure 3.20. Cumulative void size distributions in the vicinity of the spall plane near the rear surface of a soft aluminum specimen subjected to laser deposition.

plane of maximum damage is within the second zone, 50 to 100 μm from the rear of the target, where both the size and number of voids reached their maximum values. Lesser amounts of damage occurred on either side of this plane.

Post-test observations such as those described above provide only circumstantial knowledge about the processes that led to the observed damage. Questions like 'did the damage occur during the passage of the first tensile wave or a later one?' and 'what were the damage nucleation and growth rates?' cannot be answered solely on the basis of post-test observations. However, these nucleation and growth processes and the rates at which they occur are the fundamental processes that must be understood to develop methodologies for predicting the occurrence of damage under other circumstances.

3.3.3. Stress Measurements Behind Fracturing Samples

The purpose of making time-dependent measurements in spall tests is to determine the stress state at the plane of fracture. We wish to make the measurement in such a manner that the presence of the gauge does not disturb the stress waves, so that the measured stress history closely approximates the stress history that would have occurred at the gauge plane in the absence of the gauge.

In-material stress or particle velocity measurements are possible only outside the region of spall fracture. The placement of a gauge within the damage region would perturb both the stress waves and damage evolution to the extent that the measured variable could not be reliably related to its counterpart in the free field. For this reason and because in-material stress gauges cannot be used to measure tensile stress states such as those on or near the spall plane, gauges are usually emplaced outside the fracture region. The gauge usually records the information carried by stress waves from the region of fracture to the gauge plane. The character of these waves is determined by the damage at the site of maximum damage and by the lesser damage in adjacent regions through which the waves must propagate. The computed stress histories in Figure 3.21 illustrate the nature of the stress records that may be obtained at the interface with a low impedance buffer plate. Such records are used (as described later in this section) to deduce the character of the stress histories at the spall plane.

The interpretation of gauge histories is much more complex during fracture than during elastic behavior. For example, the distance between any two original points in the material (and hence the distance that waves must travel) changes during fracture in an inelastic manner. Also the wave velocities (C), densities (ρ), and hence the impedance (ρC) are altered by the presence of damage. As noted in Chapter 2, the impedance governs the amplitude of stresses and material velocities following wave interactions. In elastic wave propagation, the waves may travel completely through the sample and interact only with the boundaries. But after damage, the waves travel through layers of slightly different material because different levels of damage cause different sound velocities and densities and hence impedance (see Figure 2.5 in Chapter 2 for a sample of

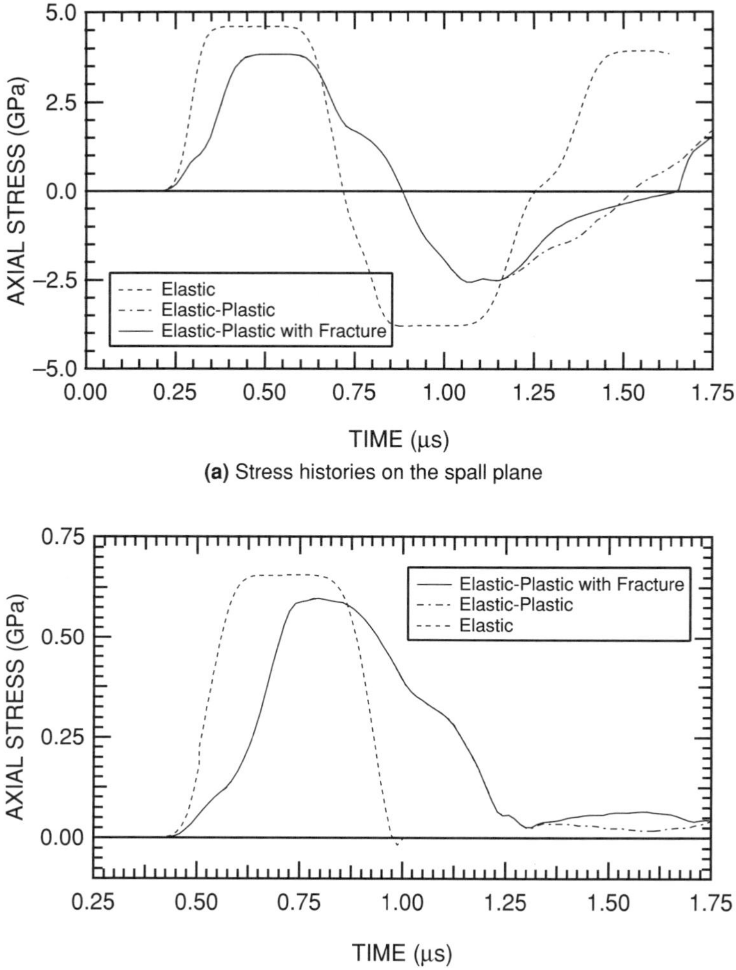

(a) Stress histories on the spall plane

(b) Stress histories on the gauge plane

Figure 3.21. Comparison of stress histories computed at the fracture plane and at a nearby gauge plane assuming elastic behavior, elastic-plastic behavior, and elastic-plastic behavior with fracture.

these wave interactions arising because of the changing damage and properties within the target). Hence, the region of fracture in the sample is a field of continuous wave interaction. No simple analytical treatment is possible for describing, in closed form, these waves and the resulting stress or velocity histories.

In the study of material behavior, it is usual to perform experiments in which a large region of the sample is at essentially the same state of stress and strain—as in a quasi-static tension test. But spall fracture is an unstable process during which there are no large regions of nearly uniform behavior. Figure 3.22 shows the simulated variation of void volume fraction with distance through an aluminum target plate. The amplitude of damage increases with impact velocity and becomes concentrated near a single plane at higher impact velocities and higher damage. This figure illustrates the phenomenon of "localization" in which regions that are partially damaged tend to attract further damage and thereby protect their less-damaged neighbors.

Experimental manganin gauge records from impacts in 1145 aluminum are presented in Figure 3.23. The important parts of these records are the recompression pulses (indicated with arrows) observed before the full release of the initial compression wave is attained.

The occurrence of the fracture signal can be explained qualitatively with the aid of the distance-time diagram in Figure 3.24(a). Along the distance (x) coordinate are the three plates: flyer, target, and epoxy buffer. The lines show the propagation of waves within these plates as a function of position and time. Initially, compression waves leave point 0 traveling left into the flyer and right into

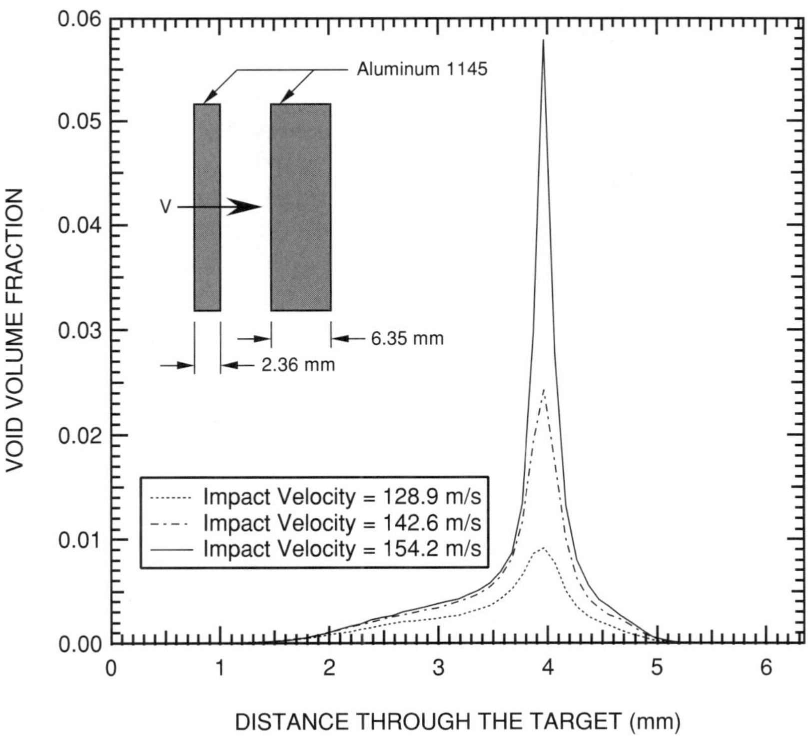

Figure 3.22. Simulated damage distributions in an aluminum 1145 target plate subjected to symmetric impact at varying impact velocities.

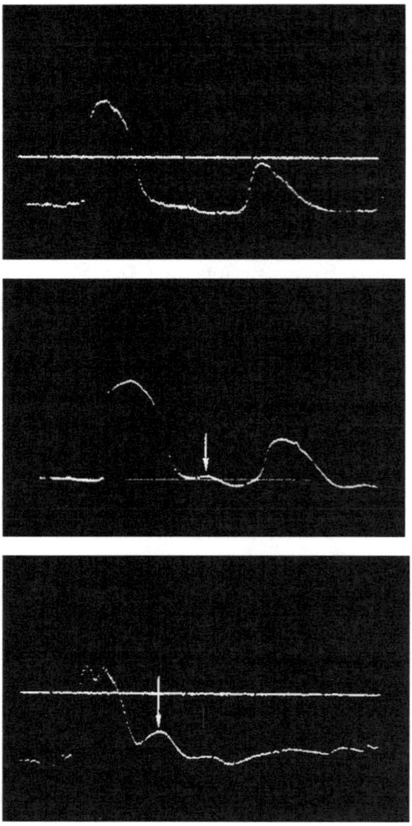

(**a**) Manganin gauge records for typical low-damage or no damage response.

(**b**) Manganin gauge record for fracture signal (arrow) arising from damage at the spall plane.

(**c**) Manganin gauge records showing a strong fracture signal (arrow), indicating high damage. No ringing is observed, indicating fracture has suppressed later loading.

Figure 3.23. Fracture signals observed in manganin gauge records (Barbee et al. [1970]).

the target, creating a compressive stress level C. At point 6 at the free surface of the flyer, the wave (0-6) is fully reflected as a rarefaction, which moves along the path 6-2-3.

The compression wave (0-1) propagating into the target is partially transmitted into the epoxy (the epoxy has lower impedance than the aluminum) and partially reflected as a rarefaction wave back into the target. The transmitted wave has amplitude C_1, which is less than C. The rarefaction from point 1 proceeds along 1-2, meeting the rarefaction from point 6, and causing tensile stresses in the region above the broken line 7-2-3. Point 2 then defines the plane of first tension and therefore the x-location of the spall plane.

When the rarefaction from point 6 reaches the epoxy at point 3, a partial rarefaction is transmitted back into the target, reducing the stress to a level C_2. For this diagram, we presume that the developing damage first becomes important at point 4 (at the same position as 2, but at some later time). The tensile stress reduction caused by the developing damage produces recompression

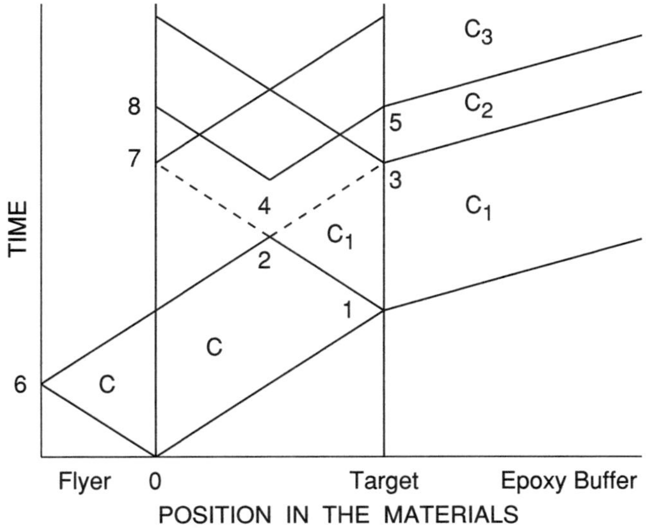

(**a**) Distance–time diagram for waves resulting from impacting an aluminum flyer onto an aluminum target backed by a thick epoxy plate.

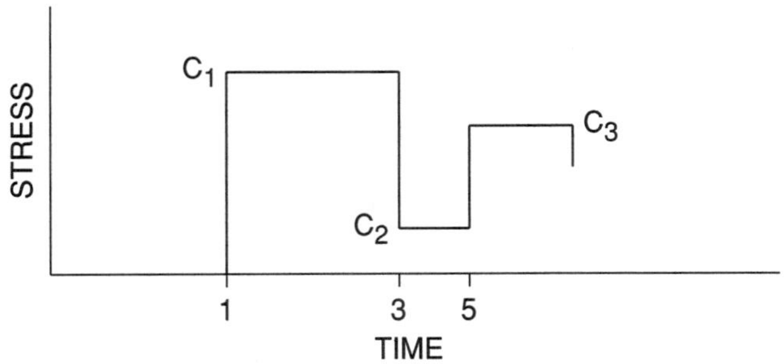

(**b**) Stress history at the interface of the aluminum target and epoxy buffer.

Figure 3.24. Effect of damage on stress history at the spall plane (Barbee et al. [1970]).

waves propagating along the lines 4-8 and 4-5. At point 5, the recompression wave is partially transmitted into the epoxy, bringing the stress up to C_3.

The history of these stresses at the target-epoxy interface is shown in Figure 3.24(b). The amplitude of the fracture signal (the difference between stress levels C_3 and C_2) depends on the strength of the recompression wave and hence on the amount of damage. The time between the main wave and the fracture signal is governed by the rate at which the damage develops in the target. Hence, the fracture signal can be used to guide us in understanding several aspects of the damage process. Measured fracture signals are important for verifying models developed to account for the effect of void or crack growth on the applied stress history because these signals are directly related to the rates at which the fracture processes are occurring.

3.3.4. Two- and Three- Dimensional Dynamic Experiments

The dynamic experiments discussed above are all one-dimensional, that is, under conditions of uniaxial strain. Of course, at later times unloading waves from the specimen edges arrive to convert subsequent behavior to three-dimensional response. Traditionally, the measurements have been considered finished at that time, and the experiments have been designed to cause minimum evolution of damage after the arrival of the edge waves, thereby making the assumption of uniaxial strain viable when interpreting the results of post-test microstructural examinations.

On the other hand, the rapid growth of computing power and three-dimensional "hydrocodes" allows two and three-dimensional dynamic experiments to be designed and interpreted. Of course, two- and three-dimensional hydrocodes have long been used to interpret experiments involving armor penetration, the dynamic response of composite materials, and similar applications. But in most of these cases, the analyses focused on specific engineering applications, and not on measuring basic material failure properties.

An example of a two-dimensional experiment to produce and quantify evolving material damage was reported by Nemat-Nasser et al. [1998]. The experiment was designed to dynamically collapse a tube of single crystal copper, and in this case the inner portion of the tube showed very large hoop strains compared to axial strains, and the assumption of two-dimensional strain is approximately fulfilled. The authors showed extremely good agreement with the localized experimental deformation and damage when performing computational simulations with the two-dimensional hydrocode DYNA2D.

In general, the availability of two- and three-dimensional hydrocodes is making it possible to perform a variety of basic material property experiments which have complicated geometries chosen to drive the material through complex stress–strain paths. We expect much more activity in this area in the future.

3.3.5. Quasi-Static Experiments

The experiments discussed above use stress waves to apply the loads, and we define such experiments to be "dynamic." Quasi-static experiments are also very useful for generating increasing levels of microstructural damage, and for aiding in construction of mesomechanical failure models. Three such experiments (tensile bars, poker chip, and creep tests) are mentioned in Chapter 1, and the various combinations of stress-strain-strain rate states that can be achieved in each test are shown in Figure 1.1. Another example of evolving microscopic damage—in the form of voids—is shown in a sectioned tensile bar in Figure 1.4.

We emphasize the term "quasi-static" instead of "static." As we discussed in Chapter 1, fracture is a dynamic process with characteristic times ranging from nanoseconds to years, and the experiments identified in Figure 1.1 were selected to produce evolving damage over as large a span in characteristic times as is practical.

Whereas the poker chip test is basically a uniaxial strain experiment, the tensile bar and creep tests usually have symmetry in two space dimensions. For these experiments, as in the two-dimensional dynamic experiments discussed above, numerical computations with two-dimensional finite element codes are used in conjunction with mesomechanical models to iteratively relate the model parameters to levels of observed damage (see Curran et al. [1987]).

4 Interpretation of Experimental Pullback Spall Signals

4.1. Estimating Spall Stresses from Experimental Data

All methods of measuring the dynamic tensile stress in materials during spalling are indirect, because it is impossible to introduce a sensor into a sample without influencing its resistance to tensile stress. Each of the indirect methods uses a different approach to determine the dynamic tensile stress; sometimes large discrepancies are obtained using different methods. This chapter describes the various methods for determining the spall threshold and explores the advantages and limitations of each method. Emphasis is placed on fracture stress determination using measurements of the free-surface velocity histories, or the particle velocity histories at the interface between the sample and a soft barrier. These methods are emphasized because they are believed to be the best available, and because they were used in obtaining most of the spall data discussed herein.

4.1.1. Wave Interactions During Spall

Measurements of spall strength are based on analysis of the one-dimensional motion of compressible, continuous, condensed media following the reflection of a shock pulse from the surface of the body. Figure 4.1 illustrates the dynamics of wave interactions during the reflection of a triangular shock pulse from the free-surface of a body under uniaxial strain conditions. Figure 4.1(a) shows a triangular loading stress history. This stress history is typical of what might be expected when an explosive is detonated in contact with the front surface of a specimen. This stress history comprises an initial peak, associated with the detonation pressure (also known as the Chapman-Jouguet, or CJ pressure), followed by a gradual decay caused by the release wave emanating from the expanding explosive gases. This release wave is commonly known as the "Taylor wave" after G.I. Taylor [1950], who first developed a theory to describe the flow that connects the CJ point to the final state in the expanding gaseous detonation products.

In the $x - t$ diagram of Figure 4.1(b), the shock front trajectory is described by the line oo'. A family of C_+ characteristics represents the unloading wave overtaking the shock front. One of these characteristics is shown to intersect the shock trajectory at point o'. Note that, in Figure 4.1(b), each unloading wavelet is represented by a single characteristic line that is a member of a fan of characteristics. When the shock front reaches the free-surface at o', the free-surface velocity undergoes a jump from zero up to $u_0 = 2u_s$, where u_s is the particle velocity behind the shock front. The unloading wave behind the shock front causes decay in the free-surface velocity as shown in Figure 4.1(d). Thus, initially (i.e., before fracture), the particle velocity history has the same triangular shape as the stress history. At the free-surface of the specimen, the shock wave is reflected as a centered unloading wave that travels backward toward the interior of the sample. This rarefaction wave is represented by the left-going C_- characteristic

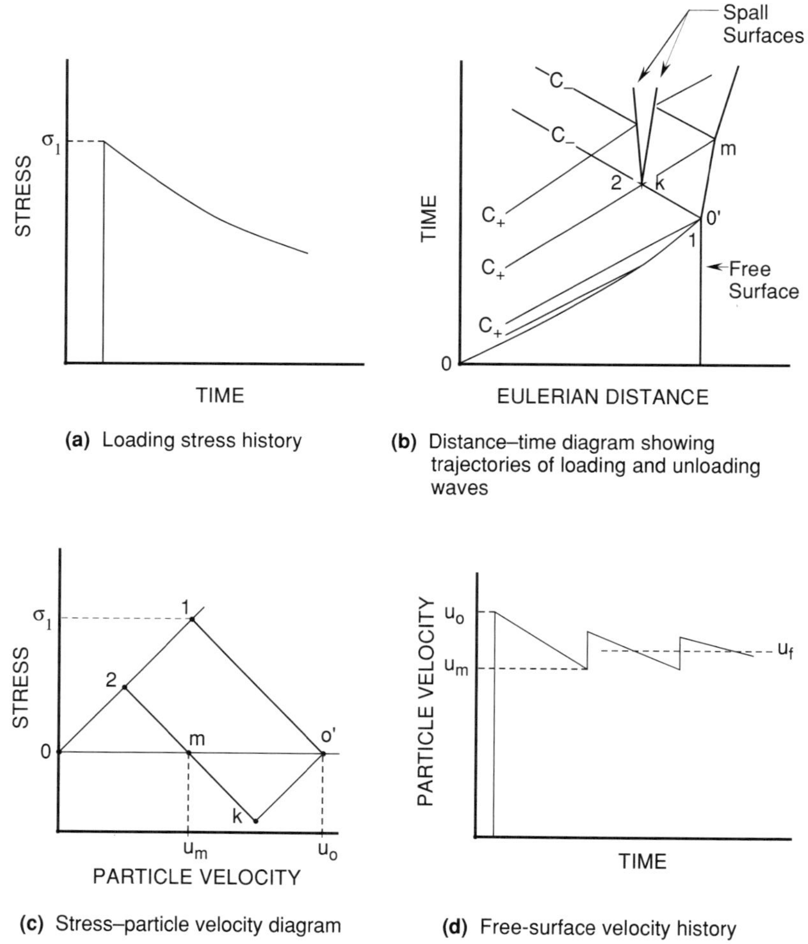

Figure 4.1. Wave interaction diagrams for the reflection of a triangular shock pulse (like that caused by an explosion) from a free-surface and formation of a fracture plane.

emanating from point o'. Here, too, each C_- characteristic is a member of a rarefaction fan centered at o'. The state of the particles must satisfy conditions on both the C_+ and C_- characteristics and is determined in the $\sigma - u$ diagram of Figure 4.1(c) by the intersection of Riemann trajectories describing states of matter along the C_+ and C_- characteristics that pass through the particle at any given time.

The maximum tensile stress is reached at each particle position as it is traversed by the terminal C_- characteristic of the centered rarefaction wave emanating from point o' on the back surface of the specimen. Therefore, the peak tensile stress at the spall plane just before fracture corresponds to the intersection of trajectories $o'k$ and $2k$ in the $\sigma - u$ plane of Figure 4.1(c). Line $o'k$ describes the change of state along the tail C_- characteristic of the centered rarefaction wave; line $2k$ represents the trajectory of the change of state along the last of the C_+ characteristics of the incident wave crossing the spall plane before the fracture.

Fracture of the material at the spall plane causes the tensile stress to decrease rapidly to zero. As a result, compression waves form in the material adjacent to the spall plane. These waves propagate to the left and right away from the spall zone. At the rear surface of the specimen, where the free-surface velocity, u_{fs}, is measured, this compression wave is manifested as a jump in velocity from u_m to some higher velocity. This signature of spall damage is referred to as the spall pulse. When the material spalls, a stress wave becomes trapped between the spall plane and the rear surface of the specimen. Later reverberations of this stress wave lead to the damped oscillations observed in the particle velocity record in Figure 4.1(d). The period of these oscillations can be used to determine the thickness of the spalled layer.

4.1.2. Determining Fracturing Stress Using Measurements of the Free-Surface Velocity History

The available methods for measuring the free-surface velocity of the sample during a spall experiment were discussed earlier, in Chapter 3. Here, we focus attention on the analysis techniques used to determine the dynamic fracture stress from the measurement of free-surface velocity histories.

The peak free-surface velocity, u_0, and the free-surface velocity just before the arrival of the spall pulse, u_m, are determined directly from the free-surface velocity profile. The tensile stress value just before spalling, σ^*, is then determined by the intersection of Riemann trajectories passing through the points ($\sigma = 0$, $u = u_0$) for C_- and ($\sigma = 0$, $u = u_m$) for C_+. Within the acoustic approach, the following linear approximation (e.g., Novikov et al. [1966])

$$\sigma^* = \frac{1}{2}\rho_0 c_0 \Delta u_{fs} \tag{4.1}$$

is used, where $\Delta u_{fs} = u_0 - u_m$ is the so-called "velocity pullback." Dynamic measurement of spall strength is based on measurement of the velocity pullback.

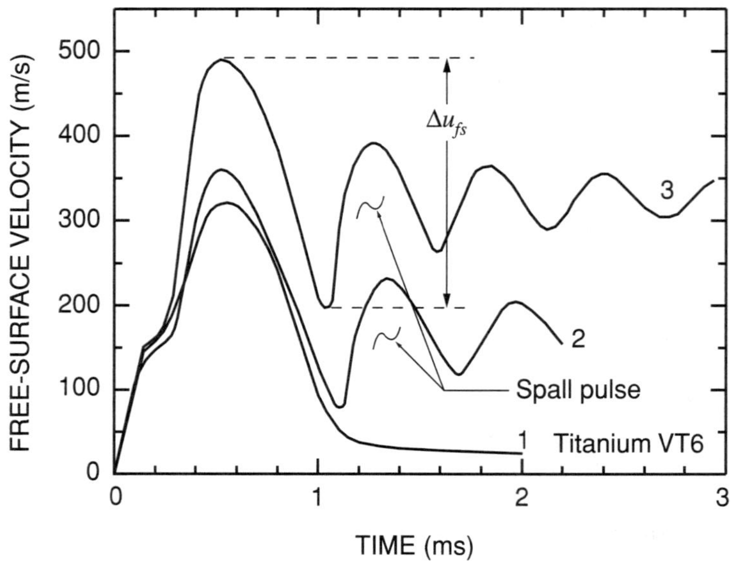

Figure 4.2. Free-surface velocity profiles for the VT6 titanium alloy at three shock wave intensities.

Equation 4.1 provides a reasonable estimate of the fracture stress so long as the density and wave speed in the material are close to their original values, ρ_0 and c_0. With increasing tensile stress, this condition becomes less accurate and knowledge of the compressibility of the material under tension becomes necessary. Generally, however, we have no knowledge of the pressure-volume curve in the tensile region except by extrapolation from compressed states under dynamic loading. If one extrapolates the material isentrope into the negative pressure (i.e., tension) region, one finds that the correction for nonlinear compressibility is on the order of 10% or less for most practical cases.

Figure 4.2 shows examples of measurements of free-surface velocity profiles for the titanium alloy VT6 (Kanel and Petrova [1981]).[1] The stress in the material at the lowest impact velocity (450 ± 20 m/s) is below the spall threshold. As a result, the free-surface velocity profile practically replicates the form of the compression pulse in the sample. The elastic-plastic compression wave and the subsequent unloading wave are both recorded. The small hysteresis in the free-surface velocity profile is due to irreversible plastic deformation during the loading-unloading cycle.

1. VT6 is a designation for a Russian titanium alloy with the following comp osition: Ti-6%Al-4%V.

The magnitude of the tensile stress that develops in the body after reflection of the compression pulse at the free-surface increases with increasing shock intensity. When the peak tensile stress reaches the spall threshold, damage begins to accumulate. The tensile stress in the damage accumulation zone decreases as the fracture develops. As a result, a compressive disturbance called a "spall pulse" appears on the free-surface velocity profile. Thereafter wave reverberation is observed within the scab between the free-surface and the damage zone. The period of velocity oscillation is a measure of the thickness of the scab. As it was discussed above, the velocity pullback, Δu_{fs}, is a measure of incipient fracture strength of the material. Experiments on many materials show that increasing the shock amplitude does not influence the magnitude of Δu_{fs} as indicated in Figure 4.2 for titanium VT6.

For most solids, the free-surface velocity profiles exhibit elastic-plastic properties. There are several wave propagation velocities in a body undergoing elastic-plastic deformation. In the case of one-dimensional flow, weak perturbations propagate with the longitudinal sound velocity, c_l, if the deformation is elastic, and with the bulk sound velocity, $c_b < c_l$, in the plastic deformation region. This, as discussed below, has a direct bearing on the procedure of calculating the tensile stress at the spall plane.

Stepanov [1976] and Romanchenko and Stepanov [1980] first discussed the influence of elastic-plastic response on the velocity pullback value. Figure 4.3 shows an axial stress-particle velocity ($\sigma_x - u$) diagram for wave interactions when a plane square compressive pulse is reflected off the free-surface of an elastic-plastic body. The process of uniaxial compression is elastic until the stress reaches the Hugoniot elastic limit (HEL). The slope of the initial elastic part of the Hugoniot below the HEL in these coordinates is $d\sigma/du = \rho c_l$. The slope in the plastic deformation region above the HEL is equal to ρc_b. Unloading from a shock-compressed state is initially elastic in both the incident and reflected waves. The elastic part of the unloading wave has a stress magnitude equal to twice the HEL. Thereafter, all expansion processes occur in the plastic region. Thus, if the shock wave amplitude exceeds twice the HEL, tension produced under interaction of rarefaction waves takes place in the plastic deformation region.

To account for the effects of elastic-plastic deformation in determining the fracturing stress, Stepanov [1976] analyzed spallation during a symmetric impact of two elastic-plastic plates. Stepanov [1976] argued that the spall pulse front should propagate with the longitudinal elastic wave speed, c_l, because the spall pulse is a compression wave that propagates through the extended material whereas the incident rarefaction plastic wave ahead of it propagates with the bulk sound velocity. Thus, to determine the fracturing stress value, Stepanov [1976] identified the state corresponding to the intersection of the terminal C_- characteristic of the reflected rarefaction wave and the C_- characteristic corresponding to the minimum velocity in the measured free-surface velocity history (point m′ in Figure 4.3). Since the spall pulse front is considered as an elastic

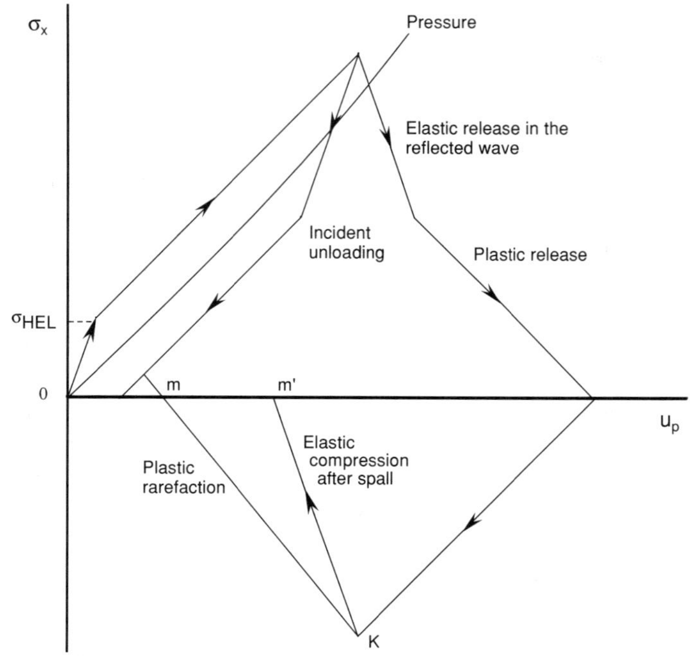

Figure 4.3. Stress–particle velocity diagram for wave interactions during the reflection of a triangular shock pulse from the free-surface of an elastic-plastic body. The diagram is drawn for the case in which the shock strength exceeds twice the HEL.

wave, Stepanov concluded that in the $\sigma - u$ plane the C_+ characteristic corresponds to the trajectory Km′ which passes through the point $\sigma = 0$, $u = u_m$ and has a slope $d\sigma/du = -\rho c_l$. On the other hand, the trajectory corresponding to states along the terminal C_- characteristic has a slope ρc_b. Within this approach, the relationship to calculate the fracturing stress is

$$\sigma_c^* = \rho_0 c_l \Delta u_{fs} \frac{1}{1 + c_l/c_b}. \tag{4.2}$$

The relationship expressed in Eq. (4.2) does not account for the effect of spall plate thickness on the wave profile. However, since the wave profile evolves as a result of interactions among stress waves that traverse the specimen at different velocities (as described above), it is reasonable to expect the spall plate thickness to play a role in shaping the wave profile. Indeed, in the case of a trapezoidal stress pulse generated by plate impact, the distortion of the wave profile increases with increasing plate thickness. Without a further correction to account for the effect of spall plate thickness, the spall strength calculated using Eq. (4.2) would decrease with increasing plate thickness even if the real strength of the material is constant. To account for the effect of spall plate thickness, of the material is constant. To account for the effect of spall plate thickness, Romanchenko and Stepanov [1980] introduced the following phenomenological relation which includes an additional correction to Eq. (4.2):

$$\sigma_c^* = \rho_0 c_l \Delta u_{fs} \frac{1}{1 + c_l/c_b} + \Delta\sigma, \tag{4.3}$$

where

$$\Delta\sigma = \frac{1}{2}\frac{d\sigma}{dt}\bigg|_{C_-} h_s \cdot \left(\frac{1}{c_b} - \frac{1}{c_l}\right),$$

$d\sigma/dt$ is the stress gradient along the tail C_- characteristic of the reflected rarefaction wave, and h_s is the spall plate thickness. Romanchenko and Stepanov [1980] did not discuss the theoretical basis that led to the choice of this functional form for $\Delta\sigma$.

Intuitively, it appears that the spall stress correction should depend on the steepness of the spall pulse front. Kanel et al. [1984] suggested the expression

$$\sigma^* = \frac{1}{2}\rho_o c_b \left(\Delta u_{fs} + \delta\right) \tag{4.4}$$

for calculating the tensile stress just before spalling, based on consideration of the measured free-surface velocity profile as a superposition of the incident wave and the spall pulse (Figure 4.4). In this equation, δ is a correction for the profile distortion due to the elastic-plastic properties of the material. Taking into account the velocity gradients ahead of the spall pulse, $\dot{u}_1$, and the gradient in its front, $\dot{u}_2$, the correction magnitude is estimated as

$$\delta = \left(\frac{h_s}{c_b} - \frac{h_s}{c_l}\right)\frac{|\dot{u}_1 \dot{u}_2|}{|\dot{u}_1| + \dot{u}_2}. \tag{4.5}$$

The Riemann trajectories in the tension region just before the initiation of spall have slopes defined by the bulk compressibility and one has to use the bulk sound velocity in Eq. (4.4) when the corrected Δu_{fs} value is used.

The three different approaches described above, which were introduced to account for the effect of the elastic-plastic response of spalling materials on the measured free-surface velocity profile, result in different values of the fracturing stress. To shed more light on the range of validity of these approaches, a more detailed analysis of the elastic-plastic wave dynamics was undertaken. Figure 4.5 shows results from a series of computer simulations of wave interactions in an elastic-plastic body. In these simulations, a triangular stress pulse was initially applied to the left boundary of a plate. After 1 μs, the linearly decreasing boundary velocity was replaced by an increasing velocity thus causing recompression in the body. The steepness of the recompression wave was varied by varying the rate of increase of the boundary velocity. The results show that the propagation velocity of the recompression wave front decreases from the longi-

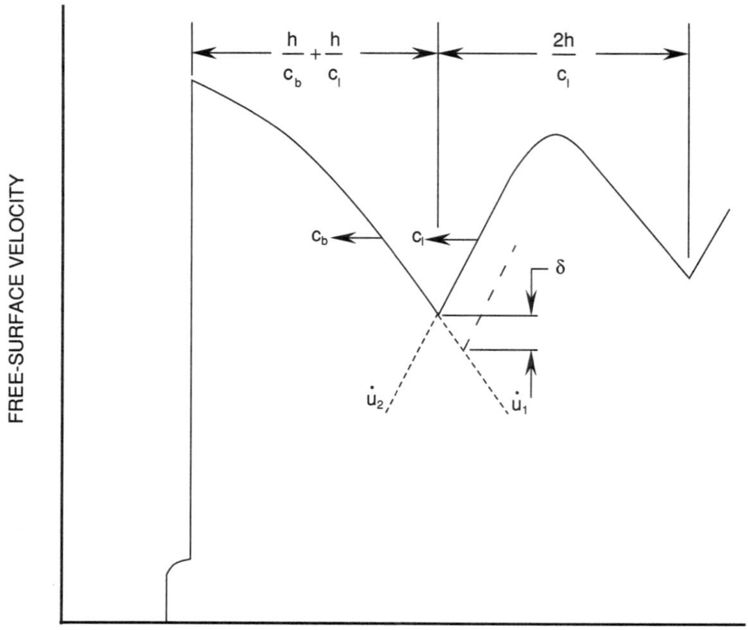

Figure 4.4. Determination of the corrected expression for calculating the tensile stress before spalling.

steepness of the shock front decreases from a shock discontinuity to zero. This result casts some doubts on the validity of Eqs. (4.2) and (4.3) and requires further discussion.

To find the front propagation speed of the second pulse, let us consider the time–distance diagram presented in Figure 4.6. The diagram shows C_+ characteristics describing a portion of the incident plastic rarefaction wave followed by an elastic recompression wave. The recompression wave front propagates with a velocity, c_F, which is in the range $c_l \geq c_F \geq c_b$. For the flow on the right side of the recompression front trajectory, F, the stress gradient along the trajectory is

$$\left.\frac{d\sigma}{dt}\right|_F = \left.\frac{\partial\sigma}{\partial t}\right|_+ + c_F \left.\frac{d\sigma}{dh}\right|_+,$$

which, using the momentum conservation equation (Eq. (2.2)), can be rewritten as

$$\left.\frac{d\sigma}{dt}\right|_F = \dot{\sigma}_+ - c_F \rho_0 \dot{u}_+,$$

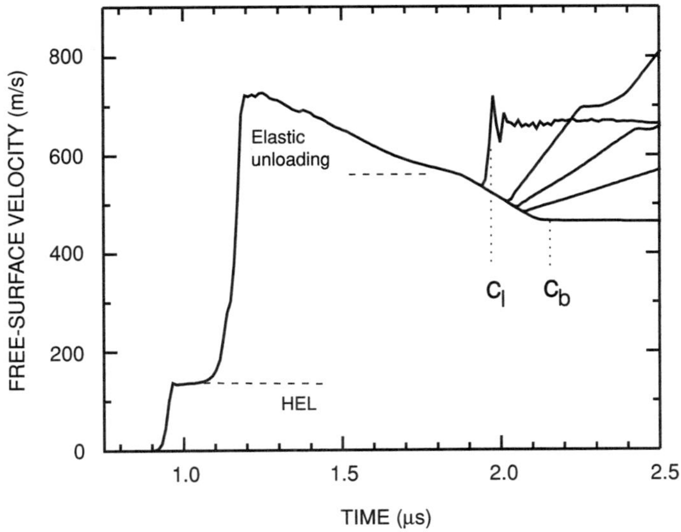

Figure 4.5. Results from a series of computer simulations of triangular compression-unloading pulse followed by a recompression wave in an elastic-plastic body. In all cases, the recompression pulse was introduced 1 μs after the beginning of the first shock compression.

where the index "+" denotes parameters on the right side of the trajectory F, and the superposed dot denotes time derivatives of the general form

$$\dot{f} = \frac{\partial f}{\partial t}.$$

Analogously, the stress gradient on the left side of trajectory F can be expressed as

$$\left.\frac{d\sigma}{dt}\right|_F = \dot{\sigma}_- - c_F \rho_0 \dot{u}_-.$$

The particle velocity gradient along the trajectory is

$$\left.\frac{du}{dt}\right|_F = \left.\frac{\partial u}{\partial t}\right|_+ + c_F \left.\frac{du}{dh}\right|_+,$$

which, using the mass conservation equation (Eq. (2.1)), can be rewritten as

$$\left.\frac{du}{dt}\right|_F = \dot{u}_+ + c_F \rho_0 \dot{V}_+.$$

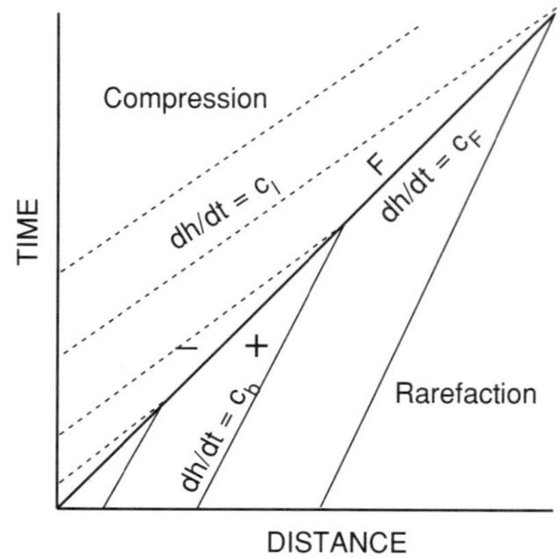

Figure 4.6. C+ characteristics of a plastic rarefaction wave followed by an elastic compression. F is a trajectory of the compression wave front.

Since the material just in front of the second wave deforms plastically, $\dot{V} = -\dot{\sigma}/\rho^2 c_b^2$, so the last equation can be presented as

$$\left.\frac{du}{dt}\right|_F = \dot{u}_+ - \frac{c_F \dot{\sigma}_+}{\rho_0 c_b^2}.$$

On the left side of trajectory F, the plastic rarefaction is replaced by the elastic compression, so the particle velocity gradient on the left side of F is

$$\left.\frac{du}{dt}\right|_F = \dot{u}_- - \frac{c_F \dot{\sigma}_-}{\rho_0 c_l^2}.$$

Since the stress and particle velocity gradients along the trajectory F should be the same on both sides of F, we come to a set of two equations for the front propagation velocity, c_F, as a function of the rate of change of the stress and the particle velocity just ahead of the recompression front and behind it

$$c_F = \frac{\dot{\sigma}_+ - \dot{\sigma}_-}{\rho_0 (\dot{u}_+ - \dot{u}_-)},$$

$$c_F = \frac{\dot{u}_+ - \dot{u}_-}{\dfrac{\dot{\sigma}_+}{\rho_0 c_b^2} - \dfrac{\dot{\sigma}_-}{\rho_0 c_l^2}}.$$

Eliminating the particle velocity gradients from the two equations above leads to the following relationship between the recompression front velocity and the stress gradients

$$c_F = c_b c_l \sqrt{\frac{\dot{\sigma}_+ - \dot{\sigma}_-}{\dot{\sigma}_+ c_l^2 - \dot{\sigma}_- c_{b-}^2}}. \tag{4.6}$$

Equation (4.6) is valid for $\dot{\sigma}_+$ and $\dot{\sigma}_-$ of different signs. According to this solution, the recompression wave front propagates with the elastic longitudinal wave velocity, c_l, when the stress gradient ahead of it, $\dot{\sigma}_+$, is zero or when the recompression wave is a shock discontinuity ($\dot{\sigma}_- \to \infty$). In the trivial case when gradual unloading is followed by a stress plateau the unloading tail propagates with the bulk sound speed.

When a spall fracture occurs as a result of reflection of a triangular compression pulse, superposition of the incident and the reflected rarefaction waves supports constant tensile stress values in each cross-section of the body until the spall pulse arrives. As a result, the spall pulse front propagates with the velocity $c_F = c_l$ independently of its steepness. In this case all characteristics of the spall pulse reach the sample free-surface and the spall strength may be calculated equivalently using both Stepanov's approach (i.e., Eq. (4.2)) or the relationship expressed by Eq. (4.4) based on consideration of states just ahead of the spall pulse with the correction

$$\delta = \left(\frac{h_s}{c_b} - \frac{h_s}{c_l} \right) \cdot |\dot{u}_1|. \tag{4.7}$$

A plate impact creates a shock compression pulse of approximately trapezoidal shape. In this case the stresses in cross-sections ahead of the spall pulse front are not constant, so $c_F < c_l$. This means that some of the characteristics of the spall pulse disappear with time as is shown in Figure 4.6, and Eq. (4.2) is not valid even for a shock-like spall pulse. In this case, only Eq. (4.4) with the correction may be used. The correction δ should be calculated using Eq. (4.6) and the propagation velocity of the spall pulse front given by c_F instead of c_l. The value of c_F is determined using $\dot{\sigma}_+ \approx \rho c_b \dot{u}_1 / 2$ and $\dot{\sigma}_- \approx \rho c_l \dot{u}_2 / 2$ near the spall plane and $\dot{\sigma}_+ \approx 0$ near the rear free-surface of the plate. Here $\dot{u}_2$ is the free-surface velocity gradient in the spall pulse front.

The computation of σ^* and δ in all four approaches is based on the assumptions that fracture is instantaneous and that damage only occurs on the spall plane. These assumptions are most likely to be satisfactory for very brittle material in which the imposed tensile stress is well above the threshold for spalling. For more ductile materials, the evolution of fracture affects the shape of the spall pulse, and measurement of the shape of the pulse provides information about the spall kinetics. There, a more complicated analysis is required, as we will describe in later chapters.

In fact, it is impossible to be sure that the extrapolation of $u_{fs}(t)$ used for determining the correction value is accurate. Actually, an undistorted profile can contain a smooth minimum near the measured value of u_m as well as a sharp minimum near its corrected counterpart with the same effect on the real free-surface velocity profile. This means that experiments must be carefully designed to provide the smallest correction possible. To this end, it can be shown that δ is on the order of 10% of Δu_{fs} in the case of an incident load pulse of triangular form and may reach 50% of Δu_{fs}, or more, in the case of a trapezoidal pulse.

Many spall experiments are performed using the flyer plate impact configuration. In this case, the stress and particle velocity profiles remain approximately trapezoidal until the distance of propagation reaches about 5 impactor thicknesses. Thereafter, the unloading wave overtakes the shock front and the load pulse begins to take on a triangular shape.[2] Thus, the optimal ratio of sample thickness to impactor thickness must exceed ~5 to have the smallest correction and smallest possible error of the spall strength value.

The influence of the ratio of sample thickness to impactor thickness on the velocity pullback, and correspondingly, on the correctness of determining the spall strength may be illustrated by the results of experiments with KhVG steel shown in Figures A.23 to A.25 of the Appendix. These experiments were essentially identical except for the noted variations in sample thickness. Whereas the strain rate is highest for the thinnest sample (Figure A.25), the pullback in the free-surface velocity profile is smallest. This behavior is not the result of abnormal spall strength dependence on strain rate. It is just the consequence of the increased distortion—associated with the decreased ratio of sample thickness to impactor thickness—of the spall signal as it propagates from the spall plane to the free-surface of the sample. Increasing the thickness ratio causes the spall strength determined using Eq. (4.2) and that determined using Eqs. (4.4) to (4.6) to approach the same value.

The validity of determining the fracturing stress using the free-surface velocity profiles as outlined above has been confirmed by many experiments with shock load intensities close to the spall strength magnitude. A spall pulse was not recorded in these experiments when the peak stress was below the spall strength, but it appeared in the velocity profiles with increasing stress above the spall strength. Then, the velocity pullback value remained practically unchanged as the shock intensity was increased. Careful comparison of free-surface velocity measurements with results of microscopic examination of impacted and recovered samples shows that the fracture nucleation threshold correlates well with a stress magnitude equal to σ^*.

2. Strictly speaking, the shock propagation distance required for the pulse to evolve from trapezoidal to triangular shape is a function of stress amplitude and material properties. the rule of thumb of having the target thickness exceed 5 times the impactor thickness is based on the experience of some of the present authors with many different materials subjected to a wide range of stress amplitudes.

4.1.3. Experiments with a Soft Buffer Plate Behind the Target

Tensile stresses can be created inside a body not only by the intersection of Taylor waves and reflections of a compression pulse from the free-surface of the body, but when the stress pulse reflects from an interface with a material of lower shock impedance. In this case, measurements of the spall strength can be based on the stress or particle velocity profiles at the interface between the sample under investigation and the lower impedance buffer plate. The experimental configuration and the stress-particle velocity diagram for this case are shown in Figure 4.7. The choice of a test configuration for use in a particular spall experiment depends on the available techniques of recording the wave profile. For

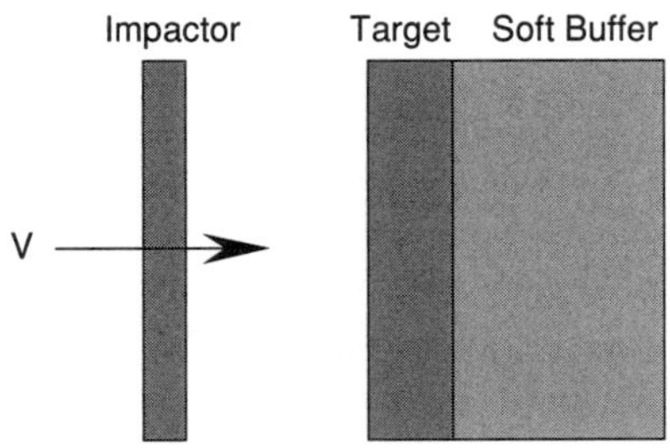

(a) Configuration of an experiment with a soft buffer behind the target

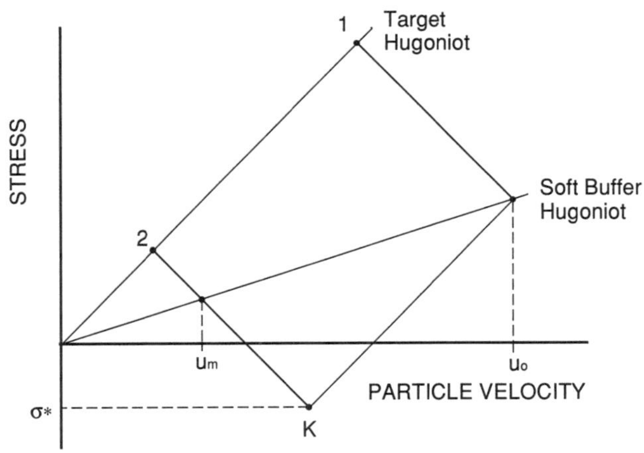

(b) Stress–particle velocity diagram for the reflection of a compression stress pulse from the interface with a softer material

Figure 4.7. Configuration and stress–particle velocity diagram for a spall experiment with a soft buffer behind the target.

example, the experimental configuration shown in Figure 4.7 is normally used when stress history measurements are desired. In this case, PMMA (Polymethyl methacrylate, often sold as "plexiglas") is most often used as the buffer material because PMMA has an impedance that closely matches the impedance of the epoxy in the stress gauge package. This minimizes the rise time of the gauge and optimizes the resolution of the stress history measurement.

Replacement of the free-surface by a low impedance buffer leads to an increase in the distance between the spall plane and the plane at which measurements are made. As a result, the potential for distortion of the wave profile also increases. An additional source of error in this case is in the equations of state used for the sample and buffer materials.

Dynamics of the fracture zone in experiments with a low impedance buffer are more complicated than in experiments without a buffer and have some specific characteristic properties due to the effect of a counterpressure from the soft barrier side. This effect was analyzed by Kanel and Utkin [1991], who used the acoustic approach to analyze cavitation in an inviscid liquid with zero strength during the reflection of a triangular compression pulse from the interface between the liquid and a low impedance material. The process is illustrated by the distance–time and the stress–particle velocity diagrams of Figures 4.8 and 4.9. The objective of the analysis is to determine how the boundary of the cavitation zone is moving and how this motion influences the velocity (and stress) history at the interface between the liquid and the softer buffer.

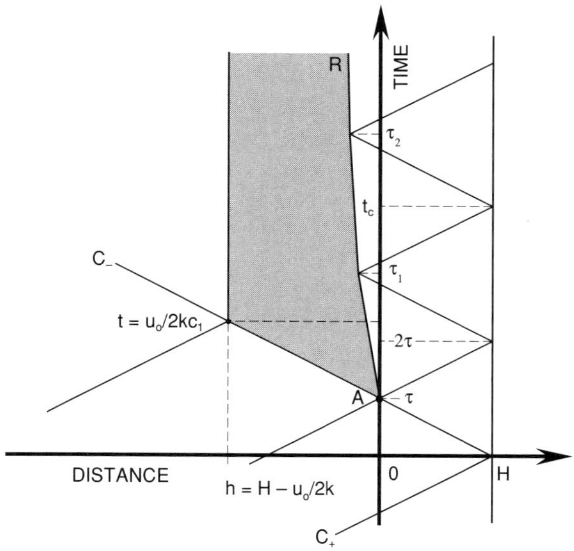

Figure 4.8. Distance–time diagram showing wave interactions during liquid cavitation caused by the reflection of a triangular compression pulse from the interface with a softer material.

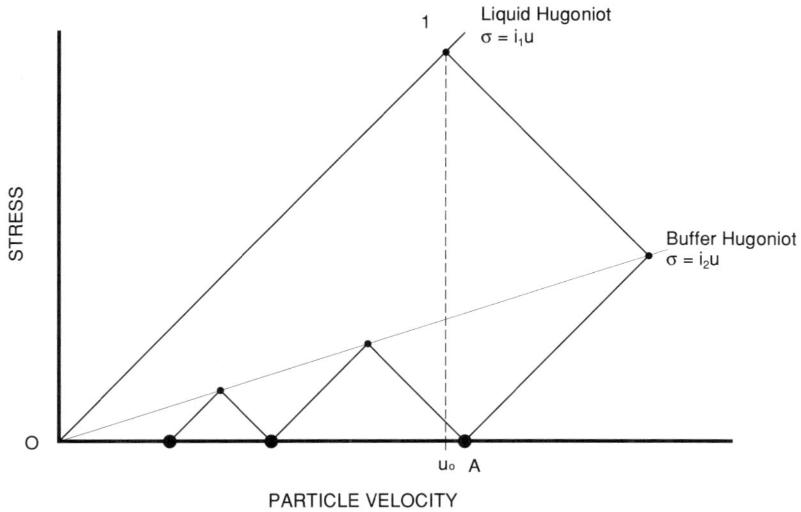

Figure 4.9. Stress (or pressure)–particle velocity diagram for cavitation of a liquid with no strength.

Let $i_1 = \rho_1 c_1$ and $i_2 = \rho_2 c_2$ be the dynamic impedances of the liquid and soft buffer, respectively. In Figure 4.8, the incident triangular compression pulse propagates to the right along C_+ characteristics. This compression pulse is reflected at the liquid-buffer interface as a centered rarefaction fan that propagates to the left along C_- characteristics. Let the velocity distribution in the incident pulse be

$$u = u_0 - k\left(c_1 t - h + H\right). \tag{4.8}$$

Cavitation occurs at $h = 0$, where the stress first reaches zero as a result of interaction between the incident compression pulse and the reflected rarefaction waves. This occurs at $t = \tau$, where τ is given by the relation

$$\tau = \frac{H}{c_1} = \frac{u_0}{kc_1}\frac{i_2}{i_1 + i_2}. \tag{4.9}$$

The right boundary of the cavitation zone (curve AR in Figure 4.8) continually shifts to the left (i.e., moves through the material) because the recompression wave approaching the boundary from the soft buffer side causes the stress to increase above zero. The particle velocity and specific volume in the inviscid liquid undergo a jump from u^- and V^- on the left side of the boundary to u^+ and $V^- = V_0$ on the right side of the boundary. Mass conservation across the boundary of the cavitation zone can be expressed by using an equation similar to the mass jump condition across a shock front as follows:

$$\frac{D-u^-}{V^-}=\frac{D-u^+}{V^+}, \tag{4.10}$$

where D is the boundary velocity. Thus, the trajectory AR of the cavitation zone boundary can now be described by the differential equation

$$\frac{dh_R}{dt}=D-u^+=\frac{V^+}{V^- - V^+}\left(u^+ - u^-\right). \tag{4.11}$$

To calculate this trajectory, we make use of the condition that pressure along AR is zero, and the particle velocity u^+ on the right side of AR is connected with the stress and particle velocity state $\bar{\sigma}$, $\bar{u}$ along the liquid-buffer interface through the Riemann invariant along C_- characteristics. Parameters on the left side of the boundary are determined by using the condition that the particle velocity in a liquid with no strength during cavitation is constant. This velocity is given by

$$u^-(t)=2[u_0-2k(H-h)] \qquad \text{for} \qquad t\geq\tau-h/c_1\,. \tag{4.12}$$

The time rate of change of the specific volume in the cavitation zone is then given by

$$\frac{dV^-}{dt}=V_0\frac{du^-}{dh}=4kV_0\,. \tag{4.13}$$

Using the parameters u^-, V^-, u^+, and V^+ calculated as described above, we finally obtain the following linear equation for the trajectory AR of the right boundary of the cavitation zone:

$$h_R=Ac_1(t-\tau) \qquad \text{for} \qquad \tau\leq t\leq\tau_1, \tag{4.14}$$

where

$$A=-1-\frac{\delta}{4}+\sqrt{1+\frac{\delta^2}{16}}, \tag{4.15}$$

$$\delta=\frac{i_1-i_2}{i_1+i_2}, \tag{4.16}$$

and

$$\tau_1=\tau\left(\frac{3+A}{1+A}\right). \tag{4.17}$$

Thus, the cavitation zone is continuously reduced with time. Its right boundary shifts to the left with a velocity Ac_1 that does not depend on the incident pulse steepness. The boundary velocity decreases with the arrival of each successive recompression wave (at time τ_1, τ_2, etc., in Figure 4.8). This motion of the cavitation boundary indicates that voids created by tensile stresses at early times could be recompacted by counterpressure from the soft buffer side at later times. This possibility must be taken into account when microscopic examinations are performed on spalled samples recovered from experiments with soft barriers.

In contrast to the case of free-surface velocity measurement, the velocity at the interface between the spall sample and a lower impedance buffer continues to decrease with time after the arrival of the spall signal at the interface. Figure 4.10 shows the interface velocity profiles calculated for $\delta = 0.5$ ($i_1/i_2 = 3$) with and without accounting for the shifting boundary of the cavitation zone.

4.1.4. Other Methods for Determining the Spall Strength

Several methods other than the two described above have been used in various studies to determine the spall threshold and the state of the material just before spall. The most straightforward of these methods relies on post-test inspection of impacted samples to determine the critical velocity that corresponds to the spall threshold. With this method, a sample of the material under investigation is impacted with a flyer plate, and the impact velocity is measured. The specimen is recovered, sectioned, and examined under a microscope for spall damage. By repeating this procedure at several impact velocities, we can determine the im-

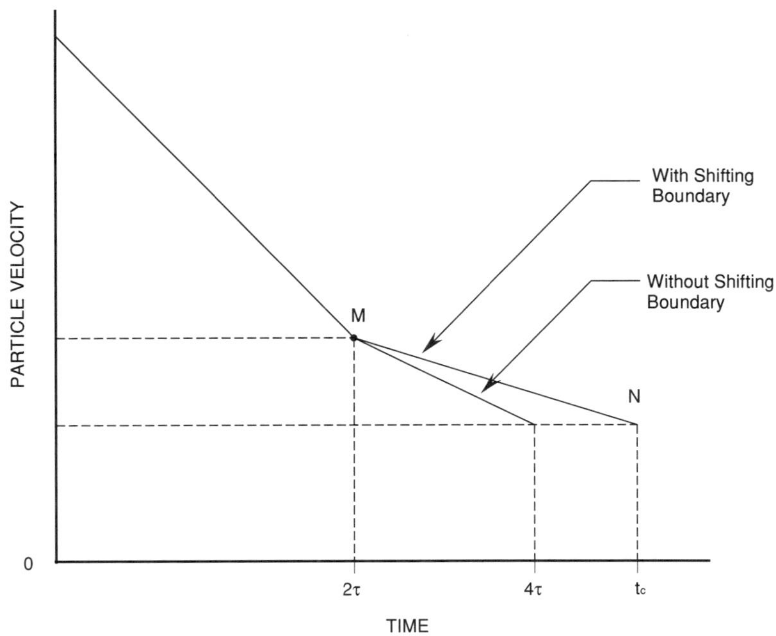

Figure 4.10. Velocity at the liquid-soft barrier interface with and without accounting for the shifting boundary of the cavitation zone.

pact velocity that corresponds to the inception of spall damage in the impacted sample. With the critical velocity so determined, the spall threshold stress under the conditions investigated in the experiment can be determined based on the known compressibility of the material.

This method has been used to determine the spall threshold in several materials including copper (Smith [1962]) and aluminum (Blinkow and Keller [1962]). Among the advantages of this method are that it requires minimal instrumentation and the experiments are relatively easy to perform. One of its disadvantages is that it is subjective. Determining the onset of damage depends on the magnification of the microscope used to inspect the sample and the adopted criteria, which in turn introduce an uncertainty into the fracture stress at the onset of spall calculated using this method.

Another method of determining the fracture stress during spall under uniaxial strain conditions is one that relies on measuring the spalled layer thickness, then performing hydrodynamic calculations to determine the maximum tensile stress experienced by the material at the spall plane (using the experimental results to constrain the calculations). This approach of determining the fracture stress was used by McQueen and Marsh [1962], among others, to study spall in copper, and by Breed et al. [1967] to study spall in aluminum, copper, nickel, and lead. Bushman et al. [1983], who surveyed several spall strength measurement studies in aluminum and copper, noted that the highest spall strength was usually obtained from experiments with relatively strong shock waves, where the strength was determined by measuring the thickness of the spalled plate, followed by hydrodynamic calculations. The spall strength in these studies is systematically overestimated because the models used in the hydrodynamic calculations neglect the effect of damage on the evolving stress waves. In general, the accuracy of this method of determining the spall strength depends, to a large extent, on the degree of sophistication built into the model used in the hydrodynamic calculations.

Yet another method of determining the spall strength involves measuring the initial velocity and average surface velocity of the spalled layer by a discrete method. The essence of this approach—which was used by Al'tshuler et al. [1966] to investigate spall in copper, and by Novikov et al. [1966] to study dynamic fracture in steel, aluminum, and copper—can be explained by referring to Figure 4.1. For instantaneous (i.e., brittle) fracture by a triangular load pulse, the final velocity of the spalled plate is close to the average value

$$u_f \cong \frac{1}{2}\left(u_0 + u_m\right), \tag{4.18}$$

and the maximum tensile stress in the spall plane can be determined using the simple expression

$$\sigma^* = \rho_0 c_0 u_f . \tag{4.19}$$

In essence, this method of estimating the spall strength is based on momentum conservation, but it does not account for dissipation due to spall damage. For this reason, Eq. (4.19) generally overestimates the spall strength and gives better estimates when the applied stress is larger than the spall strength (i.e., the material is overdriven) and when the spall process is brittle. As shown later (see Figure 7.2), if spall occurs by ductile damage growth, the velocity of the spalled plate continually decreases as damage continues to accumulate during the reverberation of the stress pulse within the spalled layer. In this case, Eq. (4.18) is a poor estimate of the average velocity. Consequently, Eq. (4.19) does not provide a reliable estimate of the spall strength. This condition also exists if complete separation at the spall plane is not achieved during the first wave reverberation (i.e., if the load duration is short). In this latter case, damage continues to accumulate during stress reverberations and the velocity continues to decrease due to energy dissipation in fracture.

The methods of spall strength determination described in this section were used primarily in early spall investigations performed during the 1960s when instrumentation and computational tools were not as developed and as sophisticated as they are today. The experimental means of continuous particle velocity and stress measurements available today were not readily available then, neither were the tools for performing detailed microscopic characterizations of impacted samples. Similarly, the computational resources and advanced damage-dependent spall models at our disposal today were not available then to aid in the interpretation of experimental data. Today, these sophisticated experimental and analytical tools are all brought to bear to provide a better understanding of spall fracture and the parameters that affect it. These tools and the manner in which they are applied to investigate spall problems are described in later chapters.

4.2. Influence of Damage Kinetics on Wave Dynamics

Unlike quasi-static loading where, for practical purposes, fracture can be considered instantaneous, the duration of spall fracture under shock wave loading conditions is comparable to the duration of the applied dynamic load. As a result, stress relaxation during fracture plays an important role in the wave process. For this reason, the damage kinetics must be taken into account during the analysis and computation of the dynamic load. The measured stress and particle velocity profiles contain some information about the damage kinetics during spall. This section is devoted to providing a better understanding of these mechanical aspects of rate-dependent fracture during spalling.

4.2.1. Evolution of the Tensile Wave

Let us begin by considering qualitatively the evolution of a tensile wave after fracture initiation in the acoustic approach (Kanel and Chernyich [1980]). (A more detailed analysis is presented in Section 4.2.3.) The total specific volume of matter, $V(p,V_v)$, is considered to be the sum of the volume of intact solid material, V_s and the volume of voids, V_v:

$$V(p,V_v) = V_s(p) + V_v(t). \tag{4.20}$$

The damage rate $\dot{V}_v$ is taken to be a function of damage itself and of the pressure, p. Using Eq. (4.20), the conservation equations for one-dimensional flow in Lagrangian coordinates take the form

$$\rho_0 \frac{\partial u}{\partial t} + \frac{\partial p}{\partial h} = 0 \tag{4.21a}$$

and

$$\frac{\partial p}{\partial t} + \rho_0^2 c^2 \frac{\partial u}{\partial h} - \rho_0^2 c^2 \dot{V}_v = 0, \tag{4.21b}$$

where the constitutive relation $dp = -c^2 dV$ is used to derive Eq. (4.21b) and, as before, ρ_0 is the initial density, u is the particle velocity, t is time, h is the Lagrangian position, and c is the sound speed in the medium, and where plastic strains are neglected.

In this case, as in the case of nonrelaxing media, characteristics are lines with slope $dh/dt = \pm c$ in the time–distance plane. Here, we assume that the sound speed c remains constant, thus neglecting the effect of void growth on the stiffness of the distended material. Derivatives of pressure and particle velocity along the characteristics are given by

$$\left.\frac{dp}{dt}\right|_{C_+} = \frac{\partial p}{\partial t} - \rho_0 c \frac{\partial u}{\partial t} = \rho_0 c \left.\frac{du}{dt}\right|_{C_+} + \rho_0^2 c^2 \dot{V}_v, \tag{4.22a}$$

$$\left.\frac{dp}{dt}\right|_{C_-} = \frac{\partial p}{\partial t} + \rho_0 c \frac{\partial u}{\partial t} = \rho_0 c \left.\frac{du}{dt}\right|_{C_-} + \rho_0^2 c^2 \dot{V}_v. \tag{4.22b}$$

Since void growth under tension means that $\dot{V}_v > 0$, trajectories of changing $p - u$ states along the characteristics deviate from the Riemann invariants given by

$$p = \pm \rho_0 c u + \text{const}. \tag{4.23}$$

toward higher pressures.

Qualitative analysis of states along the characteristics leads to some preliminary conclusions about the conditions that cause the appearance of the minimum on the free-surface velocity profile. This minimum corresponds to a confluence of trajectories of changing state along C_+ characteristics. The trajectories of changing state along C_+ and C_- characteristics have a common tangent in the confluence point because, otherwise, the conditions that

$$\left.\frac{dp}{dt}\right|_{C_-} = 0 \tag{4.24a}$$

and

$$\left.\frac{du}{dt}\right|_{C_-} = 0 \tag{4.24b}$$

must be satisfied when the C_- characteristic crosses the confluence point. Since, according to Eq. (4.22), this is possible only if $\dot{V}_v = 0$, we conclude that either the confluence of trajectories of changing state along C_+ characteristics occurs on the boundary of the fracture zone, or the trajectories of changing state along C_+ and C_- characteristics have a common tangent in the confluence point. Taking into account that

$$\left.\frac{du}{dt}\right|_{C_+} + \left.\frac{du}{dt}\right|_{C_-} = 2\frac{\partial u}{\partial t} \tag{4.25a}$$

and

$$\left.\frac{dp}{dt}\right|_{C_+} + \left.\frac{dp}{dt}\right|_{C_-} = 2\frac{\partial p}{\partial t}, \tag{4.25b}$$

we may also conclude that the trajectories of changing state along these C_+ and C_- characteristics have a common tangent with that of the particle path.

Figure 4.11 shows the $p - u$ diagram of the process calculated numerically using the acoustic approximation. The fracture process influences not only the magnitude but also the sign of the slope of the trajectories of changing state along the characteristics. It would be natural to expect that the spall signal originate at the point where the particle velocity is constant, which corresponds to a symmetric velocity distribution in the vicinity of this point. However, Figure 4.11 shows that the slope of the trajectories at the confluence point is close to the slope of the Riemann C_+ invariant. This situation and its consequences are discussed in more detail later in this chapter.

Let us now examine the evolution of the tension wave after the reflection of a triangular compression pulse from the free-surface. Superscript "+" will denote states immediately ahead of the tension jump, and "–" will denote states immediately behind the jump. The tension wave propagates in the negative direction (along C_- characteristics), and the relationship between pressure and particle velocity across the jump is

$$p^- - p^+ = -\rho_0 c(u^- - u^+). \tag{4.26}$$

Taking into account that the reflected wave is superimposed on the incident simple compression wave, where $V_v = 0$ and $dp = \rho_0 c \cdot du$, we obtain from Eqs. (4.22) and (4.26)

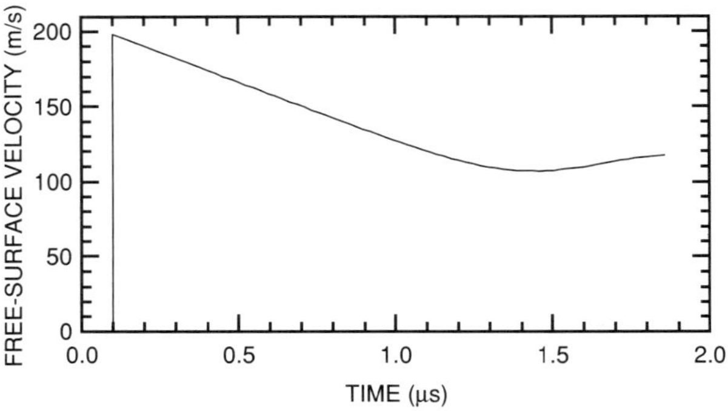

(a) Free-surface velocity profile

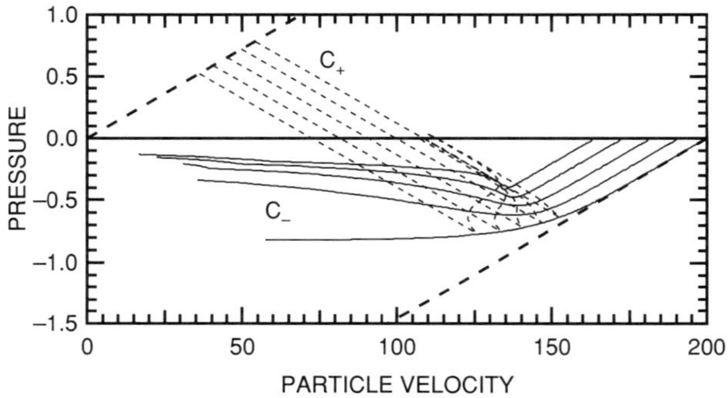

(b) Trajectories of changing states along characteristics

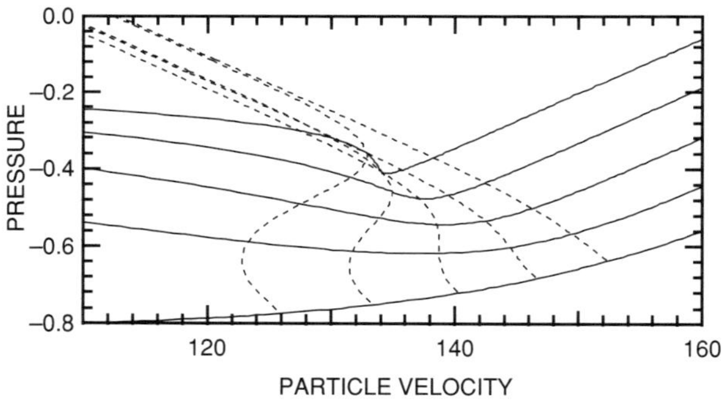

(c) Same as (b) above, magnified in the region of spall pulse formation

Figure 4.11. Numerical simulation of the evolution of spall fracture.

$$\left.\frac{dp}{dt}\right|_{C_-^-} = 2\left.\frac{dp}{dt}\right|_{C_-^+} - \left.\frac{dp}{dt}\right|_{C_-^-} + \rho_0^2 c^2 \dot{V}_v = 2\dot{p}_0 + \tfrac{1}{2}\rho_0^2 c^2 \dot{V}_{v0}, \tag{4.27}$$

where

$$\dot{p}_0 = \left(\frac{\partial p}{\partial t}\right)_0 = \frac{1}{2}\frac{dp}{dt}\bigg|_{C_-^+} \tag{4.28}$$

is the time rate of change of pressure in the incident compression pulse, and $\dot{V}_{v0}$ is the initial fracture rate immediately behind the tension jump. Within the framework of the acoustic approach, the damage does not influence the evolution of the tensile wave front when the initial damage rate is equal to zero. In the case of nonzero initial damage rate, the peak stress of the tension wave, according to Eq. (4.27), increases at a slower rate than in the case of no fracture. As an example, consider the linear dependence of the fracture rate on pressure:

$$\dot{V}_v(p) = \frac{-2Fp}{\rho_0^2 c^2}. \tag{4.29}$$

Integrating Eq. (4.27) for a triangular compression pulse ($\dot{p}_0 = \text{const}$) yields the following relationship for the peak tensile stress behind the tension jump:

$$p^- = \frac{2\dot{p}_0}{F}\left(1 - \mathrm{e}^{-\mathrm{Ft}}\right) = \frac{2\dot{p}_0}{F}\left(1 - e^{F\frac{h-h_0}{c}}\right), \tag{4.30}$$

where h_0 is the coordinate at the free-surface. The pressure asymptotically approaches the ultimate value $p^- = 2\dot{p}_0 / F$, corresponding to the condition

$$\dot{V}_v(p^-) = \frac{-4\dot{p}_0}{\rho_0^2 c^2}, \tag{4.31}$$

which also follows directly from Eq. (4.27).

4.2.2. Measurements Associated with the Peak Tensile Stress at the Spall Plane

Direct measurements of the tensile wave evolution are obviously impossible, but indirect measurements can be performed (Kanel and Glusman [1983]). The idea is as follows: When compression or rarefaction waves are reflected from an interface with a lower impedance material, the sign of the reflected wave is opposite that of the incident wave. It is possible to select a pair of materials of different compressibilities such that, after shock compression, the pressure at the interface between them remains positive (compressive) during the passage of the unloading wave from the "rigid" into the "soft" material, even if the pressure reaches a large negative (tensile) magnitude inside the "rigid" material. In this case, a pressure profile at the interface can be measured. Then, based on an analysis of wave interactions, we can indirectly recover the peak tensile stress in the interior of the "rigid" plate.

Stress measurements were performed in copper and stainless steel samples by Kanel and Glusman [1983]. Plane shock waves with approximately triangular pressure profiles were introduced into the samples through a thick paraffin layer

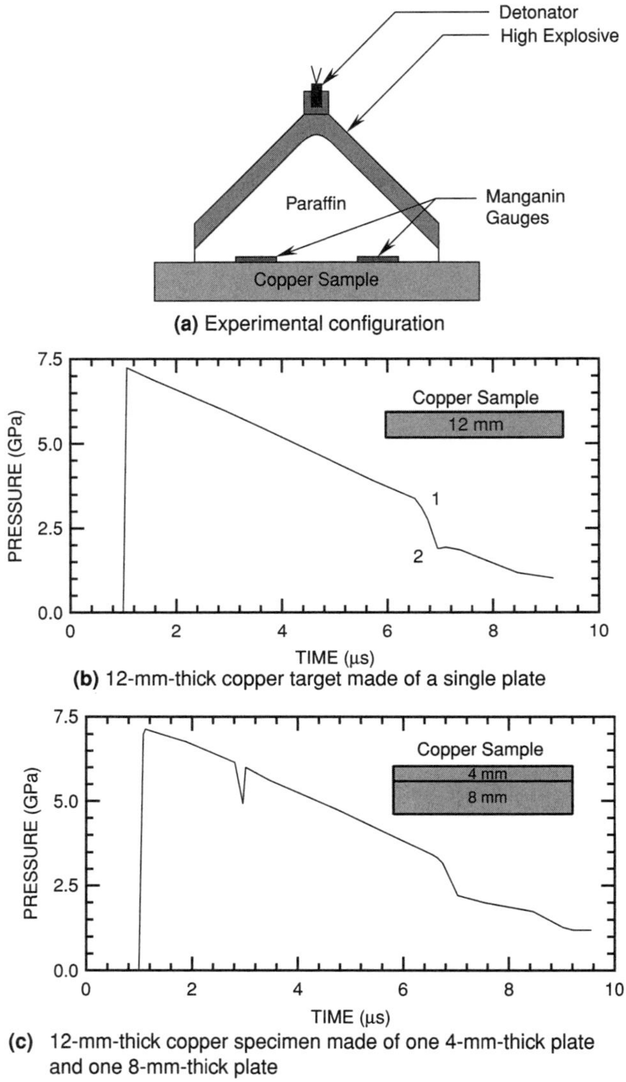

Figure 4.12. Measurements of the peak tensile stress behind the spall plane.

(Figure 4.12(a)). Pressure profiles $p(t)$ were measured at the interface between the paraffin layer and the sample using manganin gauges. The experimental oscillograms presented in Figures 4.12(b) and (c) indicate the arrival time of the shock front at the gauge location on the interface, the relatively slow pressure decay in the incident pulse, and the fast pressure drop (from 1 to 2) associated with the rarefaction wave reflected from the rear free-surface of the specimen.

The $x - t$ and $p - u$ diagrams in Figure 4.13 illustrate the methodology used to determine the pressure in the tail of the reflected rarefaction wave (i.e., the pressure corresponding to point 2 in Figure 4.12(b)). Points 1 and 2 in Figure 4.13 correspond to points 1 and 2 in Figure 4.12(b). On the $x - t$ diagram, the line 012 is a trajectory of the paraffin-sample interface, and the line 0A is the shock front trajectory. The maximum tensile pressure in the sample occurs near the paraffin-sample interface at point K in Figure 4.13. The state of the material at this point corresponds to the intersection of trajectories of changing state along characteristics 1K and K2. The location of the isentrope 1K on the $p - u$ diagram is determined by using the measured pressure, p_1, at point 1 of the ex-

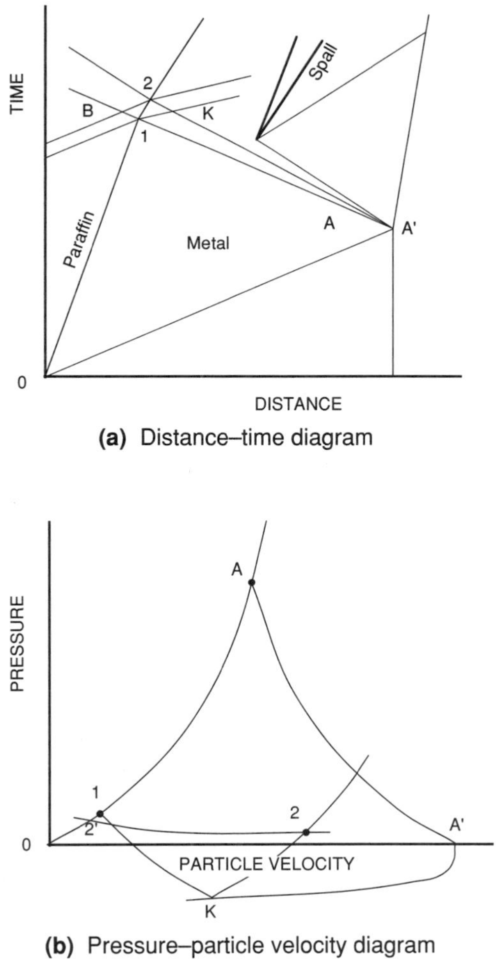

Figure 4.13. Distance–time and pressure-particle velocity diagrams corresponding to the configuration and conditions shown in Figure 4.12.

perimental pressure profile. The location of isentrope K2 is also determined from the measured pressure value, p_2. This pressure corresponds to point 2 on the isentrope of paraffin 2'2, which describes states along the characteristic B2. The location of this isentrope is determined from the measured pressure profile by extrapolating its incident part to time t_2.

Additional experiments with split targets were performed to verify the accuracy of this interpretation. Copper and steel samples in this case were composed of two plates, the combined thickness of which is the same as the total thickness of the corresponding solid sample. The ratio of the thickness of the front plate (nearest to the explosive) to that of the back plate in these experiments was designed to reduce the peak tensile stress in the front plate below the spall strength. Then the peak tensile stress near the interface between the sample and paraffin was determined from the measured pressure profile such as the one shown in Figure 4.12(c).

Table 4.1 gives the results of measurements of peak tensile stresses, p_k, behind the spall plane in stainless steel and copper samples. Also included in Table 4.1 are the spall strength, σ^*, measured for these loading conditions and estimations of the negative pressure magnitudes, p'_k, assuming no influence of fracture on the tensile wave evolution.

Table 4.1. Results of measurements of tensile wave behind the spall plane.

Material	Sample thickness (mm)	Peak pressure (GPa)	p_1 (GPa)	p_2 (GPa)	p'_2 (GPa)	p_k (GPa)	p'_k (GPa)
Copper	12	7.04	3.36	2.04	3.1	−1.22 ± 0.2	−3.5 ± 0.20
$\sigma^* = 0.8 \pm 0.1$ GPa	4 + 8	7.04	3.36	2.17	3.1	−0.66 ± 0.2	−1.0 ± 0.15
Spall thickness = 6 mm							
Stainless steel	12	10.1	4.46	2.53	3.98	−1.55 ± 0.3	−5.5 ± 0.20
$\sigma^* = 1.85$ GPa	6+6	10.1	4.46	2.54	3.98	−1.51 ± 0.1	−3.0 ± 0.15
Spall thickness = 6.9 mm	3+9	10.1	4.46	2.79	3.98	−0.49 ± 0.2	−1.0 ± 0.10

Taking into account possible errors, the peak tensile stresses at a distance of 4 to 5 mm behind the spall plane are 1.0 to 1.4 GPa for copper and 1.3 to 1.9 GPa for stainless steel. Without fracture, the estimated peak stresses near the interface are 3.5 GPa for copper and 5.5 GPa for stainless steel. In other words, the tensile stresses in stainless steel are practically limited to the spall strength (i.e., $\sigma^* = 1.85$ GPa for stainless steel). In the copper samples, a small increase of intensity of the tensile wave was observed during its propagation behind the spall plane. This small increase can be explained in terms of the dependence of spall strength on strain rate.

The main observation based on these experimental results, however, remains: the stress relaxation at fracture makes it impossible for the material to support tensile stresses in excess of some limit determined by the spall conditions. The evolution of the tensile wave is determined by the initial fracture rate.

4.2.3. Correlation Between the Spall Fracture Rate and the Free-Surface Velocity Profile

For fracture resulting from shock wave loading, the fracture time can be comparable to the load duration. This means that, for many applications, dynamic fracture must be considered as a continuous process with characteristic kinetics of damage development. In general, the events that take place inside the material affect the wave profile structure; therefore; the results of dynamic wave profile measurements contain information about the kinetics of damage evolution. For instantaneous fracture, the spall pulse should have a very sharp front. Increase of damage evolution time should cause a decrease in the slope of the spall pulse front. This section attempts to establish a general quantitative relationship between the damage rate and the structure of the wave profiles. Analysis of the flow associated with stress relaxation during spalling should provide information about the kinetics of fracture in the sense of the dependence of damage rate on stress and the degree of damage.

Let us consider the evolution of a triangular compression pulse after its reflection from the free-surface of a solid in the context of the acoustic approximation. The damage is assumed to be initiated by tensile stresses exceeding the critical value, $\sigma^* = -p_k$ and is described by evolution of the specific volume of voids, V_v. The total volume is equal, as before, to the sum of the damaged volume, V_v, and the volume of undamaged (solid) material, V_s. Let the damage kinetics depend only on the specific volume of pores and be a power function of V_v. The form of the kinetic relation is chosen to be convenient for the analysis. In general, of course, the damage rate depends on the acting stress, temperature and degree to which the damage has evolved. Because we are interested in the initial stage of spallation, the collapse of voids under compression will not be considered. In the context of this model, the behavior of the material is described by the following set of equations in Lagrangian coordinates:

$$\frac{\partial V}{\partial t} - \frac{1}{\rho_0}\frac{\partial u}{\partial h} = 0, \tag{4.32a}$$

$$\frac{\partial u}{\partial t} + \frac{1}{\rho_0}\frac{\partial p}{\partial h} = 0, \tag{4.32b}$$

$$\frac{\partial V_v}{\partial t} = \frac{(\rho_0 V_v)^\alpha}{\rho_0 \tau_\mu}, \tag{4.32c}$$

$$p = \rho_0^2 c^2 \left(\frac{1}{\rho_0} - V - V_v \right), \tag{4.32d}$$

where the characteristic damage time, τ_μ , and $\alpha < 1$ are constants, and ρ_0 is the initial density. The pressure is defined by the equation of state through the specific volume of solid material.

Figure 4.14 diagrams the flow field in the $h - t$ plane. The free-surface has a coordinate $h = 0$. Region 1 is free from the interaction of the incident load pulse with the reflected rarefaction wave. No tensile damage occurs in this region, and the dependencies of the particle velocity and pressure on time and position correspond to a triangular compression pulse described by the relations

$$u(h,t) = u_0 - k(ct - h), \tag{4.33}$$

$$p(h,t) = \rho_0 c u(h,t), \tag{4.34}$$

where u_0 is the peak particle velocity, k is a constant, and $\tau_0 = -h_0/c$ defines the length, $2h_0$, of the pressure pulse through the relation $u_0 = -2kh_0 = 2kc\tau_0$.

Tensile stresses arise in region 3, and the pressure reaches the critical magnitude, $p_k = -\sigma^*$ at the point with coordinates,

$$h = h_k = \frac{p_k}{2\rho_0 kc} \quad \text{and} \quad t = \tau_k = -\frac{h_k}{c}.$$

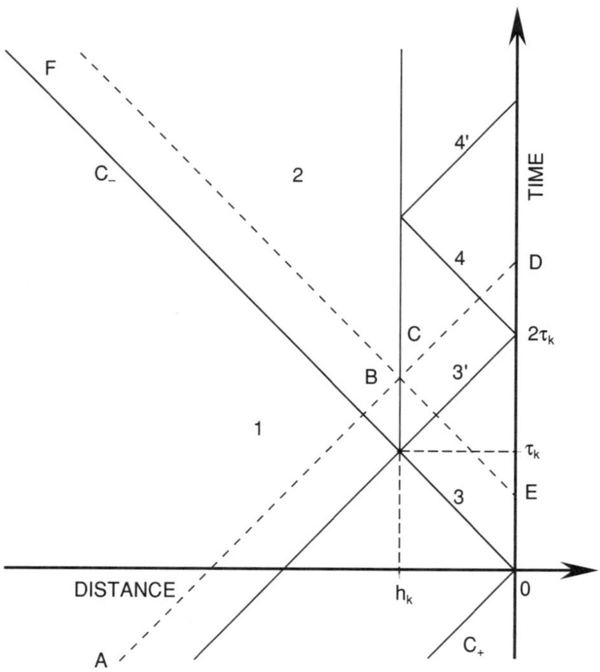

Figure 4.14. Distance–time diagram of the flow field during spalling.

The damage occurs in region 2 and the flow is determined here by solving the system of Eqs. (4.32) with boundary conditions at $h = h_k$ and as $h \to -\infty$ and initial conditions along the C_- characteristic through the J_+ Riemann invariant.

To obtain a solution of Eqs. (4.32), let us eliminate the terms V_v and V and substitute $T = t + h/c$ for t. Then, applying the Laplace transformation, we obtain a system of ordinary differential equations. The solution that satisfies the initial conditions and remains stable as $h \to -\infty$ is

$$\begin{aligned}\widehat{p}(h,s) = {} & \frac{2k\rho_0 c}{s}\left[h - h_0 - \frac{c}{2s}\left(1 - \exp\left(-2s\left(h - h_0\right)/c\right)\right)\right]\theta(h - h_0) \\ & - \frac{\rho_0 c}{2} F(s)\left(h - \frac{c}{2s}\right) + a,\end{aligned} \tag{4.35}$$

$$\begin{aligned}\widehat{u} = {} & \frac{2k}{s}\left[h - h_0 - \frac{c}{2s}\left(1 - \exp\left(-2s\left(h - h_0\right)/c\right)\right)\right]\theta(h - h_0) \\ & + \frac{1}{2} F(s)\left(h + \frac{c}{2s}\right) - \frac{a}{\rho_0 c}.\end{aligned} \tag{4.36}$$

Here, $\hat{p}$ and $\hat{u}$ are the Laplace conjugates of pressure and particle velocity, s is the Laplace variable, $F(s)$ is the Laplace conjugate of damage rate $\rho_0 V_v$, and $\theta(x)$ is the Heaviside unit function. The constant a is defined in such a manner as to keep the J_- Riemann invariant at $h = h_k$. For example, within the time interval $0 \le T \le 2\tau_k$, the constant a is given by

$$a = \frac{k\rho_0 c}{s}\left(2h_0 + \frac{c}{s}\right) + \frac{\rho_0 c}{2} F(s) h_k. \tag{4.37}$$

Equations (4.35), (4.36), and (4.37) provide the solution of Eqs. (4.33) and (4.34) in terms of the Laplace variables in region 2 of the flow field (see Figure 4.14) for $0 \le T \le 2\tau_k$. Some results can be obtained immediately from Eqs. (4.35) and (4.36) without performing the inverse transformation. The pressure history along the right side of the C_- characteristic for $h > h_0$ can be determined using the known property of Laplace transformation: $\lim_{s \to \infty} sG(s) = G(0+)$,

$$p = 2k\rho_0 ch - \frac{\rho_0 c}{2}(h - h_k)L, \tag{4.38}$$

where $L = 0$ when $0 < \alpha < 1$ and $L = 1/\tau_\mu$ for $\alpha = 0$. It follows from this equation that the pressure corresponds to the case of undamaged material if the initial damage rate is equal to zero. When $\alpha = 0$, a different situation arises. After the beginning of damage evolution at point (h_k, τ_k), the pressure continues to decrease if $\tau_\mu > (1/4k)$. It remains constant if $\tau_\mu = (1/4k)$, or it begins to increase if $\tau_\mu < (1/4k)$.

The free-surface velocity can be determined from Eqs. (4.35) and (4.36). Taking into account the conservation of J_- Riemann invariant along the C_+ characteristic in regions 3' and 4 (see Figure 4.14) and using the known formulas of inverse Laplace transformation, we obtain the following expression for the free-surface velocity within the time interval $2\tau_k \le t < 4\tau_k$:

$$\frac{u(0,t)}{2u_0} 1 - \frac{t}{2\tau_0} + \frac{c}{4u_0}\left[(1-\alpha)\frac{t-2\tau_k}{\tau_\mu}\right]^{1/(1-\alpha)} . \qquad (4.39)$$

We now analyze this solution for various values of α, the exponent in the void volume growth rate.

4.2.3.1. The Case $\alpha = 0$ (Constant Void Volume Growth Rate)

Figure 4.15 shows the free-surface velocity profiles plotted by means of Eq. (4.39) for three values of τ_μ. As can be seen, there is a critical value of relaxation time equal to $(1 / 4k)$. In this case the free-surface velocity is constant after time Δt_{spall} when information about the beginning of damage reaches the free-surface. For $\tau_\mu < (1 / 4k)$, the damage is manifested as a spall pulse in the free-surface velocity profile. For $\tau_\mu > (1 / 4k)$, the velocity continues to decrease during fracture beyond $t = 2\tau_k$. Introducing the damage rate $V_v = 1 / \rho\tau_\mu$ and the expansion rate in the unloading wave of the incident pulse, the result obtained can be stated as follows. A spall pulse on the free-surface velocity profile forms only if the initial damage rate is more than four times as great as the expansion rate in the unloading wave of the incident pulse. The slope of the spall pulse front is equal to

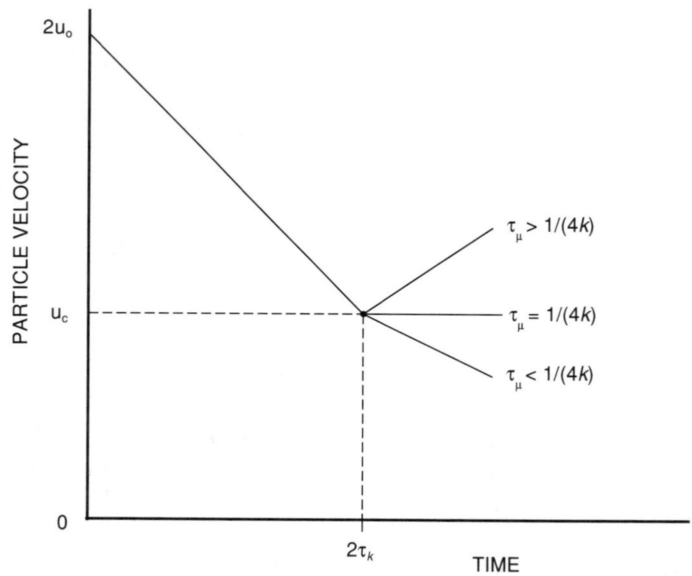

Figure 4.15. Free-surface velocity profiles for the case of constant damage rate after the spall threshold.

$$\frac{d}{dt}\left(\frac{u(0,t)}{2u_0}\right)=\frac{1}{8\tau_0}\left(\frac{\dot{V}_v}{\dot{V}}-4\right), \qquad t>2\tau_k. \tag{4.40}$$

It follows from Eq. (4.40) that the initial magnitude of the damage rate, $\dot{V}_v$, can be estimated from experimental free-surface velocity profiles.

Let us now consider the changing $p-u$ state along characteristics. The solution for the fracture zone, which follows from Eqs. (4.35) and (4.36) after inverse Laplace transformation, is

$$p(h,t)=2\rho_0 ckh+\frac{\rho_0 c^2}{4\tau_\mu}\left(t-\frac{h}{c}-2\tau_k\right), \tag{4.41a}$$

$$u(h,t)=2(u_0-kct)+\frac{c}{4\tau_\mu}\left(t+\frac{3h}{c}+2\tau_k\right). \tag{4.41b}$$

These relationships provide a constraint on the pressure and particle velocity along the C_+ characteristic on the segment BC in Figure 4.14:

$$p-p_+=\frac{\rho_0 c}{1/2k\tau_m-1}(u-u_+), \tag{4.42}$$

where p_+ and u_+ are the pressure and the particle velocity at the point of intersection of the C_+ characteristic with the straight line $h=h_k$ (the point C in Figure 4.14). In the regions $3'$ and 4 along the same characteristic C_+

$$p-p_+=-\rho_0 c(u-u_+). \tag{4.43}$$

The relationship (4.42) shows that the trajectory of changing state along the characteristic becomes vertical when $\tau_m=1/2k$. The vertical slope of the trajectory does not correspond to any spall threshold, and the damage rate value in this case is half of that corresponding to the appearance of the spall signal on the free-surface velocity profile. Figure 4.16 shows trajectories of changing state along C_+ characteristic ABCD shown in Figure 4.14 for the threshold situation when $\tau_m=1/4k$. The arrows indicate the direction of change of the state. After intersection with the tensile wave front, the pressure and the particle velocity along this characteristic are changed by a jump from point A to point B. The change from point B to point C occurs continuously, and thereafter the characteristic becomes trapped in the fracture zone.

Along the segment CD, the relationship between the pressure and the particle velocity corresponds to Eq. (4.43). The geometry of the trajectories of changing state shows that the pressure at the spall plane $h=h_k$ increases from the threshold value p_k to zero during the time $2\tau_k$. In other words, under threshold conditions the pressure on the spall plane increases at a rate equal to the unloading rate in the incident load pulse. Figure 4.16 shows also the trajectory of changing state along C_- characteristic ECF (Figure 4.14) for this threshold case.

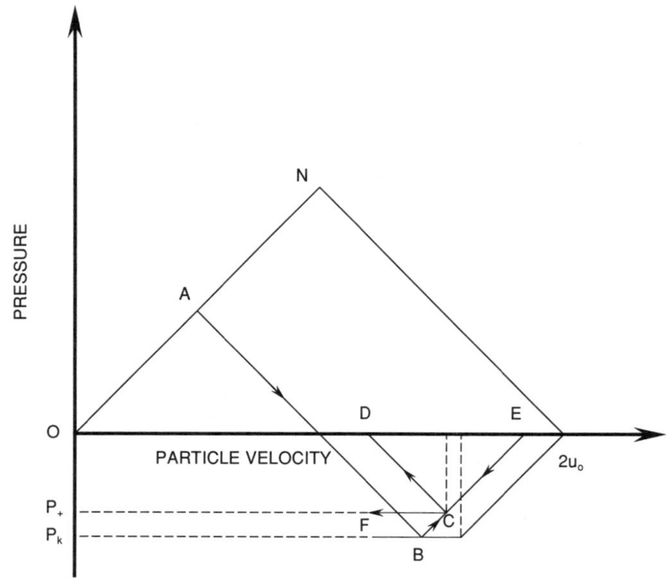

Figure 4.16. Trajectories of changing states along characteristics at a constant damage rate.

4.2.3.2. The Case α > 0 (Variable Void Volume Growth Rate)

When $\alpha > 0$, the damage evolves at an accelerating rate, beginning at an initial rate of zero. Figure 4.17 shows the profiles of free-surface velocity for this case. Curves 1, 2, and 3 correspond to increasing α or τ_μ. Unlike the case of constant damage rate, the derivative of free-surface velocity in this case is continuous at the point $t = 2\tau_k$, and a minimum is reached at $t = t_m > 2\tau_k$, where

$$t_m = 2\tau_k + \frac{\tau_\mu}{1-\alpha}(4k\tau_\mu)^{(1/\alpha)}. \tag{4.44}$$

The corresponding velocity magnitude, u_m, is derived from Eq. (4.39). In practice, the spall strength is determined through the velocity pullback $\Delta u_{fs} = 2u_0 - u_m$. For $\alpha = 0$, we have $\Delta u_{fs} = -2p_k / \rho_0 c$. In the general case, the velocity pullback also depends on the damage kinetics and the expansion rate in the incident pulse through the relation

$$\Delta u_{fs} = -\frac{2p_k}{\rho_0 c} + \frac{\rho_0 c^2 \alpha}{2\rho_0 c(1-\alpha)}(4\tau_\mu \rho_0 \dot{V})^{1/\alpha}. \tag{4.45}$$

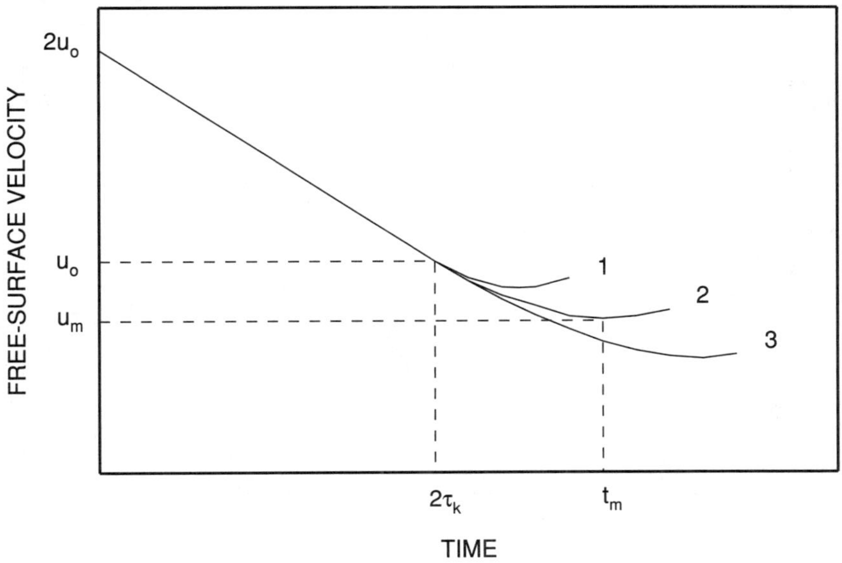

Figure 4.17. Free-surface velocity profiles for the case of accelerating damage (a > 0).

Initially, the negative pressure reaches the value $p^* = -\rho_0 c\Delta u_{fs}/2$ in the plane with coordinate $h^* = \sigma^*/(2k\rho_0 c) < h_k$. Let us estimate the damage rate value that corresponds to the minimum in the free-surface velocity profile. In terms of the model of kinetics of damage that has been used, the damage rate is largest on the plane with coordinate h_k where fracture first began. Going back from the free-surface to this plane and from time t_m, when the minimum velocity occurs, to time $t = t_m - \tau_k$, we find the damage rate to be equal to $V_v = 4k/\rho_0 = 4\dot{V}$. This result coincides with that for constant damage rate. The minimum in the free-surface velocity profile and, consequently, the beginning of spall pulse formation is observed when the damage rate on the spall plane is equal to four times the expansion rate in the unloading part of incident pulse.

Using an approach similar to the one discussed above, Utkin [1992, 1993] analyzed the wave dynamics for the case where the damage rate is assumed to be a function of pressure. In this case, a segment with horizontal slope appears on the free-surface velocity profile when the damage rate is equal to four times the expansion rate in the unloading part of the incident compression pulse.

In reality, the damage rate is a function of both tensile stress and degree of damage. As a result, the threshold damage rate that corresponds to the appearance of a minimum in the free-surface velocity profile can be reached at many times during the development of fracture. This time interval decreases with increasing tensile stress while the reflected rarefaction wave propagates from the free-surface into the body. Figure 4.18 shows the threshold line in the

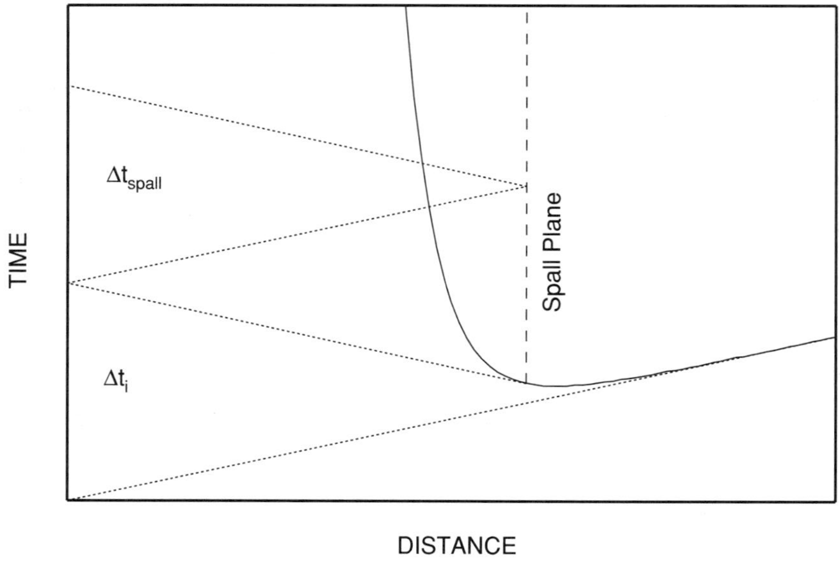

Figure 4.18. Time–distance diagram for the spall process caused by the reflection of a triangular compression pulse from the free-surface (represented here by the time axis) for the case where the damage rate is a fun ction of tensile stress and damage.

time–distance diagram, along which the condition $\dot{V}_v = 4\dot{V}$ is satisfied. The spall signal arrives at the sample surface from the point on the this line where the slope is

$$\frac{dt}{dx} = -\frac{1}{c}. \tag{4.46}$$

As a result, the duration, Δt_i, of the first velocity pulse on the free-surface velocity profile exceeds the periods, Δt_{spall}, of later (after beginning of fracture) velocity oscillations. The difference between these time intervals is interpreted as an apparent delay time of the spall fracture. Obviously, development of the fracture to the left of the spall plane is suppressed by the compression wave created as a result of the relaxation of the tensile stress in the spall plane.

4.3. Estimating Spall Fracture Kinetics from the Free-Surface Velocity Profiles

Fracture needs to be predicted in many applications ranging from micrometeorite impact and pulsed laser attacks to large-scale impacts and explosions. The fracture model should be efficient over a wide range of load durations. Many fracture models based on approaches ranging from microstatistical to empirical have been developed to describe damage and fracture kinetics under dynamic loading conditions. The nucleation and growth (NAG) modeling approach, described in detail in later chapters, is a well-known example of the microstatistical approach. NAG models are usually validated by comparing the measured damage distributions and the wave profiles predicted by the model with those measured in spall experiments. This comparison requires many laborious tests accompanied by careful post-test examinations of the impacted samples.

For some applications, however, this approach may not be necessary. Constitutive models simpler than those developed using the NAG approach can be constructed and verified based only on information derived from experimentally measured free-surface velocity profiles. The development of such simple empirical models is facilitated if the model formulation is guided by an analysis of a series of free-surface velocity profiles. The analysis can provide preliminary information about the damage kinetics and may permit the estimation of some parameters of the chosen constitutive relationship.

Section 4.2 described an acoustic analysis of the spall process in an attempt to correlate the free-surface velocity profiles with the rate of fracture at the spall plane. We now use these results to formulate empirical constitutive relationships for fracture damage under spall conditions.

The analysis of the spall process presented in Section 4.2 permits us to make the following observations about the initial stages of spall fracture in a specimen loaded by a triangular stress pulse. The fracture process sets a limit on the growth of the peak tensile stress behind the spall plane when the reflected tensile wave propagates into the body. Assuming the damage rate, $\dot{V}_v$, depends linearly on the pressure, p (see Eq. 4.29), the ultimate tensile stress is reached at the damage rate

$$\dot{V}_v = -\frac{4\dot{p}_0}{\rho_0^2 c^2}, \tag{4.47}$$

where $\dot{p}_0$ is the unloading rate in the incident compression pulse. Thus, the ultimate magnitude of the reflected tensile pulse corresponds to a void growth rate equal to four times the unloading rate in the incident compression pulse. This same damage rate leads to the appearance of the minimum (i.e., spall signal) on the free-surface velocity profile. The spall signal is formed only if the damage rate is more than four times as great as the expansion rate during unloading in the incident compression pulse.

Under the threshold conditions represented by Eq. (4.47), the pressure at the spall plane increases at a rate exactly equal in magnitude to the unloading rate in the incident load pulse. In other words, the appearance of the spall signal on the free-surface velocity profile means that the damage rate is increasing so rapidly with the development of fracture, that this increase compensates for the relaxation of the tensile stress. Therefore,

$$\frac{\partial \dot{V}_v}{\partial V_v} \geq -\frac{\rho_0^2 c^2}{4} \frac{\partial \dot{V}_v}{\partial p}. \tag{4.48}$$

The pressure and void volume vary continuously during fracture; as a result, the damage rate also varies. The damage rate after damage initiation is related to the rise time of the spall pulse front. The slope of the spall pulse front is

$$\frac{du_{fs}}{dt} = \frac{\dot{p}_0}{8\rho_0 c}\left(\dot{V}_v / \dot{V}_0 - 4\right), \tag{4.49}$$

where $\dot{V}_0$ is the expansion rate in the incident pulse.

It is reasonable to assume that the damage rate is a function of both tensile stress and degree of damage. As a result, the threshold damage rate that corresponds to the appearance of a minimum in the free-surface velocity profile can be reached at many times during the development of fracture. This time interval decreases with increasing tensile stress, while the reflected rarefaction wave propagates from the free-surface into the body. Figure 4.18 showed the threshold line in the time–distance diagram, along which the condition $\dot{V}_v = 4\dot{V}$ is satisfied. The spall signal arrives at the sample surface from the point on this line where the slope is

$$\frac{dt}{dx} = -\frac{1}{c}. \tag{4.50}$$

As a result, the duration, Δt_i, of the first velocity pulse on the free-surface velocity profile exceeds the period, Δt_{spall}, of later (after beginning of fracture) velocity oscillations. The difference between these time intervals is interpreted as an apparent delay time of the spall fracture. Obviously, development of the fracture to the left of the spall plane is suppressed by the compression wave created as a result of the relaxation of the tensile stress in the spall plane.

As an example, let us consider the spall fracture rate for the Al - 6% Mg alloy. Figure 4.19 shows results of measurements of the spall strength σ^* as a function of the unloading expansion rate in the incident shock pulse. The dashed line in Figure 4.19 is a fit to the power function

$$\sigma^* = 0.12 \cdot \left(\frac{\dot{V}}{V_0}\right)^{0.18} \text{ GPa}, \tag{4.51}$$

where V_0 is the initial specific volume of the material. This empirical relationship reflects the dependence of the damage rate on the applied tensile stress. Using the results of the analysis discussed above, we may conclude that the damage rate depends on the tensile stress as

$$\dot{V}_v = 4V_0 \left\{\frac{\sigma}{0.12}\right\}^{\frac{1}{0.18}}. \tag{4.52}$$

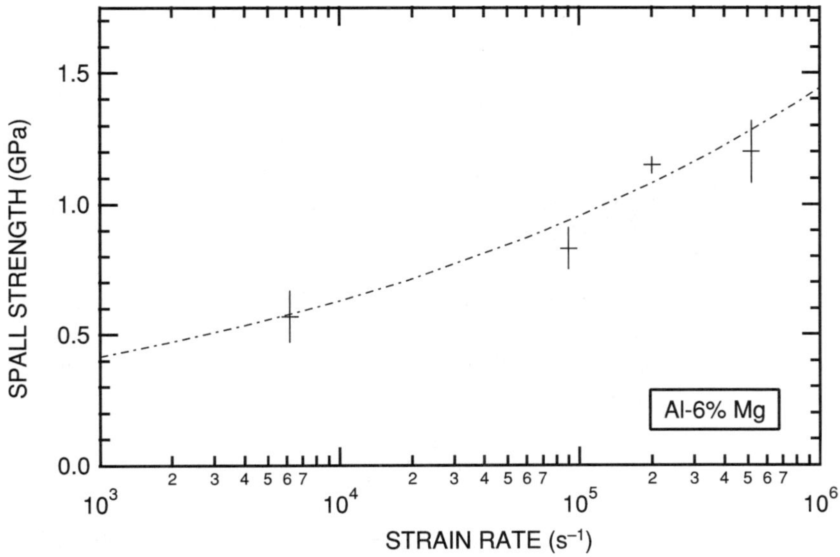

Figure 4.19. Dependence of the spall strength of the Al-6% Mg alloy on strain rate.

Experimental profiles for this alloy do not indicate any notable delay of the fracture. Within the experimental error of the measurement, the period of the velocity oscillations after the beginning of spall fracture corresponds to the duration of the first velocity pulse with allowance for the difference between the propagation velocities of the spall pulse front (longitudinal sound velocity c_l) and the incident unloading wave ahead of it (bulk sound velocity c_b). Additionally, free-surface velocity profiles for this alloy do not show any notable stress relaxation ahead of the spall signal. Thus, the expression for $V_v(\sigma)$ obtained above describes the initial, or near initial, damage rate. Experimental profiles show also that the steepness of the spall pulse front is always proportional to the velocity gradient in the incident unloading wave.

In other words, a faster initial damage rate is accompanied by a proportionally faster damage rate on the following phases of the fracture process. The initial damage rate and the damage rate at later times seem to be controlled by the same parameter that appears as a multiplier in the constitutive relationship. This multiplier can represent, for example, the number of damage sites. We cannot determine the concentration of the damage sites from the free-surface velocity profiles, but it is reasonable to assume that this concentration is determined, for example, by the ultimate tensile stress at which damage is activated.

A simple constitutive relationship consistent with our observations can be expressed in the following form:

$$\frac{\dot{V}_v}{V_0} = \frac{\sigma}{\tau\sigma_n}\left(\frac{\sigma_{\max}}{\sigma_n}\right)^{\alpha-1}\left(\frac{V_v}{V_0}\right)^{\beta}, \tag{4.53}$$

where $\sigma_{\max}$ is a point function representing the peak tensile stress experienced by the material during the passage of the rarefaction wave, constants σ_n and α are taken from the empirical relationship (4.52), and the time factor τ and the parameter β are yet to be determined. The relation (4.53) carries the implication that all damage nucleation sites are activated simultaneously when the peak tensile stress is reached.

For preliminary estimations, let us begin by finding the threshold line for the spall process after reflection of a triangular compression pulse from the free-surface. Let the pressure gradient at unloading in the incident pulse be $\dot{p}_0 = -\frac{1}{2}kc$, which corresponds to the expansion rate of

$$\dot{\overline{V}} = \frac{-\dot{p}_0}{\rho_0 c^2} = \frac{k}{2\rho_0 c}. \tag{4.54}$$

The ultimate tensile stress increases linearly as a function of the propagation distance of the reflected rarefaction wave:

$$\sigma = kx. \tag{4.55}$$

As a first approximation, we consider only the initial stage of the fracture development, assuming that small initial increments of voids do not lead to substantial relaxation of stress. In this case, the condition (4.47) gives

$$\dot{\overline{V}}_v = A\sigma^{\alpha}\overline{V}_v^{*\beta} = \frac{2k}{\rho_0 c}, \tag{4.56}$$

where $\overline{V}_v = V_v/V_0$ and $A = 1/\tau\sigma_n^{\alpha}$. Solving for the volume of voids, we obtain

$$\overline{V}_v^* = \left(\frac{2k}{\rho_0 c}\frac{1}{A\sigma^{\alpha}}\right)^{\frac{1}{\beta}}. \tag{4.57}$$

Another expression for the void volume can be obtained by integrating the kinetic relationship (4.53):

$$\overline{V}_v^* = \left[(1-\beta)A\sigma^{\alpha}\Delta t\right]^{\frac{1}{1-\beta}}, \tag{4.58}$$

where Δt is the time interval needed to reach V_v^*.

The last two relationships can be combined to obtain the following expression for Δt:

$$\Delta t = \frac{1}{A(1-\beta)}\left(\frac{2k}{A\rho_0 c}\right)^{\frac{1-\beta}{\beta}}\sigma^{-\frac{\alpha}{\beta}}, \tag{4.59}$$

and the threshold line in the time–distance diagram is obtained as

$$t = \frac{x}{c} + \Delta t,$$

or

$$t = \frac{x}{c} + \frac{1}{A(1-\beta)} \left(\frac{2k}{A\rho_0 c} \right)^{\frac{1-\beta}{\beta}} (kx)^{-\frac{\alpha}{\beta}}. \tag{4.60}$$

The spall signal is formed at the point where the slope of this curve is $-1/c$. This condition is satisfied at the distance

$$x^* = \frac{1}{k} \left[\frac{c}{2} \frac{k\alpha}{A\beta(1-\beta)} \left(\frac{2k}{A\rho_0 c} \right)^{\frac{1-\beta}{\beta}} \right]^{\frac{\beta}{\alpha+\beta}}. \tag{4.61}$$

Figure 4.20 shows threshold lines for the damage kinetics (4.53) calculated with different values of β for the same x^*. In these calculations the time factor τ was increased with decreasing β. The apparent delay of the spall is almost linearly proportional to the value of β. Let us now check the condition (4.48). For the fracture kinetics (4.53), this condition is

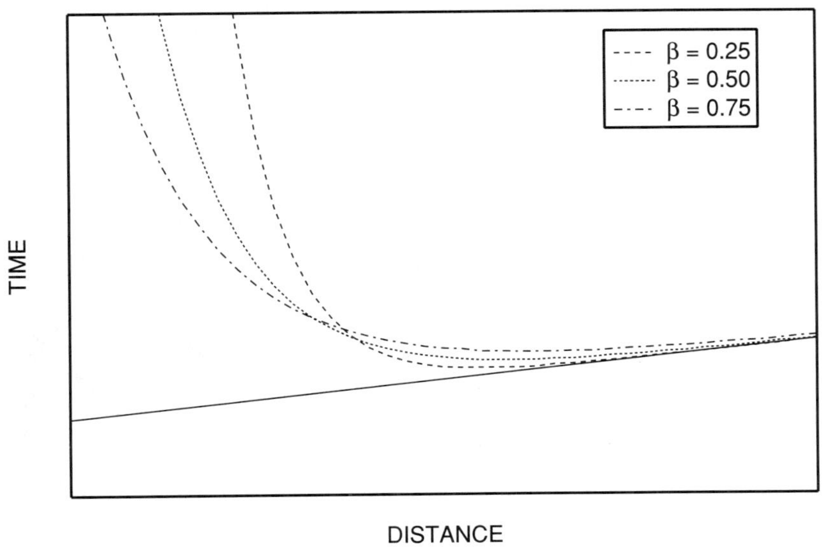

Figure 4.20. Threshold lines calculated with the constitutive relatio nship (4.53).

$$\frac{\sigma}{\overline{V}_v} \geq \frac{\rho_0 c^2}{4\beta}. \tag{4.62}$$

In the case of smaller β, the condition (4.48) ceases to be satisfied at smaller porosity, which means decreasing amplitude of the spall signal. At some small β, the condition (4.48) is satisfied only at $\overline{V}_v \leq \overline{V}_v^*$ and a spall signal cannot form. The apparent delay of the fracture seems inevitable with the assumed fracture kinetics in the form of Eq. (4.53).

Calculations of the threshold line are much simpler than complete computer simulations of the spall process and they provide an effective tool for obtaining preliminary estimates of the constitutive model parameters.

Figure 4.21 compares experimental free-surface velocity histories with the results of computer simulations of the spall experiments, using the constitutive relationship (4.53). The model parameters used in the simulations are $\sigma_n = 0.12$ GPa and $\alpha = 5.65$, as follows from Eq. (4.52), and $\beta = 0.5$ and $\tau = 4.2\,\mathrm{x}\,10^{-2}$ s. In the calculations, complete fracture was assumed to correspond to a void volume equal to 25% of the initial volume, or $0.25\,V_0$. The com-

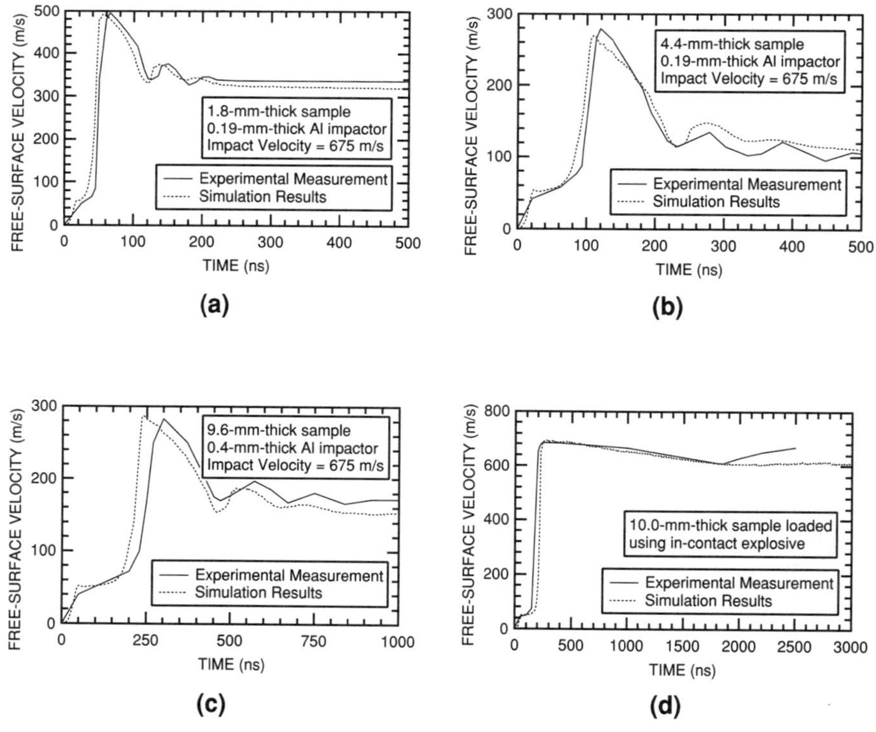

Figure 4.21. Comparison of measured free-surface velocity profiles with those calculated using the constitutive relationship (4.53) for the Al-6% Mg alloy.

puter simulations were done with the one-dimensional Lagrangian code EPIF. The elastic-plastic properties were described by the structural Marzing model, which represents each elementary volume of the body as a set of parallel elements with different yield strengths.

In this section, we have demonstrated how a simple constitutive model for describing spall damage can be established based only on free-surface velocity profiles. The model is empirical, and it does not attempt to relate damage in the material to underlying micromechanisms. This model can be improved, but that will require additional experimental and theoretical information and will lead to a more complicated and less computationally efficient set of equations, which is beyond the scope of this simple analysis. Chapters 6 and 7 provide a detailed description of the theoretical, experimental, and numerical aspects of the nucleation-and-growth (NAG) modeling approach, an approach used to develop microstructurally based constitutive models for describing the dynamic behavior of ductile and brittle materials.

4.4. Summary of the Information Obtained from the Spall Signal

As discussed above, the free-surface velocity histories measured in spall experiments contain more information about spall fracture than simply the velocity pull-back. The shape and amplitude of the spall signal, as well as the frequency and decay of the free-surface velocity oscillations in the later phases of spalling also contain useful information about the process parameters. The rise time of the spall signal front is determined by the fracture development rate, and a higher peak velocity at the peak of the spall signal may be expected for more rapid completion of the fracture process. Additionally, the dependence of the experimentally determined spall strength on the load pulse duration can be related to the dependence of the incident fracture rate on stress. The relationship of the first period of the free-surface velocity oscillation to the periods of subsequent oscillations in the $u_{fs}(t)$ profile is associated with the delay time of the rapid fracture onset. Also, decay of the particle velocity oscillations during the wave reverberation inside the spall plate can be related to the damping characteristics of the fracture surface layer. If the spalling process does not culminate in complete fracture during the first wave reverberation, an overall deceleration of the spall plate should be observed at some later time in the free-surface velocity record.

These qualitative observations can be quantified through the use of analytical and modeling techniques that aid in the interpretation of the spall signal. To make effective use of these techniques, it is important to establish bounds within which certain analysis tools are appropriate, and others are not. Below is a brief discussion that identifies some of the uncertainties associated with various spall analyses.

Most analytical tools are founded in continuum mechanics and, as such, they assume continuous flow through uniform material. The continuum mechanics approach, being quite fruitful in general, is, nevertheless, limited by the nature of real fracture processes, which proceeds through damage development at numerous but separate and discrete sites. This distinction may become essential when measurements are made with a high degree of space resolution using laser velocimetry techniques. In such cases, the discrete nature of the damage development process causes scatter in the measured velocity pullback and in the spall signal shape. On the other hand, the localized nature of fracture results in geometrical dispersion of the rarefaction and compression waves recorded in the free-surface velocity history, which complicates the analysis and interpretation of the record. Decay of the velocity oscillations may also result from geometrical dispersion on the rough fracture surface as well as from bulk viscous dissipation in porous damaged layers.

The next source of uncertainty in the interpretation of the spall signal is distortion, resulting primarily from nonlinear material compressibility and elastic-viscoplastic behavior, that influence the spall pulse as it propagates from the spall zone to the free-surface of the specimen. Since the sound speed increases with increasing pressure, the spall signal has a tendency to become steeper as it propagates from the spall zone to the sample surface. In contrast, viscoplastic effects cause wave dispersion, and hysteresis associated with the elastic-plastic deformation cycle causes a decrease in the amplitude of the spall signal thus making it difficult to compare the spall behavior of materials having different yield strength. The influence of these nonlinear effects on the spall pulse increases with increasing propagation distance, making it nearly impossible in some cases to recover the incident stress history in the vicinity of spall plane from the velocity history measured some distance away on the rear surface of the specimen. In experiments with a relatively large ratio of impactor thickness to sample thickness, the competing effects of these nonlinear sources of distortion may combine to produce the same spall signal in samples that may have experienced differing kinds and levels of damage.

In carefully designed experiments, measurements of the free-surface velocity profiles provide rather unambiguous information about the fracture stress, the spall plate thickness, the spall fracture delay, and the corresponding initial fracture rate. Extracting information about the fracture mechanisms and the fracture kinetics requires more interpretation. In order to reveal the kinetic parameters of fracture development, computer simulations of the phenomenon are performed. These code simulations are also subject to various limitations. The material model used in the simulations is of critical importance. Constitutive models usually contain a number of free parameters. If the model and/or the parameters are not constrained, it is possible to obtain the same result from simulations using different constitutive models or even using various combinations of parameters of one model. The uncertainty can be reduced substantially by choosing an appropriate model and by calibrating the model using experimental data that span a wide range of load conditions. Since the elastic-viscoplastic properties of the material influence the shape of both the incident load pulse and the spall pulse, it is important that these properties be properly represented in the simulations.

The formation of the spall signal is a result of a precipitous drop in the ability of the material to resist tension, which, in turn, may be the result of different combinations of microfracture events. Unfortunately, the existing experience of investigations of the spall phenomena by means of measurements of the free-surface velocity profiles does not permit one to infer an unambiguous relationship between the fracture mechanism and the shape of the spall signal. For example, it cannot be determined, solely on the basis of pullback signal, whether the fracture proceeds by means of cracking (i.e., brittle behavior) or by means of nucleation and growth of nearly spherical voids (i.e., ductile fracture). Whereas there is a tendency for more brittle materials to produce steeper spall signals, there is also experimental evidence of face-centered cubic metals producing very steep spall signals. Thus, two types of experimental investigations are required to span the full spectrum of experimental spall studies: pullback signal measurements and metallurgical examinations. Neither approach is in itself conclusive. Rather, the two approaches complement each other and, together with analytical and computational modeling techniques, they provide effective tools that can be used to develop a comprehensive understanding of spall based on physical damage and deformation mechanisms.

5
Spallation in Materials of Different Classes

In this chapter, we summarize the results of instrumented measurements of the resistance of materials of different classes to dynamic fracture. The results discussed here were obtained from wave propagation experiments conducted under uniaxial strain conditions, and consist primarily of free-surface velocity measurements. The recorded motion of the free-surface of the plate does not provide any direct information about cracks, voids, or other physical characteristics of spall damage, but it does provide the most direct and most reliable information about the fracturing stress magnitude and the stress relaxation at fracture. The experiments discussed here investigated the influence of the peak shock pressure, temperature, load duration and orientation, and the structure of the materials on the resistance to spall fracture for materials of different classes including commercial metals and alloys, ductile and brittle single crystals, glasses, ceramics, polymers, and elastomers.

These experiments were performed by Genady Kanel and his co-workers at the Russian Academy of Sciences. Thus, the materials investigated are those readily available in the Former Soviet Union (FSU), and in some cases, those materials are somewhat different from those available elsewhere. Table 5.1 gives the composition of the various materials tested by Kanel et al. to facilitate comparison of the Russian data provided in this document to spall data from other sources. Whenever appropriate, Table 5.1 also provides the U.S. equivalent of the FSU alloys.

This chapter includes only a fraction of the results—that fraction needed to show trends and discuss specific aspects of the behavior of the various materials investigated. A compilation of all the data is provided in the Appendix. The Appendix provides a comprehensive, self-contained summary of each of 148 experiments, and includes (1) a description of the material investigated including its density and elastic properties; (2) a schematic diagram of the experiment; (3) the dimensions and conditions of the material investigated; (4) the technique used to perform the measurement and the associated experimental error; and (5) the experimental results which in all cases take the form of a particle velocity history recorded at the free-surface of the sample or at the interface between the sample and a softer material.

Table 5.1. Western equivalents of FSU metal alloys.

Material	FSU alloy designation	U.S. alloy designation	Composition	
Aluminum	AD1	1100		
	AMg6M	2017		
	D16	2024		
Magnesium	Ma1	MTA		
Titanium	VT5-1	Ti-5Al-2.5Sn (alpha phase)		
	VT6	Ti-6Al-4V (alpha plus beta)		
	VT8	Ti-7Al-4Mo (alpha plus beta)		
Steel	3		C	0.14%–0.22%
	(Low carbon)		Mn	0.3%–0.5%
			Si	<0.07%
			P	<0.045%
			S	<0.055%
	45		C	0.42%–0.5%
	(Structural carbon steel)		Mn	0.5%–0.8%
			Si	0.17%–0.37%
			P	0.04%
			S	0.04%
			Cr	0.25%
			Ni	0.25%
	KhVG		C	0.9%–1.05%
	(Doped tool steel)		Mn	0.8%–1.1%
			Si	0.15%–0.35%
			Cr	0.9%–1.2%
			W	1.2%–1.6%
	Kh18N10T		C	<0.12%
	(High-doped stainless steel of the austenite class)		Mn	1.0%–2.0%
			Si	<0.08%
			P	0.035%
			S	0.02%
			Cr	17%–19%
			Ni	9%–11%
			Ti	0.5%–0.7%
	35Kh3NM†		—	—
	EP836‡		—	—

† The composition of 35Kh3NM steel was not available in the public domain at the time of publication.
‡ The composition of EP836 steel was not available in the public domain at the time of publication.

5.1. Metals and Metallic Alloys

Metals are the materials most thoroughly investigated in research on spall phenomena. Figure 5.1 shows results of experiments with the strong Fe-Cr-Ni-Mo steel 35Kh3NM (Glusman et al. [1985]). The samples were cut from a bar workpiece in two orientations. Some of the tests were performed under shock loading in the rolling direction, whereas others were performed under shock loading in the transverse direction. In both cases, the free-surface velocity profiles were recorded.

Those profiles, shown in Figure 5.1, display the influence of load direction on the dynamic strength. The resistance to dynamic fracture is 4.0 to 4.4 GPa for loading in the rolling direction and 10% to 15% less for loading in the lateral direction. The faster decay of the velocity oscillations in the latter case is due to a more highly developed fracture surface. For samples loaded in the rolling direction, rather uniform, fine-grained, and light fracture surfaces were observed. In the samples loaded in the transverse direction, fracture occurred mainly along intergranular boundaries, and consequently, fracture surfaces demonstrated more pronounced texture. The fracture surfaces were dark and much less uniform. The nonuniformity of the fracture surface accelerates the dispersion and decay of the wave reflected from the spall plane.

We can estimate the average velocity of the spall plate from the free-surface velocity profiles shown in Figure 5.1 simply by calculating the shift of the surface during one velocity oscillation and then dividing this shift by the duration of this oscillation. Such estimations indicate that the average spall plate velocity

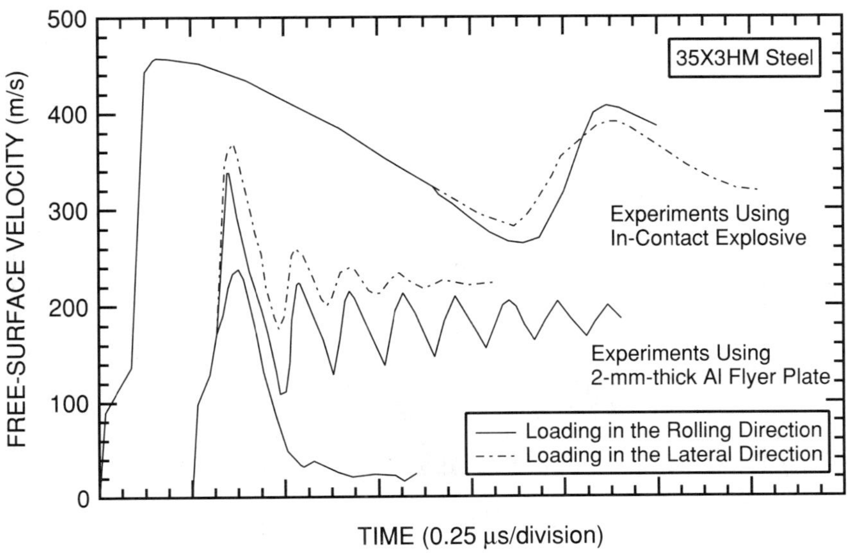

Figure 5.1. Free-surface velocity histories at spalling in 35Kh3NM steel (Fe-Cr-Ni-Mo).

is practically constant. We may then conclude that the fracture time for this steel is less than half a period of the free-surface velocity oscillations, which is less than 10^{-7} s. Therefore, fracture of this steel is relatively brittle in nature and is completed by the end of the first half-period of velocity oscillations.

A protracted, slowly evolving spall fracture was observed for more viscous materials such as stainless steel (Kanel [1982a] and [1982b]) (Figure 5.2). In this case, the spall plate continued to decelerate for a long time after the initial appearance of the spall pulse, indicating the continuous evolution of fracture. Similar behavior was also observed for other materials when the incident pulse duration was shorter than the time required for complete fracture.

Like the quasi-static strength, the spall strength of steels varies depending on the grain size and the phase state of the material. Low-pressure experiments with 4340 steel after various heat treatments (Butcher [1967]) exhibited an increase in the incipient spall threshold from 2.55 GPa to 4.4 GPa with increasing hardness of the samples from 15 HRC to 54 HRC. This behavior correlated with the low-strain-rate true strength values of 1.33 and 2.66 GPa for the 15 HRC steel and the 45 HRC steel, respectively. The influence of heat treatment on the spall strength of Russian 40Kh (40X in Russian letters) steel was studied by Golubev and Novikov [1984]. Microscopic examination of samples impacted at different velocities showed that the spall threshold increases from ~3 GPa at 20 HRC initial hardness to ~5 GPa at 55 HRC. The damage nucleated on the sulfide in-

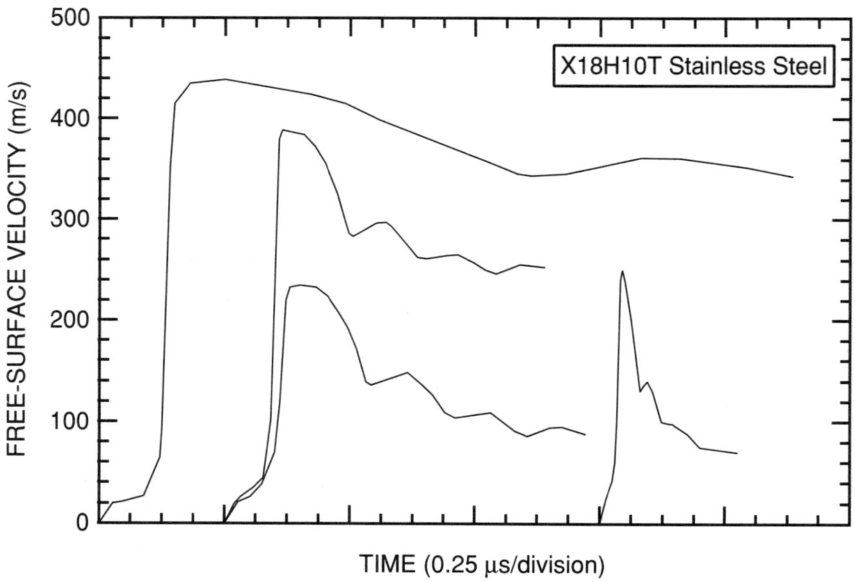

Figure 5.2. Free-surface velocity profiles at spalling in stainless steel. Note the continued deceleration after the arrival of the spall signal at the free-surface.

clusions or on the boundaries of martensite blocks. Work hardening as a result of shock loading was observed for all initial states of the samples.

The effects of microstructure on spall fracture were studied for 1008 carbon steel (Zurek and Follansbee [1990]; Zurek et al. [1990]) in more detail. The measurements indicated that there is no essential grain size dependence in the range of 25 to 45 μm. However, there is a substantial dependence of the spall strength on the number of carbides per unit volume. The spall strength of 1008 steel increases from 1.2 to 1.8 GPa at decreasing carbide size distribution from 1.5 x 10^{-3} carbides/μm^3 to 3.7 x 10^{-3} carbides/μm^3. In other words, because the carbides serve as nucleation sites for the brittle fracture, the material with the larger spacing between the damage nucleation sites exhibited the higher spall strength. However, 1007 carbon steel of 5-μm grain size and submicron carbides (Zurek et al. [1990]) exhibited the higher spall strength (2.4 GPa) and transition to ductile fracture.

Table 5.2 gives the results of measurements of spall strength for several metals and alloys at various load durations. These measurements showed that the resistance to fracture under these impact conditions increases with increasing characteristic rate of expansion. No correlation is apparent between the total time of fracture and the fracture stress dependence on the characteristic load duration. This can be verified, for example, by comparing the experimental data for stainless steel with similar data for other materials. The dependence of fracture stress on characteristic load duration appears to be determined rather by the intrinsic properties of existing damage nucleation sites and their initial growth velocity. The total time of the fracture process depends on the rate of later development and subsequent coalescence of voids.

To evaluate the effect of load duration on the measured strength, we next compare the spall strength measured under impact loading conditions with the corresponding strength determined under quasi-static conditions. Standard quasi-static tensile tests may be interpreted in terms of engineering stress (peak load divided by original cross section area), σ_b, which characterizes the maximum load withstood by the sample, or in terms of the true stress $S_k > \sigma_b$ found with due regard for the change in the cross-sectional area of the sample in the neck region.

For ductile materials, the discrepancy between S_k and σ_b may be very large. Because σ_b is a measure of the loss of stability of the sample (but not of the material), it cannot be used as a basis for comparing the strength properties determined under different loading conditions (i.e., uniaxial stress for the quasi- static test and uniaxial strain for the dynamic spall test). The true tensile breaking stress S_k provides a more objective basis for comparison. Comparison of the spall strength, obtained at load durations on the order of 1 μs, with the quasi-static true breaking stress shows that the spall strength usually exceeds the static strength by a factor of 1.5 to 2.0.

Since the fracture time under shock wave loading conditions is comparable to the duration of the load, it is important to determine the stage of fracture that

Table 5.2. Spall strength of metals and alloys determined based on measurements of the free-surface velocity history.

Material	Skl (GPa)	Strain rate (s^{-1})	Spall strength (GPa)	Spall thickness (mm)	Reference
Aluminum alloy		6.2×10^3	0.57 ± 0.1	4.7	Kanel, Razorenov, and
AMG6m		8.9×10^4	0.83 ± 0.08	0.61	Fortov [1984a]
(Al-6%Mg)		2×10^5	1.15 ± 0.05	0.34	
(sheet)		5.1×10^5	1.2 ± 0.12	0.18	
Aluminum P1C		10^5 to 10^6	0.9 to 1.2	0.25 to 0.9	Speight et al. [1973]
Aluminum alloy 2024	0.4 to 0.6	3×10^4	1.15 ± 0.1	2.75	Taylor [1968]
Aluminum alloy V95			1.6 to 1.8	4 to 10	Romanchenko and Stepanov [1980]
Copper M2		2.8×10^3	0.8 ± 0.1		Kanel, Razorenov, and
		3.4×10^4	1.1 ± 0.1	0.5	Fortov [1984b]
		2.5×10^5	1.64 ± 0.1	0.2	
Copper OFHC			1.2	0.8 to 3	Cochran and Banner [1977]
Magnesium		3×10^4	0.8	0.62	Kanel, Razorenov, and
Ma1		1.1×10^5	0.88	0.29	Fortov [1984b]
Magnesium AZ31B			0.8	3	McQueen et al. [1970]
Titanium alloy	1.6	7.3×10^4	3.7 ± 0.2	1.7	Kanel and Petrova [1981]
VT6		5.7×10^4	3.5 ± 0.2	1.8	
		8.3×10^3	3.4 ± 0.2	9.8	
Titanium alloy Ti 6Al 4V			4.1 to 5		Me-Bar et al. [1987]
Titanium alloy VT8		7.5×10^4	4.65 ± 0.3	1.65	Kanel, Razorenov, and Fortov [1987]
Uranium			2.4 ± 0.1	0.8 to 3	Cochran and Banner [1977]
Tantalum		8×10^4	4.6 ± 0.2	1.15	McQueen et al. [1970]
Tantalum			4.4	0.8 to 3	Cochran and Banner [1977]
Tungsten W-2			0.74		Rosenberg [1987]
Tungsten			0.42 to 0.66		Asay et al. [1980]
Armco iron	1.2	2.9×10^4	1.65 ± 0.1	1.1	Kanel and Shcherban
		6×10^3	1.4 ± 0.1	4.2	[1980]
Austenite	1.6 to 1.8	1.1×10^5	2.3 ± 0.1	0.5	Kanel [1982a]
Stainless steel		4.5×10^4	2.0 ± 0.1	1.6	
Kh18N10T		2×10^4	1.9 ± 0.1	3.75	
		5.3×10^3	1.85 ± 0.1	6.9	
Stainless Steel 304		2.5×10^4	2.1	3.4	McQueen et al. [1970]
				0.8 to 3	
Austenit steel		3.2×10^4	2.9	4	McQueen et al. [1970]
Steel 4330V			4.8	0.8 to 3	Cochran and Banner [1977]

corresponds to the fracture stress determined by a given method. The damage rate is approximately equal to the product of the concentration of damage nucleation sites times their average growth rate. This damage growth rate cannot be arbitrarily large. Consequently, significant overstressing of the material is possible if the load is applied at a high-rate. However, several thresholds of spalling (incipient fracture, intermediate stage, and the main crack formation) can be observed in samples after testing with load pulses of various durations (Smith [1962]). The existence of several thresholds is a consequence of evolution of the fracture process in time. The question arises: Which stage of fracture is related to the spall strength determined from the free-surface velocity profile?

The rate of damage that can be observed in the wave profiles depends on the decompression rate in the incident load pulse. Actually, the spall pulse can be seen on the free-surface velocity profile if the fracture kinetics are fast enough. A slower fracture process leads to an increased rise time of the spall-pulse front. At some damage rate, we can obtain the spall pulse with zero gradient in its front. In this regime the compression wave produced by stress relaxation during the relatively slow fracture process compensates for the rarefaction in the incident release wave. Obviously, the damage rate that leads to this condition will be different for different load parameters.

The situation with wave compensation is realized at the minimum point of the $u_{fs}(t)$ profile ahead of the spall pulse. The acoustic analysis presented in Chapter 4 shows that a minimum in the free-surface velocity profile appears when the rate of damage is equal to four times the decompression rate in the incident load pulse. In other words, each spall stress value corresponds to some specific damage rate:

$$\dot{V}_v^* \approx \frac{4\dot{u}_1}{2\rho c_b}, \tag{5.1}$$

where $\dot{u}_1$ is the surface-velocity gradient in the decompression part of the incident load pulse. Comparison of the duration of the first velocity pulse and the periods of the following velocity oscillations (after fracture initiation) on the $u_{fs}(t)$ profile can give the delay time for achieving this damage rate. The period of oscillations must be shorter than the first velocity pulse if any delay exists. The effect of the elastic-plastic properties of the material must be taken into account. Similar analysis for metals shows that the damage rate discussed usually corresponds (practically without any delay) to the time at which the tensile stress on the spall plane reaches σ^*. Thus, an empirical dependence $\sigma^*(\dot{u}_1)$ or $\sigma^*(\dot{\varepsilon})$, where $\dot{\varepsilon} = V = -\dot{u}_1 / 2c_b$, should reflect the initial damage rate dependence on the applied tensile stress, that is, $V_v(\sigma)$.

It can be noted that irregularities of the fracture surface become finer with decreasing load pulse duration. The shortening of the load pulse is accompanied by an increase in the overstress in the sample. As a result, it becomes possible to nucleate fracture at finer defects. Smaller and more numerous damage nucleation sites are activated as the peak stress increases during propagation of the tensile wave into the sample, leading to an increase of the initial damage rate. Near the surface of the sample, the tensile stresses are not so high and fracture is initiated at coarser defects. Because the number of coarse defects is relatively small, the resulting fracture is not fast enough to provide substantial relaxation of tensile stresses. Development of fracture near the surface of the specimen is arrested by unloading because of the fast fracture in deeper layers of the sample.

Thus, the measured fracture stress determines the conditions of spall initiation. Comparing the spall strength deduced from free-surface velocity measurements with thresholds of spallation determined through metallographic examination of recovered samples shows that the spall strength values agree with the incipient fracture threshold or, in some cases, are even less than this threshold. Two other thresholds, corresponding to intermediate and complete fracture, occur at higher shock intensities. This does not necessarily mean that fracture occurs under higher tensile stresses. Indeed, stress relaxation just after fracture initiation limits the growth of tensile stresses as the shock wave intensity is increased. Full separation of the sample at the spall plane occurs much later than the time instant when maximum tensile stress is reached. The development and completion of the fracture process occur at lower stress and require additional consumption of energy.

Since the fracture time under shock wave loading conditions is comparable to the duration of the load, the resistance to spall fracture should depend on the temporal parameters of load pulse. Figure 5.3 shows the results of experimental data for five metals plotted in terms of σ^* as a function of V/V_0. The decom-

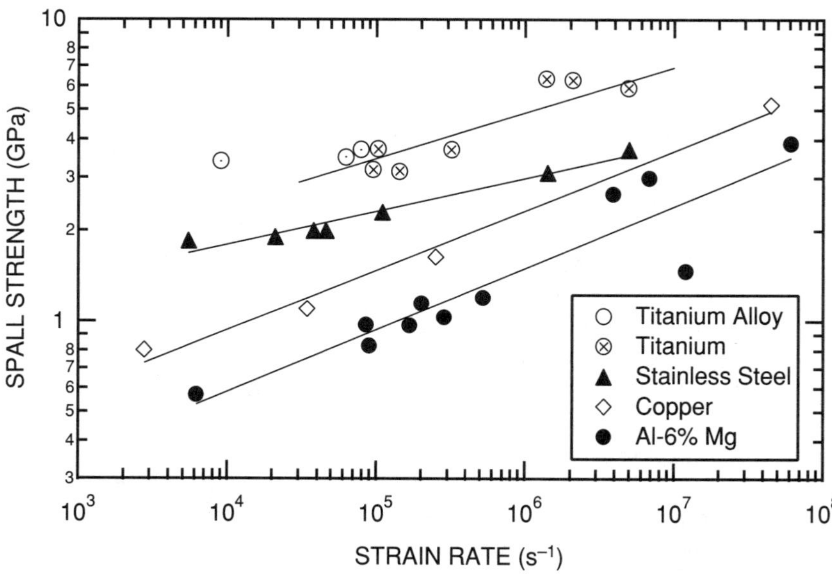

Figure 5.3. Spall strength as a function of unloading strain rate in the incident load pulse.

pression rate in the load pulse, $\dot{V}/V_0$, has been chosen as a parameter characterizing the time dependence of the spall strength because such a dependency can be used to construct kinetic relationships according to the analysis presented in Chapter 4. The resistance to spall fracture can be represented as a power function of the decompression rate of the form

$$\sigma^* = A(\dot{V}/V_0)^m. \quad (5.2)$$

The constants A and m in this equation are presented in Table 5.3 for stainless steel, titanium, copper, aluminum, and magnesium.

Table 5.3. The spall strength as a function of the strain rate for selected materials.

Material	A(GPa)	m
Stainless steel 35Kh3NM	0.650	0.110
Titanium	0.390	0.190
Copper	0.150	0.200
Aluminum-6%Mg alloy, sheets	0.088	0.210
Aluminum AD1 (1100), rod	0.635	0.059
Magnesium Mg95, casting	0.390	0.072

Under conditions of shock wave testing, materials undergo fast compression and heating before the beginning of tension and fracture. Thus, the effect of temperature on the resistance to spall must be evaluated to properly interpret experimental results. Kanel et al. [1987a] studied the effect of prestrain and irreversible heating in the shock wave on the spall strength of titanium alloys in the peak shock pressure range of 2 to 90 GPa. Samples of VT8 titanium alloy were impacted by 2-mm-thick aluminum flyer plates at velocities of 660 m/s, 1900 m/s, and 5300 m/s. This range of impact velocities led to peak shock pressures ranging from 6.5 to 77 GPa, a variation of more than an order of magnitude.

The results, in the form of free-surface velocity profiles, are shown in Figure 5.4. These measurements show that the spall strength of the alloy is 4.16 ± 0.06 GPa under the conditions investigated, it remains practically constant, and it does not depend on the peak shock pressure before the fracture. Note that the residual temperature of samples before spall fracture occurred reached as high as 1100K as a result of irreversible heating in the shock front, and the total deformation in the complete loading-unloading cycle was as high as 60%. Furthermore, these experiments, as well as experiments involving variation of the load direction, show that, because deformation in the shock front and shock-induced changes in the material microstructure do not influence the resistance to spalling, damage nucleation sites are relatively coarse preexisting defects, such as inclusions, micropores, and grain boundaries. It is natural to assume that higher stresses are necessary for damage nucleation on a dislocation level.

Figure 5.5 shows experimental data on the spall strength of aluminum AD1 (Al 1100) over a wide range of peak pressures and load durations (Kanel et al.

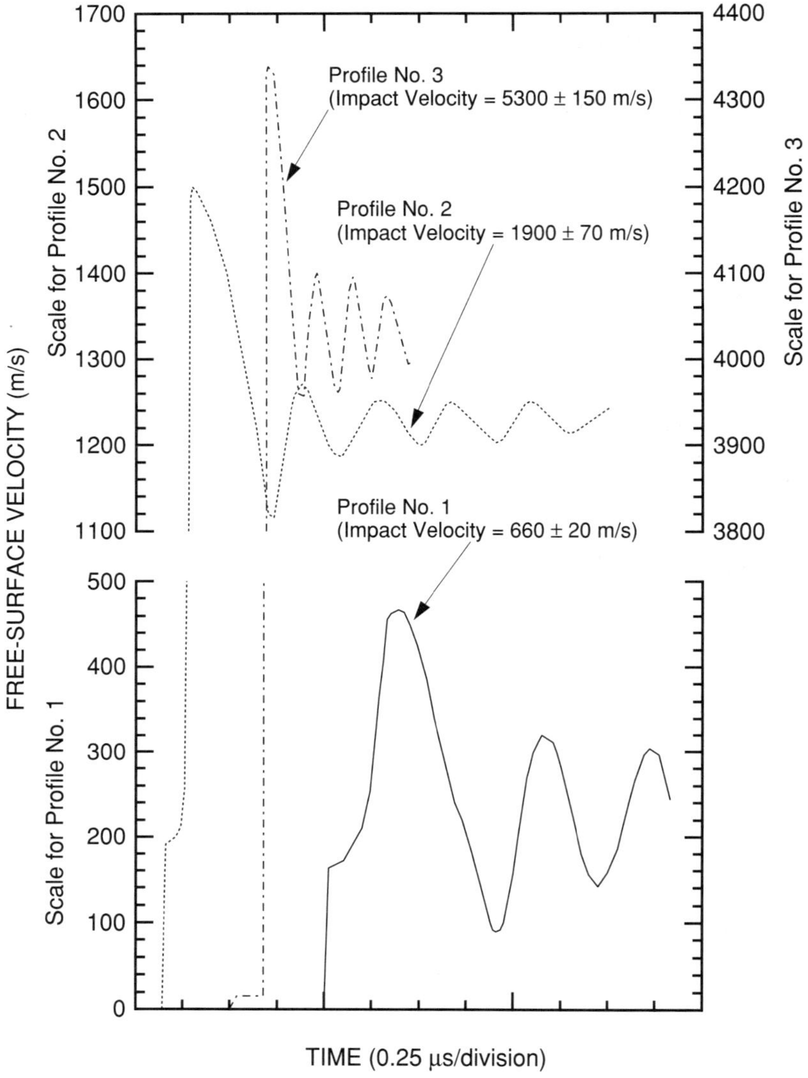

Figure 5.4. Free-surface velocity profiles for the VT8 titanium alloy obtained using the plate impact experimental configuration.

[1996a]). The spall strength, in fact, does not depend on the peak shock pressure in a range up to 35 GPa. Two experiments were performed at a peak pressure of 50 GPa. Measurements at this pressure were not very accurate, but in general the results show approximately a 20% decrease in the spall strength. According to the equation of state of aluminum (McQueen and Marsh [1960]), the residual

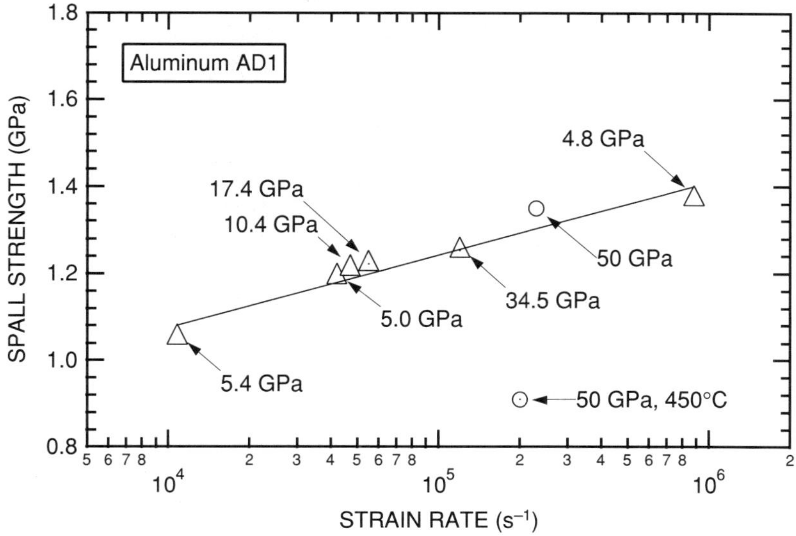

Figure 5.5. Spall strength of aluminum AD1 (Al 1100) as a function of strain rate and peak applied pressure.

temperature of the sample after unloading from a peak shock pressure of 50 GPa is about 450ºC (723.2K), or about 80% of the 933.3K melting temperature of aluminum.

These results suggest that the spall strength of metals decreases with increasing temperature near the melting point. Further evidence of this trend is illustrated in Figures 5.6 and 5.7, which show the dependence of spall strength on temperature for tin and lead, respectively (Kanel et al. [1996b]). These measurements show a substantial drop in the spall strength when the peak shock pressures approach values high enough to achieve melting after the shock wave compression and unloading process.

Relatively few spall studies have been conducted on moderately preheated materials (Bless and Paisley [1984], Dremin and Molodets [1990], Duffy and Ahrens [1994]), and in contrast to results obtained under quasi-static loading conditions, these results show that the dynamic strength of metals is usually not affected significantly by moderate increases in temperature. Free-surface velocity profiles obtained in experiments with copper at 300K and 700K initial temperatures (Bless and Paisley [1984]) did not display any essential variations of the spall strength. Experiments with pure molybdenum preheated to 1700K (Duffy and Ahrens [1994]) resulted in a spall strength of 2.4 GPa, only slightly different from the value of 2.31 GPa reported for room temperature molybdenum by Chhabildas et al. [1990] at the same peak pressure and from values of 1.3 to 2.4 GPa measured for commercial grade molybdenum at room temperature (Kanel et al. [1993]). Similar experiments with Armco iron (Dremin and

Molodets [1990]) over an initial temperature range from 77K to 540K showed an insignificant decrease of spall strength with increasing temperatures that did not exceed the experimental error.

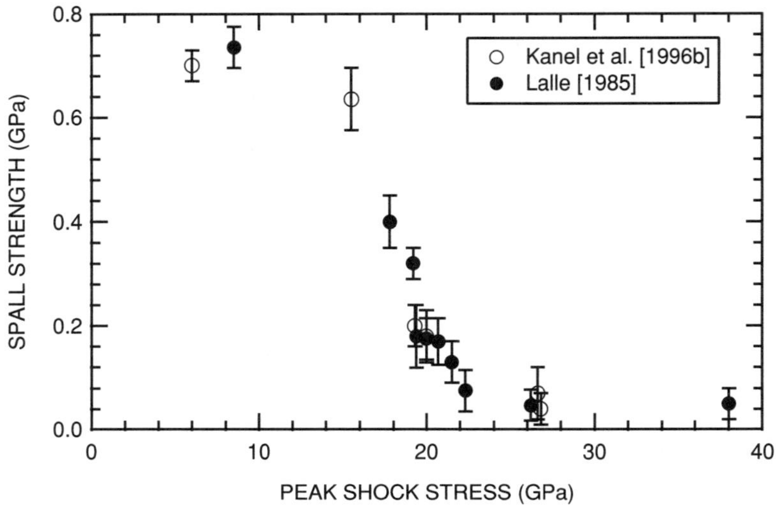

Figure 5.6. Spall strength of tin as a function of peak shock stress.

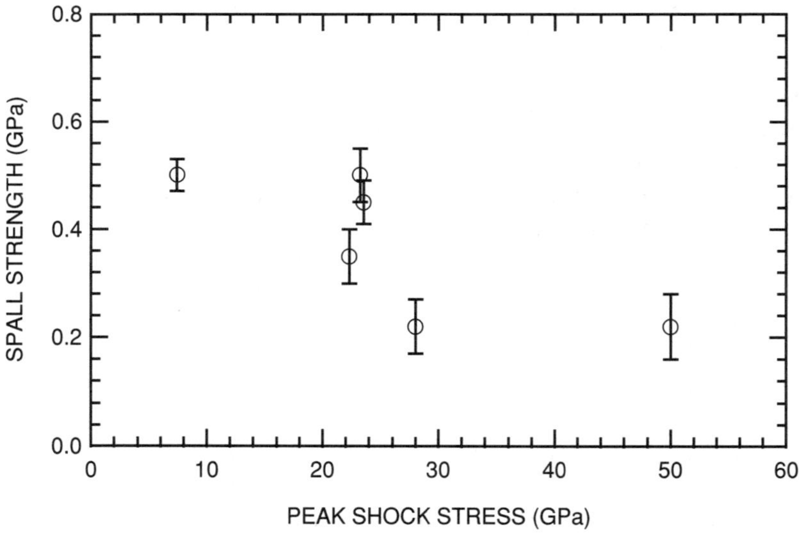

Figure 5.7. Spall strength of lead as a function of peak shock stress.

Metallographic examination of recovered samples tested by impact at various impact velocities and different temperatures provide additional insight into the effect of temperature on spall. Experiments with OFHC copper at room temperature and at 425°C (Bless and Paisley [1984]) indicate that increasing the temperature causes a slight increase in the stress required to initiate damage. The voids are more spherical in room temperature copper, and more small voids are present at elevated temperature. The observed void distributions are consistent with the hypothesis that voids grow slower at elevated temperature. According to Golubev et al. [1985], the incipient fracture threshold for copper decreases from ~1.8 GPa at an initial temperature of –196°C to ~1.2 GPa at 800°C. The room temperature spall strength of copper deduced by Cochran and Banner [1977] and by Kanel [1982a] from free-surface velocity profiles at the same load conditions is also 1.2 GPa, so there is a probability of over-estimating the spall strength values by the post-test inspection method at lower temperatures. For nickel the incipient fracture threshold decreases from ~3.5 to ~1.5 GPa over this same temperature range (Golubev et al. [1985]); the spall strength value measured instrumentally at room temperature is 1.5 GPa. For aluminum alloys from –196°C to 600°C, the damage nucleates at approximately the same stress between 1 and 1.2 GPa (Golubev et al. [1983]). According to Golubev et al. [1988], the incipient fracture threshold of Al-6% Mg alloy decreases from 1 GPa at 0°C to 0.7 GPa at 500°C for rod samples and is approximately constant at ~0.7 GPa over this temperature range for sheet samples.

Spall measurements on samples preheated over the full temperature range

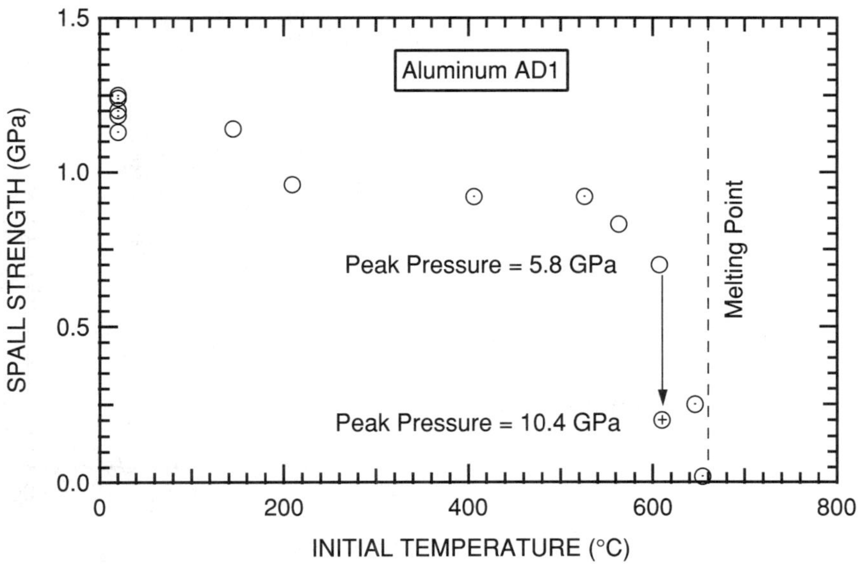

Figure 5.8. Spall strength of aluminum AD1 as a function of initial temperature at a peak shock pressures of 5.8 and 10.4 GPa.

from ambient to near melting were performed by Kanel et al. [1996a]. Figure 5.8 shows the dependence of the spall strength on the initial temperature for aluminum AD1. All but one of these measurements were done at a 5.8 GPa peak shock pressure. In general, the spall strength decreases slightly with increasing initial temperature up to ~0.8 of the melting temperature. A sharp drop in the spall strength is associated with heating above 550° to 600°C. One experiment in this series was done with peak shock pressure of 10.4 GPa and an initial temperature of 610°C. Comparisons of spall strength data at different shock intensities and similar initial temperatures shows that, as a result of the additional heating of the material by the shock wave, the spall strength is a strong function of the peak shock pressure when the initial temperature is near the melting point.

Figure 5.9 shows spall strength measurements for magnesium Mg95 as a function of initial temperature. As for aluminum, a precipitous reduction in spall strength was observed for magnesium as temperatures neared the melting temperature. The sharp drop initiates at temperatures exceeding approximately 500°C.

An additional complication in the dynamic response of iron and steels is the reversible phase transformation that occurs at approximately 13 GPa. The influence of shock pre-strain and peak pressure on the spall strength of pearlitic 4340 steel was studied by Zurek et al. [1992]. They found that increasing the shock wave amplitude from 5 to 10 GPa led to a decrease in the spall strength from 3.1

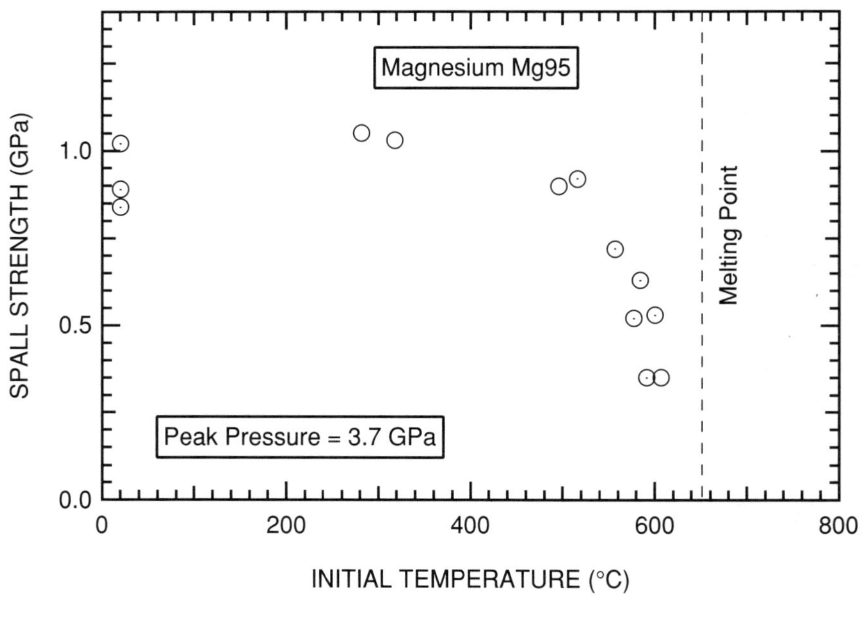

Figure 5.9. Spall strength of magnesium Mg95 as a function of initial temperature at a peak shock pressure of 3.7 GPa.

to 2.6 GPa. At 15 GPa peak pressure, which is above the 13-GPa transition pressure, an increase in the spall strength up to 4.8 GPa was observed for annealed samples. Pre-strain decreased the spall strength down to 2.9 to 3.9 GPa at above the 13 GPa transition pressure and had a small effect at 10 GPa peak pressure.

Figure 5.10 shows the free-surface velocity profiles for 40Kh steel samples in two initial states. The 40Kh is a chromium-doped structural steel with the following chemical composition (wt %): C - 0.4, Si - 0.3, Mn - 0.6, Cr - 1. The material was tested in two conditions: "as received" with a hardness of 17 to 19 HRC; and quenched, with a hardness of 45 to 54 HRC. In the "as received" state, the steel had a ferrite-pearlitic structure with a ferrite grain size of ~20 to 30 μm and pearlitic wafers of about 0.5 to 1.5 μm. The carbide content was about 10%. Quenched steel samples had a martensite structure with isolated grains of the residual austenite (about 2%) and chromium carbide (about 1%). The martensite needles were ~0.5 μm in diameter and ~15 μm long; the austenite grain size was ~1.5 to 2.5 mm. The heat treatment did not result in a large HEL increase, but had a much larger effect on the spall strength values.

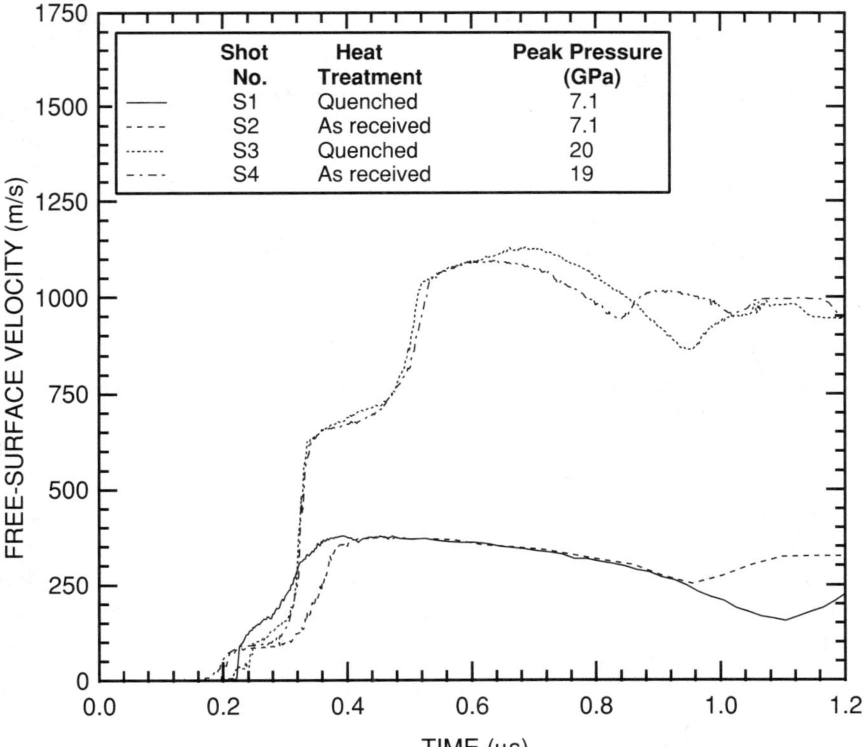

Figure 5.10. Free-surface velocity profiles for chromium-doped structural steel 40Kh subjected to different heat treatments and shock pressures.

Table 5.4. Results of the spall strength measurements for the chromium-doped structural steel 40Kh at different heat treatment and peak shock.

Shot no.	Sample condition	Configuration a (mm)	b (mm)	c (mm)	V (km/s)	Peak stress (GPa)	Strain rate ($10^5\,s^{-1}$)	Spall thickness (mm)	Spall strength (GPa)
S1	Quenched	2	3.17	6	0.7	7.1	0.73	1.85	4.2
S2	As received	2	3.95	6	0.7	7.1	0.54	1.50	2.3
S3	Quenched	2	3.17	—	1.9	20	1.7	0.53	4.7
S4	As received	2	3.95	—	1.9	19†	1.6	0.50	2.9
S7	Quenched	2	3.17	6	5.3	~80‡	2.4	0.38	4.75
S8	As received	2	3.95	6	5.3	~75‡	2.1	0.37	4.0

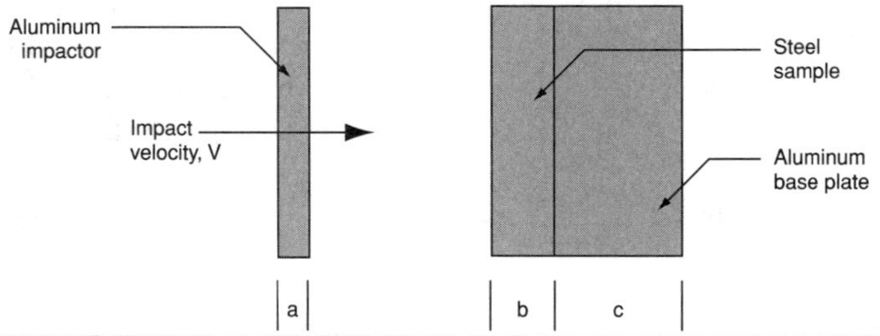

† Spall fracture occurred at the boundary between the material in incident phase and transformed matter.
‡ A decay of the shock wave occurred. The peak stress was estimated using computer simulation of the experiment.

Results of the spall strength measurements are summarized in Table 5.4. Note that the periods, Δt_{sp}, of velocity oscillations after spall fracture and the time interval between the first and the second plastic waves in shot S4 are both equal to ~0.18 μs, indicating that the spall fracture occurred near the boundary between the material in the pretransition state and the transformed matter.

At low peak stresses, the spall strength values are in a good agreement with the incipient spall threshold that was determined by microscopic examination (Golubev and Novikov [1984]). Measurements show the strength increases as a result of reversible phase transformation for both states of the steel. It is known that the reversible phase transformation causes reduction in the grain size, produces numerous defects, and, as a result, provides a sharp increase in the hardness. These effects hinder the growth of microcracks, which can be the reason for the increase in spall strength. Quenched samples initially have a fine-grain martensite structure; therefore, the strengthening effect of the reversible phase transformation is lower in these samples.

5.2. Metal Single Crystals

The study of spall phenomena in single crystals yields information about favorable conditions of damage nucleation at a structure level close to that of the ideal crystal structure. Single-crystalline materials of high purity are free from such relatively coarse defects as grain boundaries and inclusions. The largest nonuniformities of the structure of single crystals, which can serve as stress concentrators for damage nucleation, are formed as a result of interactions of dislocations during plastic deformation.

To shed light on the behavior of single crystals under shock wave loading conditions, Kanel et al. [1992a, 1994a] measured the spall strength of copper and molybdenum single crystals. These two metals differ from each other in many ways including crystal structure (copper has a face-centered cubic structure and molybdenum has a body-centered cubic structure), ductility, elastic limits, and other characteristics of strength. The small lateral dimensions of available single-crystalline samples required the use of shock wave generators with short load duration. Foil impactors or radiation from a high intensity ion beam were used to produce shock loadings of submicrosecond and nanosecond durations. The influence of crystal orientation, as well as that of the intensity and duration of the applied load, on the resistance to the spall fracture was investigated.

Figures 5.11 and 5.12 show the free-surface velocity profiles of copper single crystals and commercial-grade copper. Differences between the response of single crystals and polycrystalline material are immediately apparent from comparison of these profiles. The spall strength of single crystals is approximately three times larger than the strength of polycrystalline copper (see also Figure 5.13). For single crystals, the spall pulse is more clearly pronounced. It has a steeper front and higher amplitude than for polycrystalline copper. It is possible to conclude that fracture of single crystals is more brittle in the sense that there is a more pronounced threshold and more rapid evolution of the damage.

Another essential difference between the behaviors of single crystals and polycrystalline material is the rate of decay of the velocity oscillations in the $u_{fs}(t)$ profiles. The slowest decay was observed in single crystals loaded in the <100> direction. Visual observations of shock loaded specimens showed that single crystal specimens loaded in this direction had a smoother fracture surface than the other samples. Comparison of the duration of the incident load pulse with that of the velocity oscillations after spall suggests a delayed fracture. Annealing of single crystals at 900°C for two hours had a small effect on the Hugoniot elastic limit and no effect on spall strength, as shown in Figure 5.12.

The strain-rate dependencies of the spall strength of copper presented in Figure 5.13 show that the strength of polycrystalline copper becomes equal to the measured strength of single crystals when the strain rate reaches 3 to 7×10^7 s^{-1}. This strain rate corresponds to a load duration of ~0.5 ns or to a load pulse length of several micrometers. This result suggests that several micrometers is the characteristic distance between damage nucleation sites that have the same nature in a polycrystalline body and in single crystals. The same size determines the diameter of copper whiskers, where tensile strength values close to those measured in our experiments are realized.

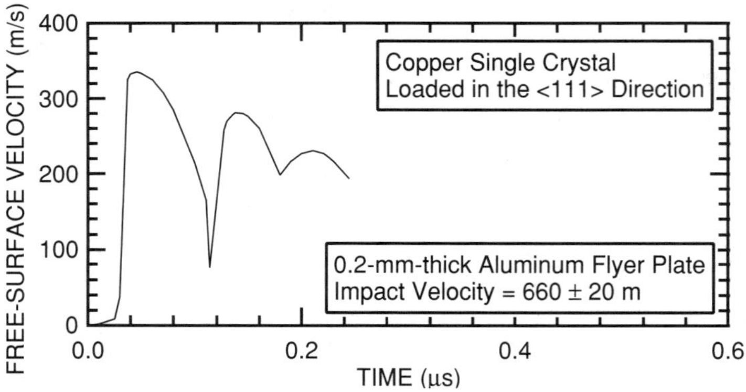

(a) Single crystal (0.2-mm-thick aluminum flyer plate)

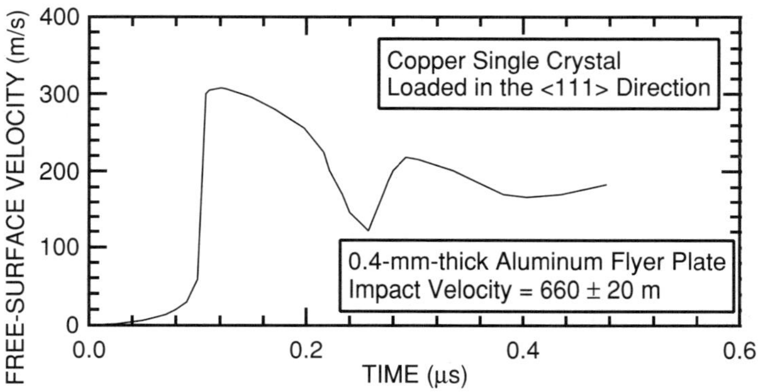

(b) Single crystal (0.4-mm-thick aluminum flyer plate)

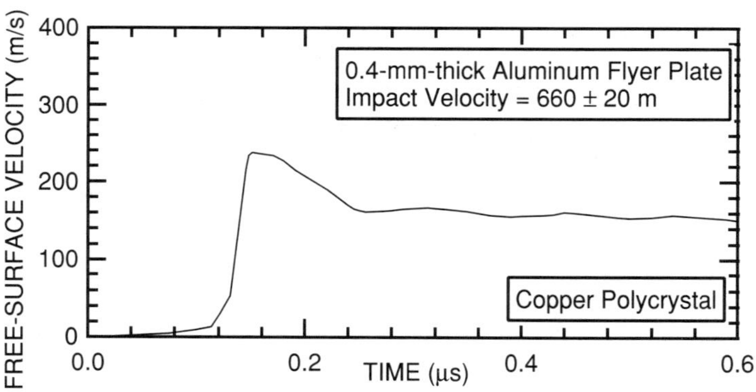

(c) Polycrystal (0.4-mm-thick aluminum flyer plate)

Figure 5.11. Free-surface velocity profiles for copper samples.

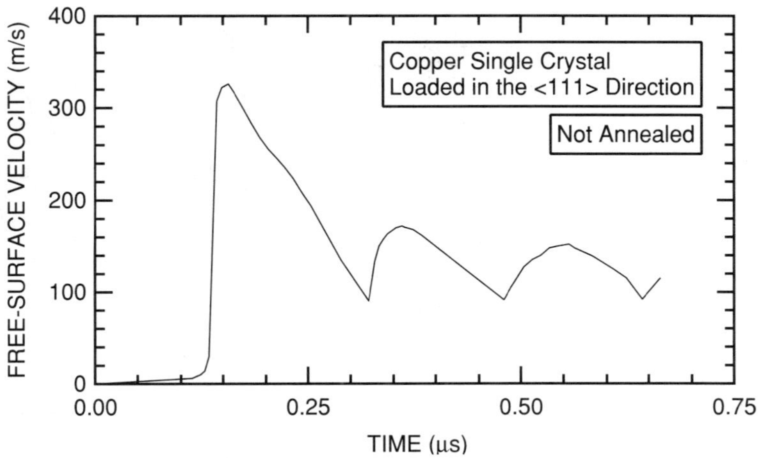

(a) Untreated single crystal loaded in the <111> direction

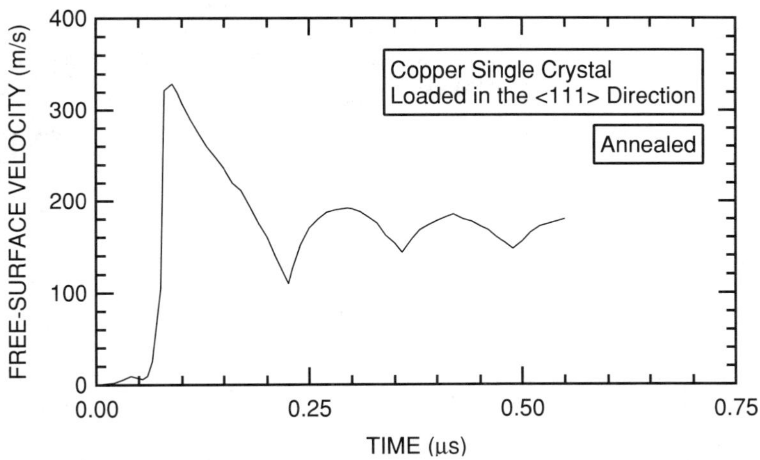

(b) Annealed single crystal loaded in the <111> direction

Figure 5.12. Effect of heat treatment on the free-surface velocity profiles for copper single crystals.

Figures 5.14 and 5.15 show typical free-surface velocity profiles for undeformed and deformed molybdenum and niobium single crystals and for polycrystalline molybdenum. Results of measurements of the spall strength as a function of strain rate are summarized in Figure 5.16. As for copper, the dynamic strength of the molybdenum single crystals is much higher than that of polycrystalline samples. Some small but remarkable delay of fracture was observed in shots with single crystals of all orientations.

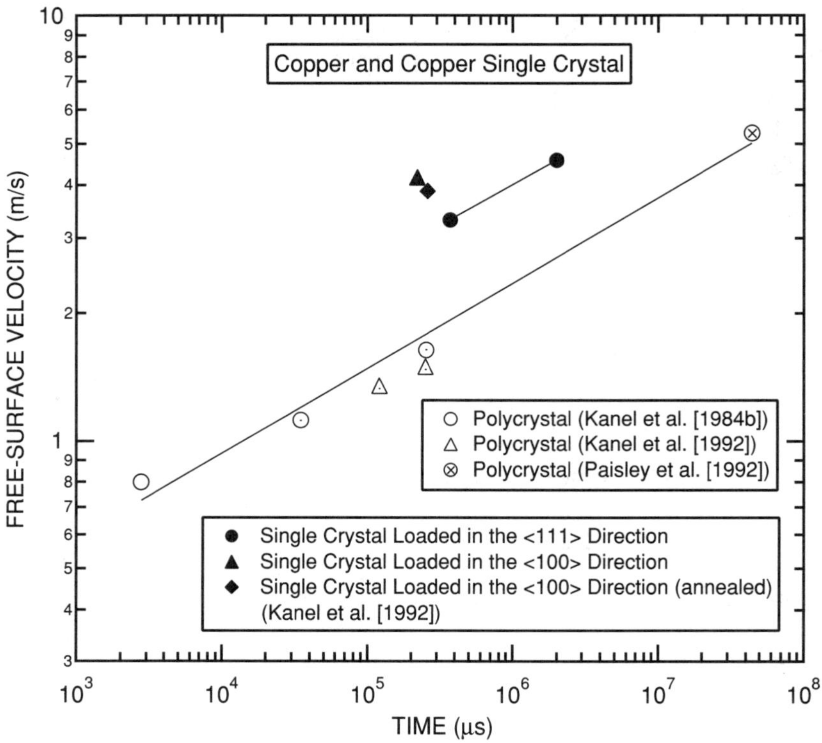

Figure 5.13. Spall strength of commercial grade copper and copper single crystals as a function of strain rate.

The spall strength of molybdenum single crystals is approximately twice the strength of polycrystalline material. As mentioned above, for copper this difference amounted to a factor of 3. The difference between deformed and undeformed single crystals is not so large. However, the amount of dislocations generated by high-speed plastic deformation at the shock front is comparable to the initial dislocation density ($\sim 10^{10}$ cm^{-2}) in the deformed single crystals. The measured spall strength was not reproducible for the copper and molybdenum single crystals in the experiments performed. Because the scatter of experimental data is large, it is difficult to reveal unambiguously the influence of the orientation on the resistance to spall fracture. However, cubic crystals are highly isotropic, and under our load conditions the stress tensor is nearly spherical. Due to these reasons, anisotropy of strength probably does not exceed the experimental scatter. The reason for the scatter in the strength of single crystals is not clear, but it could reflect a scatter in the initial state of the material. Experiments with more thoroughly characterized samples should lead to a better understanding of the nature of damage nucleation on the microstructural level.

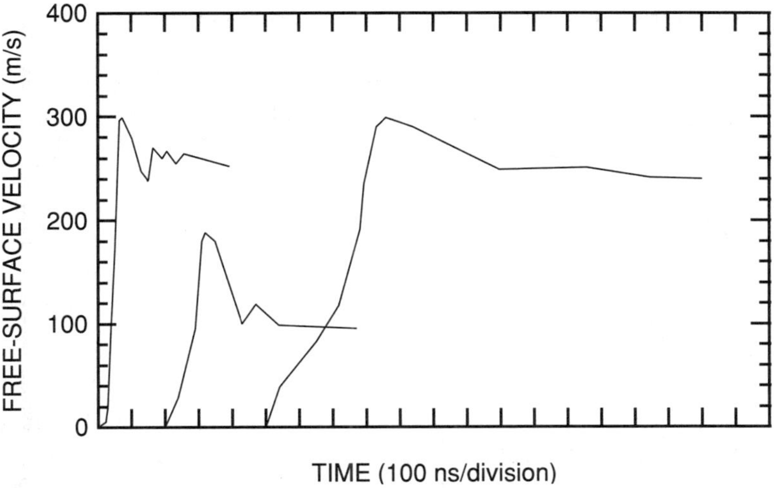

(a) Untreated single crystal loaded in the <111> direction

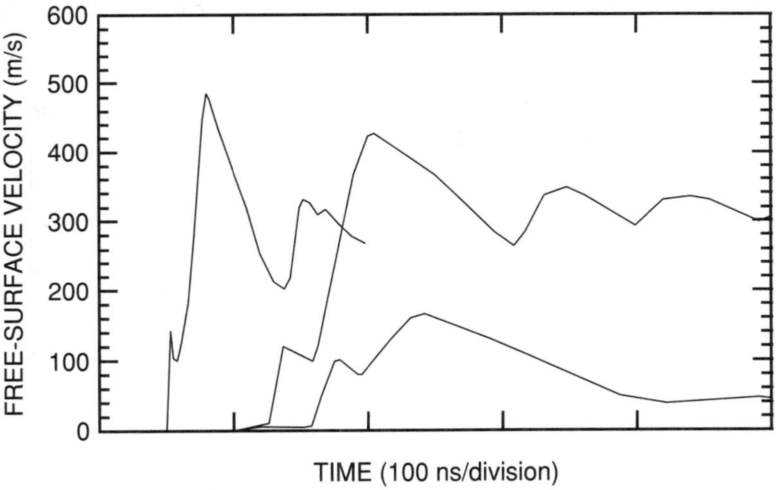

(b) Annealed single crystal loaded in the <111> direction

Figure 5.14. Spallation in molybdenum and molybdenum single crystals.

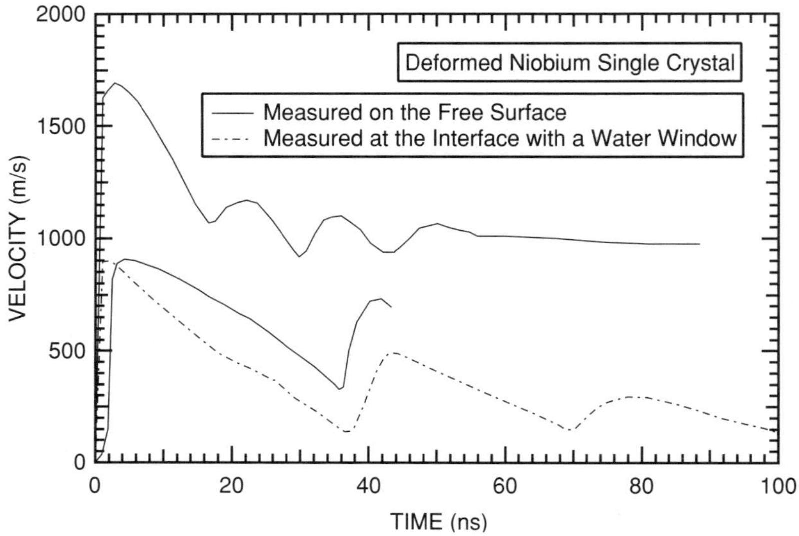

Figure 5.15. Results of spall measurements for deformed niobium single crystals.

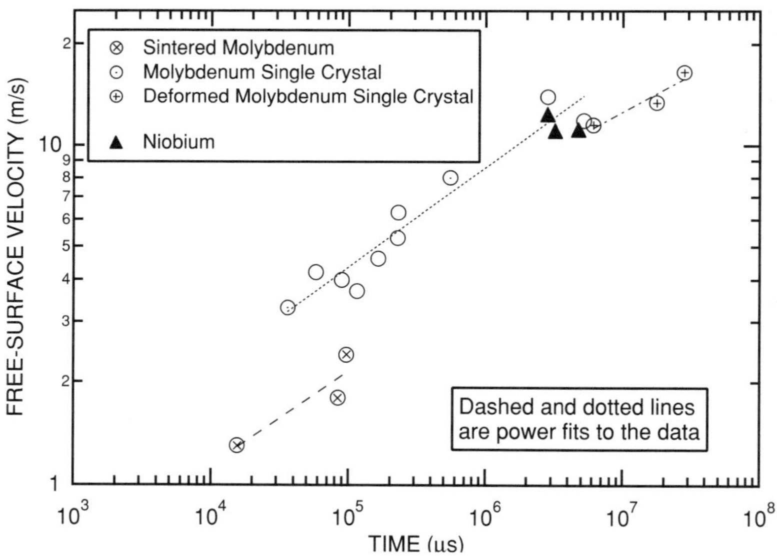

Figure 5.16. Spall strength as a function of strain rate for molybdenum and niobium.

In pure single crystals, damage nucleation may result from dislocation interactions during plastic deformation. In this case the concentration of damage nucleation sites must increase with plastic deformation, that is with increasing shock load intensity. Therefore, the absence of a significant influence of shock wave amplitude on the spall strength of single crystals is unexpected. Nevertheless, we can conclude that damage nucleation at the dislocation level requires relatively high tensile stresses and will occur in single crystals only. In polycrystalline materials, the fracture can be initiated on coarser defects, and as a result of stress relaxation, tensile stresses do not have time to reach the necessary magnitude to activate finer defects.

It is interesting to compare the measured spall strength values with the ultimate theoretical strength. Estimates of the ultimate tensile strength are on the order of $K/10$ to $K/6$, where K is the bulk elastic modulus. The ultimate tensile strength can also be estimated from the minimum of the $p(V)$ Hugoniot curve as

$$S_{th} = \frac{\rho_0 c_0^2}{4s}, \tag{5.3}$$

where c_0 and s are coefficients of the linear relation between the shock front velocity and the particle velocity. Such estimations give S_{th} = 55 GPa for molybdenum, 36 GPa for niobium, and 24 GPa for copper. The corresponding experimentally measured values are 16.5, 12.4, and 4.6 GPa. In other words, up to 35% of the theoretical strength is reached in single crystals subjected to loading pulses on the order of a nanosecond. Extrapolation of the measured spall strength data shows that the ultimate strength of molybdenum single crystals can be reached at a strain rate of about 5×10^8 s^{-1}, which corresponds to a load duration of 0.5 to 1 ns. This value is only one order of magnitude smaller than the shortest shock compression pulses realized in the experiments performed, and exceeds the period of atomic oscillations in solids by three to four orders of magnitude.

5.3. Constitutive Factors and Criteria of Spall Fracture in Metals

Experiments with steel specimens of various orientation as well as experiments with single crystals show that spall strength is a structure-sensitive parameter. Owing to the finiteness of the fracture rate, the dynamic strength of metals increases appreciably with increasing strain rate. The resultant excess stresses initiate fracture at increasingly smaller and more numerous damage nucleation sites. From the ratio between the fracture stresses for single crystals and polycrystalline metals, it follows that fracture initiation at a dislocation level occurs only in single crystals. Here, the tensile strength is two to three times larger than the fracture stress for the polycrystalline material.

Shock wave loading is accompanied by supplementary factors that can influence the fracturing process. Extremely high strain rates develop in shock waves. The shock wave loading is adiabatic, and it is associated with material heating.

It is important to understand the factors that govern the fracture of metals under these conditions. Experiments show that the fracturing stresses are much less sensitive to the shock-induced temperature rise and plastic deformation. That the effect of shock heating on the resistance to high-rate fracture is small may be explained in terms of a change of plastic deformation mechanisms in the vicinity of flaws. Transfer from thermally activated dislocation slip to thermal overbarrier dislocation slip may occur under large overstressing in the material (Merzhievsky and Titov [1986]). The fast drop in dynamic strength near the melting point is probably explained by spontaneous nucleation of large amount of vacancies in the premelting state. Coalescence of these vacancies to micropores forms new damage nucleation sites.

As for the effect of plastic deformation on fracture, we note the results of studies in large plastic flow and fracture performed by Bridgman [1964] under quasi-static conditions. The strain was varied by an order of magnitude or more in these experiments, but in fact; rupture occurred when stresses reached the ultimate strength value independent of prestraining. Thus, the total deformation itself is not a critical influence on a material's resistance to fracture.

However, experiments by Chhabildas and Asay [1992] on tantalum indicate a 27% increase in spall strength for samples precompressed to 60 GPa quasi-isentropically compared with samples precompressed to only 19 GPa. Thus, we cannot exclude the possibility of simultaneous but opposite effects of the plastic strain and temperature on the spall strength. These contributions can be separated by measurements on preheated samples.

The "spall strength" obtained in a given experiment defines only the conditions for damage nucleation; it does not exhaustively characterize the material response. Development and completion of the spall process occur at lower stress, but require consumption of additional energy for the growth of flaws and the associated plastic deformation of material around them. When the initial load pulse is short, fracture, once started, may not proceed to complete separation of the body into distinct parts.

An energy criterion (Grady [1988], Ivanov [1975], Kanel [1982a,b]) defines the possibility of complete rupture through comparison of the work of fracture and the amount of energy stored in the body. The work of fracture is the energy expended per unit cross-section area of the body during fragmentation. In reality, the dissipation of energy due to fracture takes place in some layer of finite thickness; therefore, the work of fracture, generally speaking, increases with increasing thickness of the failed zone.

Estimation of the energy dissipated in a fracture process can be based on either of two kinds of experiments. The first type of experiment determines the critical impact velocity that produces spall for given thicknesses of impactor and sample. Using the critical impact velocity thus determined and the parameters of the experiment, the energy dissipated is determined through consideration of the balance of energy and momentum as

$$E_d = \frac{\rho h_i v^2}{2}\left(1 - \frac{h_i}{h_t}\right), \tag{5.4}$$

where v is the impact velocity, and h_i and h_t are thicknesses of the impactor and target, respectively. The dissipated energy thus calculated is an upper bound for the magnitude of the work of fracture. The second kind of experiment involves measurements of a free-surface velocity profile. The loss of kinetic energy of the spall plate as it decelerates during the spall process can be inferred from this profile and can be used for estimating the work of dynamic fracture.

The fracture stress and work of fracture describe the strength properties of the materials when subjected to one-dimensional dynamic tension. However, these two parameters are not enough to predict the occurrence of complete separation of the scab or to estimate its velocity after the separation.

Figure 5.17 shows the evolution of spalling in a limited area of a plane body (Razorenov and Kanel [1991]). The incident shock is attenuated not only by the axial unloading waves but also by the lateral release. Upon reflection of this

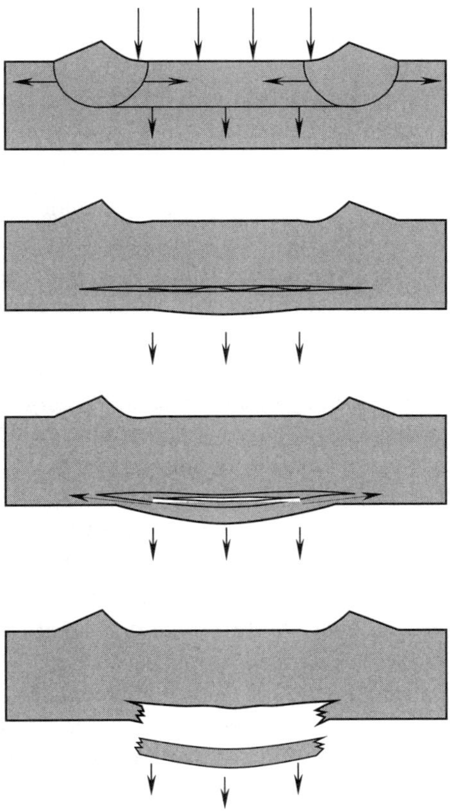

Figure 5.17. Edge effects at spallation.

attenuated load pulse from the free-surface, the tensile stresses and energy stored, which are sufficient for the complete fracture, are realized only near the axis. Thus, the fracture of the body by the plane wave is limited to some inner region. The later evolution of the process is determined by the kinetic energy stored in the spalled layer.

The inertial motion of the spalled layer is decelerated by bonding forces in the periphery of the scab. The scab motion can be decelerated, and even arrested, by the edge effects. Additional work is necessary for plastic deformation and fracture to take place along the edge of the scab. The work of edge deformation and fracture is proportional to the length of the spall element perimeter, whereas the value of energy stored in the element is proportional to its area. The ratio of these two values increases with decreasing spalled layer radius. For a small radius, development of spalling can be stopped at some intermediate stage and the spalled layer remains connected to the main body. Investigation of spalling with spall layers of different radii allows us to determine the critical stored energy value required for complete separation to take place. This variation of the spall layer radius can be arranged by varying the radius of the impact area or by placing a limiting ring on the rear surface of the sample.

5.4. Brittle Materials: Ceramics, Single Crystals, and Glasses

Hard ceramics contain many stress concentrators including pores, microcracks, and grain boundaries. A localized fracture can nucleate at the sites of these inhomogeneities even in the region of elastic deformation of the material as a whole. Microfracture in such brittle materials can appear during the compression phase. The degree of fracture increases with increasing load intensity, and the damage that occurs under compression decreases the capacity of the brittle material to resist the tensile stresses that follow the initial compression phase.

This behavior can be seen in Figure 5.18, which presents results of spall strength measurements as a function of normalized peak stress for alumina (see Fig. 5.18(a)) (Rozenberg [1992], Dandekar and Bartkowski [1994]); titanium diboride (see Fig. 5.18(b)) (Grady [1992], Winkler and Stilp [1992], Dandekar [1994]); silicon carbide (see Fig. 5.18(c) (Grady [1992], Kipp and Grady [1992], Winkler and Stilp [1992]); and boron carbide (see Fig. 5.18(d)). Figure 5.18(c) indicates that the spall strength of silicon carbide in the elastic region increases with increasing peak pressure. This probably occurs because silicon carbide is the most "ductile" of these brittle ceramics. However, even for silicon carbide, a sharp drop is observed in the tensile strength when the peak stress becomes larger than the HEL. The data for boron carbide in Figure 5.18(d) do not exhibit any loss of strength because the peak stress reached is lower than the HEL.

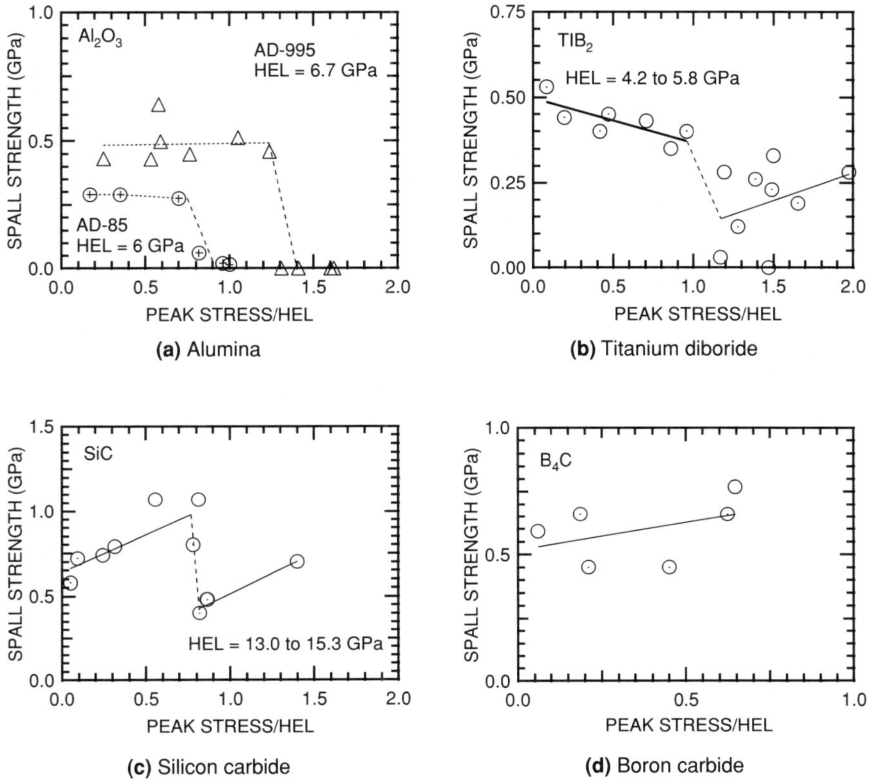

Figure 5.18. Spall strength of various ceramic materials as a function of the peak shock stress.

Staehler et al. [1994] tested vacuum-hot-pressed alumina samples to shock pressures ranging from 1.3 to about three times the HEL. They found that the spall strength undergoes a transition, first decreasing near the HEL, then increasing with increasing pressure above the HEL. This transition was attributed to a change in the dominant inelastic deformation mechanism from microcracking near the HEL to dislocation activities at higher peak stresses.

The spall strength of ceramic materials has been shown to increase with decreasing initial porosity and grain size of the ceramic material (see, e.g., Nahme et al. [1994] and Bourne et al. [1994]).

Figure 5.19 shows free-surface velocity profiles of samples of a ceramic consisting of titanium carbide particles bonded with nickel (Kanel and Pityulin [1985]). The mass fraction of titanium carbide in the composite is 80%. The free-surface velocity records do not show a well-pronounced elastic precursor in the wave profile of this ceramic. This is due to wave dispersion caused by multiple reflections of the stress waves between constituents with different dynamic

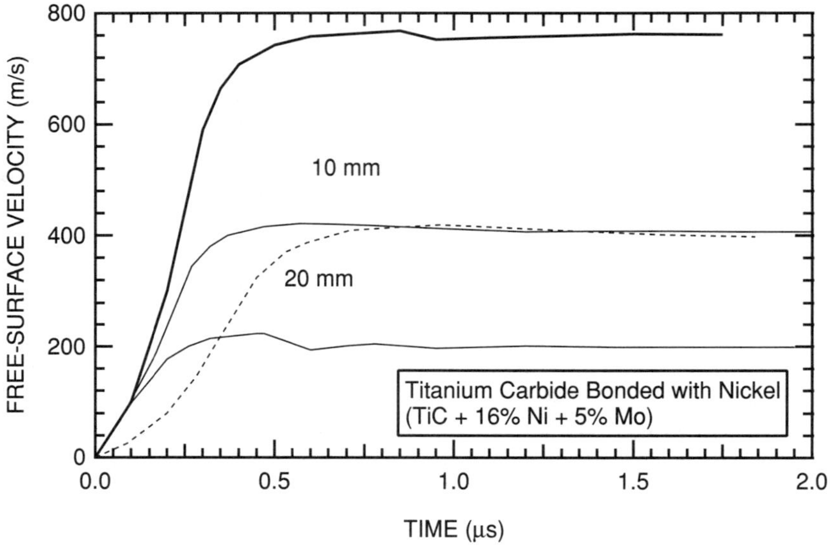

Figure 5.19. Free-surface velocity profiles for titanium carbide bonded with nickel.

impedances. The spall strength is 0.4 to 0.55 GPa and it decreases with increasing peak stress. Because the spall strength of nickel is three to four times higher than the values measured for the titanium carbide-nickel ceramic, the results confirm that fracture nucleates in the brittle titanium carbide phase.

Realization of the ultimate tensile strength in the practically undistorted initial structure is possible for single crystals with high Hugoniot elastic limits, because, for such materials, generation of large tensile stresses without any plastic deformation is possible below the HEL. In this case, fracture initiates at a structural level near that of the ideal crystal lattice.

Figure 5.20 shows free-surface velocity profiles measured for x-cut quartz samples (Kanel et al. [1992b]). Shock pulse amplitudes near the rear surface were 2.8, 4.6, and 5 GPa. According to Graham [1974], the HEL of x-cut quartz is 6 GPa.

The free-surface velocity profile corresponding to a shock pulse amplitude of 2.8 GPa in Figure 5.20 replicates the form of the compression pulse inside the sample and does not display any symptoms of spallation. Increasing the pulse amplitude to 4.6 GPa causes spall damage, as evident from the spall pulse in the free-surface velocity profile. A further increase in pulse amplitude to 5 GPa modifies the free-surface velocity profile dramatically. The unloading part of the pulse does not pass to the surface, which means that the tensile strength is practically zero in this case.

These measurements show that the dynamic tensile strength of x-cut quartz reaches 4 GPa for shock loading below the HEL and falls to zero for loading

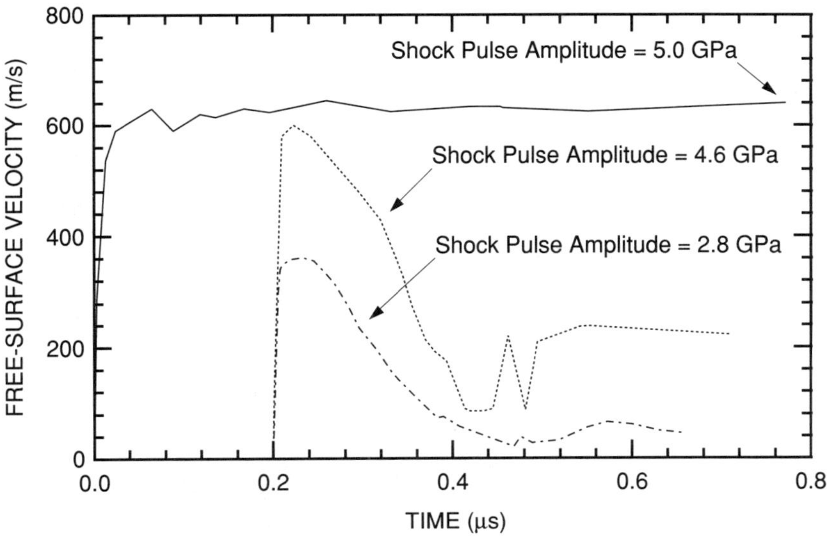

Figure 5.20. Free-surface velocity profiles for x-cut quartz at different peak shock amplitudes.

close to the HEL. Cracking in the brittle single crystals under compression is a plausible explanation for the diminishing tensile strength near the HEL.

Experiments analogous to those just described were conducted with ruby (Razorenov et al. [1993]) and sapphire (Kanel et al. [1994a]). Measurements were made at shock wave intensities below the HEL, which is 14 to 20 GPa for alumina single crystals (Graham and Brooks [1971]). Figure 5.21 shows experimental profiles for z-cut sapphire and ruby crystals. The spall pulse in this case has a drastically steep front—a strong indication of a damage evolution process with fast kinetics. The spall strength of ruby was found equal to 8.6 GPa at 15.1 GPa peak shock stress and 10 GPa at 13.5 GPa peak stress. No spallation was observed in one of two shots with sapphire, where the shock intensity was 23 GPa and the peak tensile stress reached 20 GPa. In the other shot, the peak shock intensity was 24 GPa and spallation was observed at a tensile stress of 10.4 GPa. Thus, the high homogeneity of sapphire allows it to sustain higher tensile stresses than those measured in any other alumina.

In contrast to the behavior of hard single crystals, the spall strength of glasses is not sensitive to shock intensity at levels near the HEL. Figure 5.22 shows free-surface velocity profiles for K19 glass (Razorenov et al. [1991]). Spall was not observed in these shots, which means that the spall strength of the glass exceeds 4.2 GPa both below and above the HEL. An irreversible densification occurs in glasses compressed beyond the elastic limit, both under static and dynamic conditions (Sugiure et al. [1981]). Densification can play the role

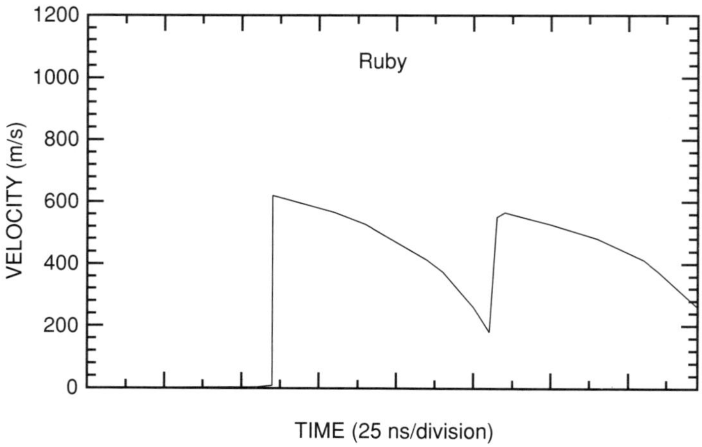

(a) Free-surface velocity profile in ruby

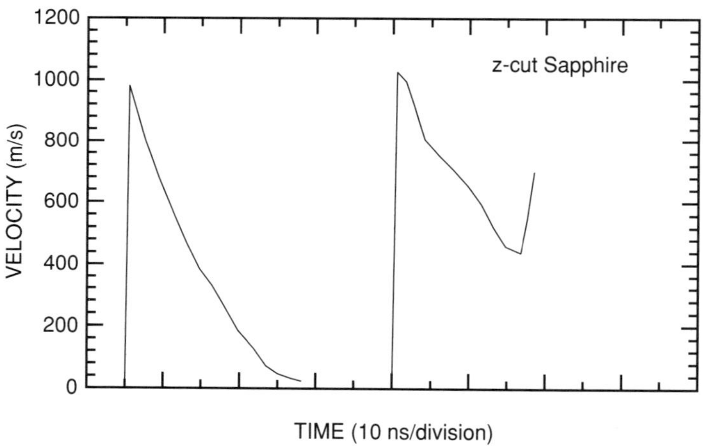

(b) Free-surface velocity profile in z-cut sapphire

Figure 5.21. Spallation in alumina single crystals.

of a plastic deformation mechanism for glasses (Ernsberger [1968]), which could suppress cracking under compression.

Shock wave loading of glass can be accompanied by the formation of failure waves. The possibility of fracture wave formation imposes certain constraints on the design of experiments with homogeneous brittle materials. The failure wave was observed in the following experimental configuration (Razorenov et al. [1991]). A triangular, long duration, 4.5-GPa shock pulse was introduced into a plane sample through a thick copper base plate, and the free-surface velocity profiles were measured. In experiments with fused quartz

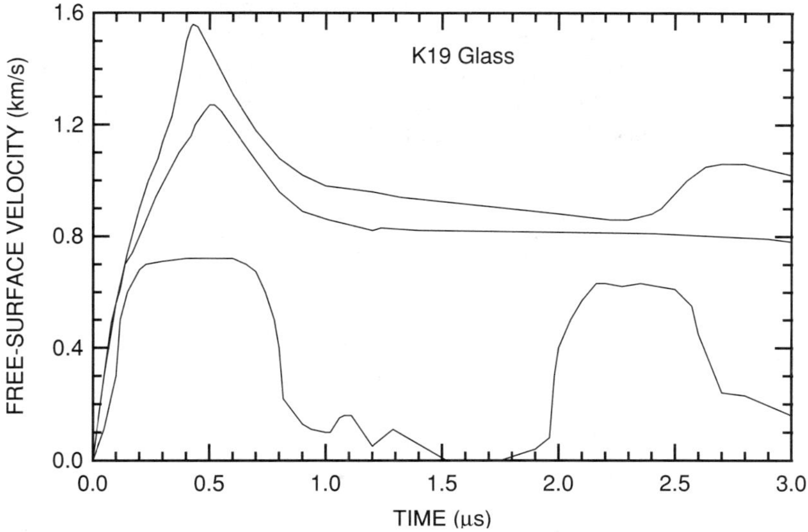

Figure 5.22. Free-surface velocity profiles for K19 glass impacted by aluminum flyer plates at different impact velocities.

samples, $u_{fs}(t)$ profiles contained short negative velocity pullbacks as shown in Figure 5.23(b). This pullback was a result of rereflection of the rarefaction wave at the sample-base plate interface (Figure 5.23(a)). The base plate had a higher shock impedance than the sample, which caused the rarefaction wave to be reflected back into the sample as a rarefaction wave. This rereflection produced a short tensile pulse in the unloaded sample, which propagated through the sample and produced a negative velocity pullback when it reached the rear surface.

No rereflected tensile pulses were observed in experiments with K19 glass (Figure 5.23(c)). Instead, a small velocity rise was noted on the free-surface, and the arrival time of the second velocity jump at the free-surface was less than the elastic wave reverberation time for the sample. This modification of the wave process can be explained in terms of a failed layer formation near the sample-base plate interface. This layer has a lowered impedance and zero tensile strength. Probably, this layer is a network of cracks initiated by uniaxial compression. Experiments with samples of different thicknesses have shown the thickness of this failed layer to increase with time. This process can be interpreted as a failure wave propagation, and the velocity of the failure wave decreases with propagation distance. Experiments with short load pulses have shown that unloading will arrest the failure wave propagation.

Failure waves in glass were observed at stress levels near, as well as below, the HEL. Compression above the HEL causes irreversible densification of glasses, which plays the role of a plastic deformation mechanism. Brar et al. [1991] in their spall experiment showed that, behind the failure wave, the tensile strength drops and the transverse stress increases, indicating a decrease in shear strength. Raiser and Clifton [1994] confirmed this observation. They also found that surface roughness does not appear to play a significant role in the formation of a failure wave.

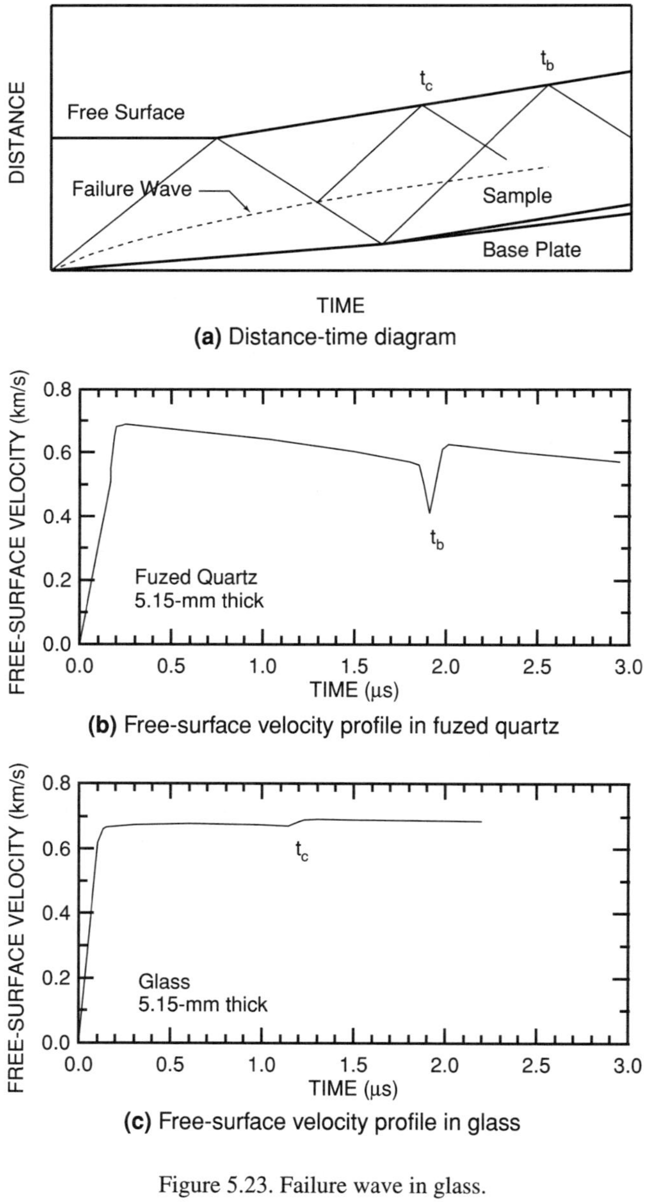

Figure 5.23. Failure wave in glass.

5.5. Polymers and Elastomers

Figure 5.24 shows the free-surface velocity profiles of polymethylmethacrilate (PMMA) samples at various peak pressures and durations of the shock load. In the low-pressure range, PMMA behaves as a brittle material and is fractured through the growth of penny-shaped cracks nucleated at preexisting solid and

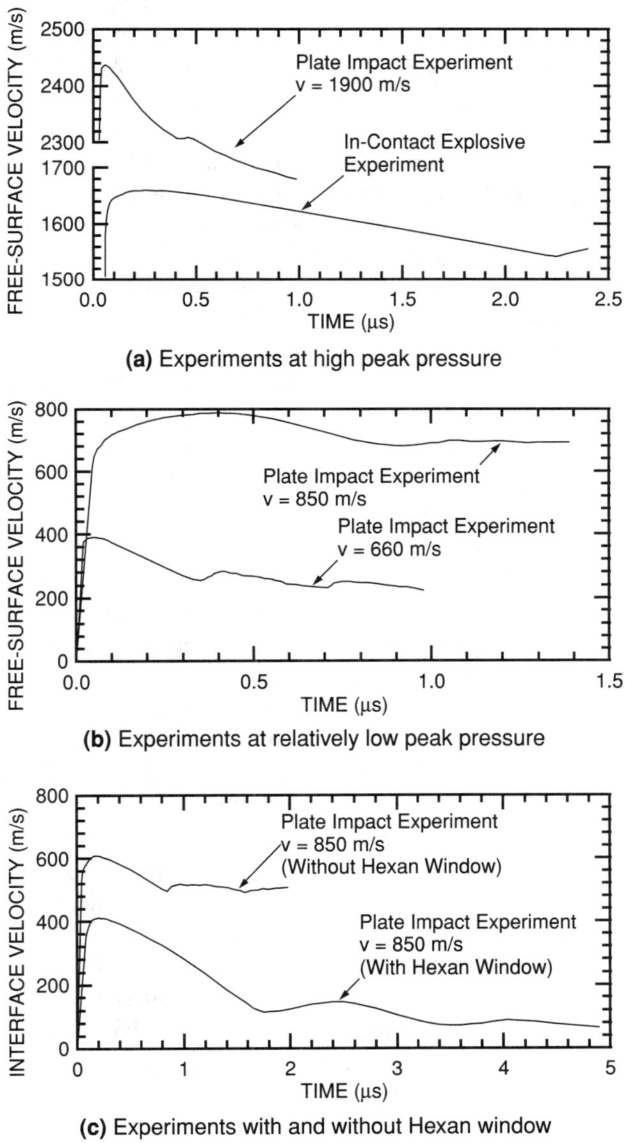

Figure 5.24. Free-surface velocity profiles for PMMA at different peak pressures.

gas inclusions in the intact material. The form of the wave profiles at low intensities of shock load is typical for solids. The appearance of small-scale oscillations on the spall pulse is a peculiarity of this material.

The character of spalling changes with increasing peak pressure. The fracture becomes more viscous, manifested as a protracted deceleration of the spall plate as observed in the free-surface velocity profiles. The most plausible reason for this trend is the heating and plastification of the material in the shock wave. Examination of samples recovered after tests at different temperatures (Golubev et al. [1982a]) showed an increasing zone of local plasticity near the crack tips, accompanied by a reduced crack propagation velocity. We cannot exclude the possibility of collapse of the incident microvoids in the material by shock pressure, which reduces the number density of the fracture nucleation sites, thereby reducing the rate of the fracture. Measurements have showed that, in spite of the varying fracture characteristics, the spall strength of PMMA is practically independent of peak shock pressure over the range of 0.6 to 6 GPa and depends only mildly on strain rate, increasing from 0.17 GPa at ~10^4 s^{-1} to 0.21 GPa at 10^5 s^{-1}. The spall strength of epoxy under the same conditions is about 0.3 GPa. It seems that spall strength values of 0.1 to 0.3 GPa are typical for homogeneous polymers (Golubev et al. [1982b]).

Elastomers, like rubber, are characterized by an ability to undergo large reversible deformations and for this reason, they constitute a unique class of polymer materials that deserves special attention. The behavior of these materials at spall fracture is different from that of other solids, including polymers.

Experimental results for rubber (Kalmykov et al. [1990]) are presented in Figure 5.25. The dashed lines in the figure represent the free-surface velocity profiles calculated from results of window tests with the assumption that rubber is not damaged in the negative pressure region. If the material does not have any appreciable resistance to tension, its surface velocity behind a shock front would be constant. The measured free-surface velocity profile (curve 3) takes some intermediate position between these two extreme cases. A small initial part of the measured free-surface velocity profile corresponds to the incident loading pulse, followed by a weak spall pulse registered at point s. Then, a slow protracted deceleration of the surface velocity is observed. Visual examination of the recovered samples did not reveal any remarkable damage. The measured value of the spall strength of rubber is 27 ± 3 MPa. The true breaking stress S_k under quasi-static tension has been found equal to 88 MPa.

Figure 5.26 presents the free-surface velocity histories for a butadienenitrile caoutchouc-based filled elastomer propellant simulant (Kanel et al. [1994b]). The filler content was 75% by mass, including 61.6% KCl. Two compositions with different filler particle sizes were tested: a coarse-dispersed composition with 160- to 200-μm KCl particles and a fine-dispersed composition with 20- to 50-μm particles. The initial density of the material was 1.6 g/cm^3, and its sound velocity at ambient conditions was 1.85 km/s. The measured free-surface velocity profiles for these propellant simulants are similar to those for rubber. Meas-

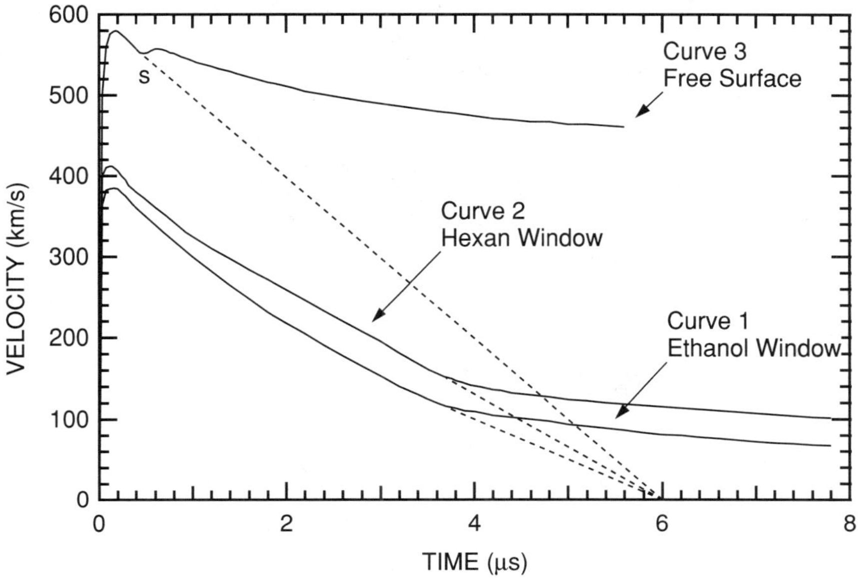

Figure 5.25. Wave profiles under spall conditions in rubber.

ured values of the spall strength are 24 to 30 MPa for fine-dispersed samples and ~15 MPa for coarse-dispersed samples. The samples from shots 1 and 2 were recovered for post-test inspection. Like rubber, these propellant simulant samples did not show evidence of a distinct spall plane even though the peak shock pressure in shots 1 and 2 was equal to ~150 MPa, which exceeds the measured spall strength value by a significant amount.

It is known that void formation precedes the rupture of elastomers. These voids nucleate under much lower stresses than is required to break the sample. Origination of microdiscontinuities is not yet a destruction event itself. Thus, in tests of natural vulcanizates under triaxial tension (Gent and Lindley [1959]), voids formed at stresses of 1 to 3 MPa and insignificant deformation occurred. Then, after this stage, samples underwent further deformation, reaching several hundred percent. According to these findings, it seems reasonable to assume that the measured spall strength values of 15 to 30 MPa characterize the nucleation of microdiscontinuities in elastomer, but do not characterize its failure.

The velocity deceleration behind the spall pulse is determined by the resistance of the material to tension. Among the elastomers tested, the resistance to tension is highest for rubber (curve 3 in Figure 5.25), lower for the elastomer with a finely dispersed filler (curves 1, 3, and 4 in Figure 5.26), and lowest for the coarsely dispersed elastomer (curve 2 in Figure 5.26). The observed free-surface deceleration is inversely proportional to the size of the filler particles.

For the finely dispersed filler, the velocity deceleration is 3.5 times as high as that of coarsely dispersed sample. Filler particle size also influences the shock front rise time. The rise time for coarsely dispersed composition is twice as long as for the finely dispersed sample.

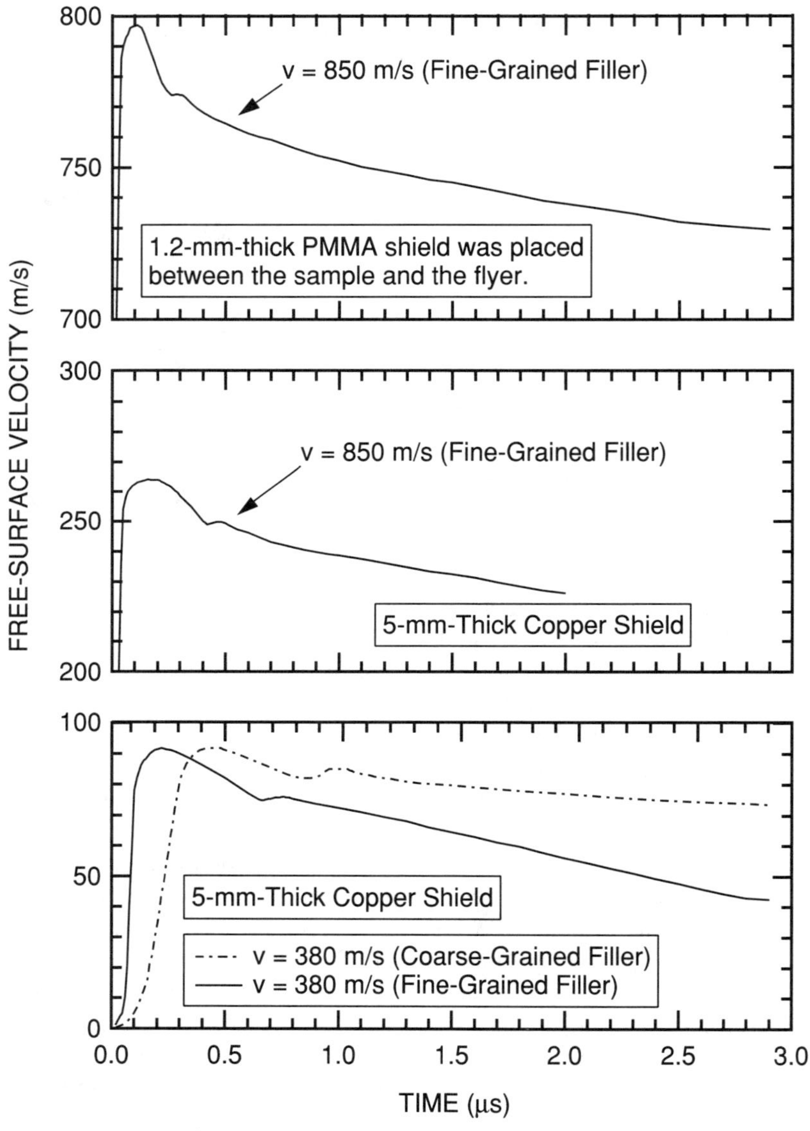

Figure 5.26. Free-surface velocity profiles for rocket propellant simulants impacted by PMMA flyer plates at impact velocities of 380 and 850 m/s.

5.6. Dynamic Strength of Liquids

Liquids, like solids, have some resistance to multidimensional tension, which can be treated as a bulk strength property of the liquid material. The ultimate bulk strength value for liquids can be estimated from intermolecular bonding forces (which, for example, gives a value for water on the order of 1 GPa) or by extrapolation through the minimum of the isotherm in the negative pressure region (which gives a value for water on the order of ~300 MPa).

The strength of liquids can be measured experimentally if the experiment is arranged to prevent the lateral narrowing of the sample under tension (Kornfeld [1951], Trevena [1967]). Static measurements have shown that the bulk tensile strength of liquids is one to two orders of magnitude lower than the ultimate theoretical estimations. In part, this may be explained in terms of the small bubbles of gas or vapor that are always present in liquids. The resistance to growth of a bubble of finite dimensions is determined by the surface tension forces, which are much lower than intermolecular forces. Even thermal fluctuation in liquids can create sites of nucleation of voids under tension.

Application of plane shock waves provides the necessary conditions for measuring the bulk strength of liquids, because the motion is one-dimensional under these conditions and because the tension is realized inside the body where surface effects do not influence the rupture process. Typical values of the spall strength of liquids have been obtained for several liquids including water and glycerol. According to Erlich et al. [1971], the spall strength of glycerol is ~25 MPa, which corresponds to the initial embryonic bubble size of ~0.01 μm. The dynamic strength of water at load duration of $\sim 10^{-4}$ s was estimated by Couzens and Trevena [1969] to be 0.85 MPa. Deionization of water increases its bulk strength up to 1.5 MPa. In the microsecond load duration range, Marston and Unger [1986] measured the spall strength of water to be 3.9 to 11.5 MPa.

6 Constitutive Modeling Approaches and Computer Simulation Techniques

To develop a material model to represent fracture processes under high-rate loading, we must begin by facing the fact that we will use the model in a computer program to simulate fracture-causing events. With this assumption, we develop in this chapter many requirements for the model and then proceed to discuss the connection of the model to a computer code. We begin the chapter with an outline of some of the fracture modeling approaches that have been used. This is followed by a discussion of elastic-plastic models to provide the nomenclature we use later in the text, and also because elastic-plastic models are part of fracture models. The discussion includes an outline of physical and thermodynamic requirements that ensure stability and well posedness of constitutive models.

The process of connecting the constitutive model to a finite-element code introduces many more requirements. Here we outline some major types of finite-element codes and we discuss the special code features that must be taken into account in constructing the material model. We suggest means for facilitating implementation of the model in any host code. With some finite-element codes, material is advected from one element to another: advection presents special problems for models, so we outline methods for dealing with these problems.

6.1. General Constitutive Modeling Approaches

6.1.1. Response of Intact Material

In this section, we provide an overview of the modeling of elastic-plastic behavior as an introduction to the more complex topic of modeling fracture processes. Here we introduce the nomenclature that will also be used in later sections dealing with fracture modeling. More complete treatments of modeling are available in Malvern [1969] and Brannon et al. [1995], for example.

In the range of pressures and temperatures that is typical for spall experiments, the elastic-plastic properties of solids have a significant effect on the shock-wave structure. The state of solids in this case is described by two tensors:

the stress tensor σ_{ij} and the strain tensor ε_{ij}, where subscripts i and j represent the coordinate directions x, y, and z of the orthogonal Cartesian coordinate system. The stress σ_{ij} is the force per unit area in the body along the direction i, acting on an area with a normal oriented along the j axis. The components σ_{xx}, σ_{yy}, and σ_{zz} are the normal stress components and $\sigma_{xy} = \sigma_{yx}$, $\sigma_{yz} = \sigma_{zy}$, $\sigma_{xz} = \sigma_{zx}$ are the tangential or shear stress components. The normal stresses are also represented by the singly subscripted symbols $\sigma_x = \sigma_{xx}$, $\sigma_y = \sigma_{yy}$, and $\sigma_z = \sigma_{zz}$; and the shear stresses by $\tau_{ij} = \sigma_{ij}$.

A body is said to have undergone strain when a state of deformation causes a change in the relative positions of two points in the body. Deforming materials may undergo two types of strain: longitudinal and shear. Longitudinal strain can be expressed in terms of the change in length of a line segment relative to the original length of the segment. Mathematically, and referring to Figure 6.1, this relationship may be expressed as

$$\varepsilon = \frac{\left|\overline{A'B'}\right| - \left|\overline{AB}\right|}{\left|\overline{AB}\right|}. \tag{6.1}$$

The longitudinal strain ε is positive when the deformational state causes the length of the line segment to increase, in which case the strain is called *elongation*. In contrast, the longitudinal strain is negative when the deformational state causes the length of the line segment to decrease, in which case the strain is called *contraction*.

Shearing strain may be visualized in terms of the change in the right angle initially formed by two intersecting line segments as shown in Figure 6.2. Thus,

$$\gamma = \frac{\pi}{2} - \alpha. \tag{6.2}$$

The shearing strain γ is positive when the deformational state causes the angle α

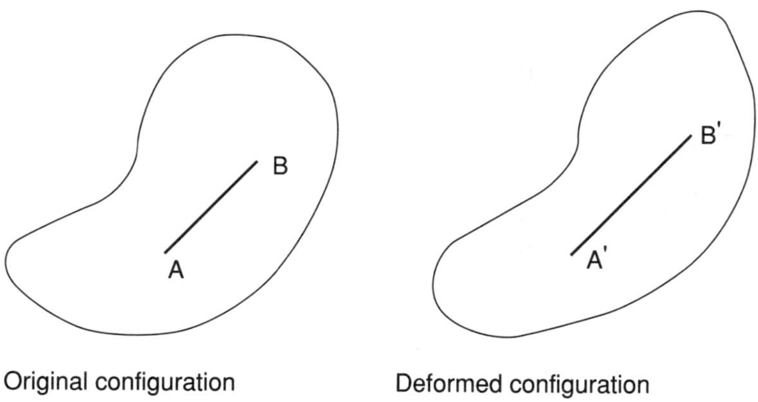

Figure 6.1. A line segment in a deforming body before and after deformation.

to be less than 90°, and it is negative when the angle α is greater than 90°.

For small strains, the strain components derived from Eqs. 6.1 and 6.2 are called engineering strains. These strains are not components of a tensor (e.g., Malvern [1969]). An equivalent measure of infinitesimal strain with tensorial properties may be expressed in terms of the displacement vector **u** (with components u_i) as follows:

$$\varepsilon_{ij} = \frac{1}{2}\left\{\frac{\partial u_i}{\partial x_j} + \frac{\partial u_j}{\partial x_i}\right\}. \tag{6.3}$$

For finite strain (when the components of the deformation gradient are not small compared to unity), Eq. (6.3) is modified to include a higher-order term and is cast either within a Lagrangian framework, where quantities are expressed in terms of material coordinates in the undeformed configuration, or a Eulerian framework, where quantities are expressed in terms of spatial coordinates in the deformed configuration. For infinitesimal strains, the difference between Lagrangian and Eulerian descriptions is negligible and the distinction is often ignored.

Because the infinitesimal strain tensor is symmetric ($\varepsilon_{ij} = \varepsilon_{ji}$), it only has six independent components. The strain components ε_{xx}, ε_{yy}, and ε_{zz} are the normal strains describing elongations (or contractions) along the coordinate axes. These longitudinal strain components are equal in magnitude to the longitudinal engineering strains one might obtain from Eq. (6.1). Tangential or shear strains are represented by $\varepsilon_{xy} = \varepsilon_{yx}$, $\varepsilon_{yz} = \varepsilon_{zy}$, and $\varepsilon_{xz} = \varepsilon_{zx}$, and they are related to the engineering shearing strains (Eq. 6.2) by the expression

$$\gamma_{ij} = 2\varepsilon_{ij}. \tag{6.4}$$

Like other tensor quantities, the strain tensor is invariant with respect to coordinate transformations. For this reason, it is often used in constitutive models

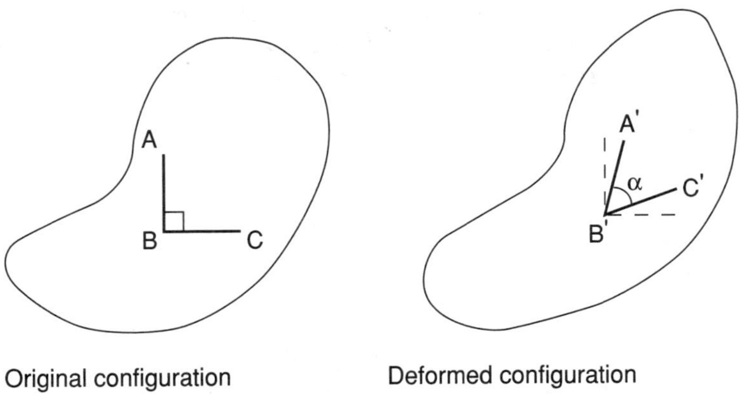

Figure 6.2. Two normal line segments in a deforming body before and after deformation.

(as opposed to the engineering strain) and in computational codes where frame invariance is an important requirement.

At each point and for every admissible state of stress in the solid, there exist three mutually perpendicular planes on which the shear stresses are zero. The stress components acting on these planes are known as the principal stresses, and the orientations along which the principal stresses act are known as the principal directions. The principal stresses are designated by the symbols σ_1, σ_2, and σ_3 where $\sigma_1 \geq \sigma_2 \geq \sigma_3$. At least two of the principal stresses have maximum and minimum magnitudes for all possible normal stresses in all orientations. The maximum shear stress, τ_{max}, acts on the plane with a normal vector that bisects the angle between the maximum and minimum principal stresses. The magnitude of the maximum shear stress is equal to half the difference between the maximum and minimum principal stresses and is therefore given by the relation

$$\tau_{max} = \frac{1}{2}(\sigma_1 - \sigma_2). \tag{6.5}$$

The normal strains $\varepsilon_1 \geq \varepsilon_2 \geq \varepsilon_3$ acting along the principal directions also have maximum and minimum magnitudes in one-dimensional problems. Filaments lying along the principal axes of stress and strain can change their lengths but they do not rotate. The maximum shearing strain occurs in the direction intermediate between the directions of the maximum and minimum normal strains and is given by

$$\gamma_{max} = \varepsilon_1 - \varepsilon_3. \tag{6.6}$$

Small relative changes in the specific volume V are equal to the sum of the relative elongations in any three orthogonal directions.

$$\frac{dV}{V} = d\varepsilon_{xx} + d\varepsilon_{yy} + d\varepsilon_{zz}. \tag{6.7}$$

For an isotropic material, it is useful to separate the stress and strain tensors into spherical (or hydrostatic) and deviatoric (or distortional) components so that the volumetric and shearing aspects of the material behavior can be treated separately. The spherical component of the stress tensor is the hydrostatic pressure p (with sign reversed):

$$p = -\frac{1}{3}(\sigma_{xx} + \sigma_{yy} + \sigma_{zz}). \tag{6.8}$$

The negative sign reflects the common convention that stresses are positive in tension, whereas pressure is positive in compression. The deviatoric components of the stress tensor σ'_{ij} are computed by subtracting the mean stress or pressure from the stress tensor:

Normal components $$\sigma'_{ij} = \sigma_{ij} + p\delta_{ij}, \quad i = j \tag{6.9a}$$

Shearing components $$\sigma'_{ij} = \sigma_{ij}, \quad i \neq j \tag{6.9b}$$

where δ_{ij} is the kronecker delta and has a value of either 1 or 0 depending on whether $i = j$ or $i \neq j$. The deviatoric stress components are primarily concerned with shear behavior and with yield phenomena. On the other hand, the spherical component is more closely related to hydrostatic phenomena.

The components of the deviatoric strain tensor, ε'_{ij}, are related to the components of the strain tensor by the equation

$$\varepsilon'_{ij} = \varepsilon_{ij} - \frac{1}{3} e_v, \tag{6.10}$$

where $e_v = \varepsilon_{11} + \varepsilon_{22} + \varepsilon_{33}$ is the volumetric strain, or dilatation, and it represents the change in volume of an element. On the other hand, the deviatoric strain components represent the change in shape of the element.

In the linear theory of elasticity, the stress and strain increments are related by Hooke's law, which can be expressed in the following incremental form:

$$\text{Normal} \qquad d\left(\sigma_{ij} + p\delta_{ij}\right) = d\sigma'_{ij} = 2G\left(d\varepsilon_{ij} - \frac{1}{3}\frac{dV}{V}\delta_{ij}\right), \quad i = j, \tag{6.11a}$$

$$\text{Shear} \qquad d\tau_{ij} = d\sigma_{ij} = d\sigma'_{ij} = Gd\gamma_{ij} = 2Gd\varepsilon_{ij}, \quad i \neq j, \tag{6.11b}$$

where G is the shear modulus. The increments in the spherical components of stress and strain are related as follows:

$$dp = -K\frac{dV}{V}, \tag{6.12}$$

where K is the bulk modulus.

The yield condition or limiting elastic state may be defined by many criteria. The purpose of these criteria is to use a standard test to define the conditions under which plastic flow occurs for given load conditions. For example, according to the criterion of Coulomb and Guest (see Hill [1950]), the yield condition is reached when the maximum shear stress reaches the value corresponding to the yield strength, Y, in simple tension:

$$\left|\tau_{\max}\right| = \frac{Y}{2}. \tag{6.13}$$

Another commonly used criterion is that of von Mises in which the yield limit is reached when the so-called effective stress, $\bar{\sigma}$, is equal to the yield strength. The effective stress, an invariant of the deviatoric stress tensor, is defined by

$$\begin{aligned}\bar{\sigma}^2 &= \frac{3}{2}\sigma'_{ij}\sigma'_{ij} \\ &= \frac{3}{2}\left\{\left(\sigma'_{xx}\right)^2 + \left(\sigma'_{yy}\right)^2 + \left(\sigma'_{zz}\right)^2\right\} + 3\left\{\left(\sigma'_{xy}\right)^2 + \left(\sigma'_{yz}\right)^2 + \left(\sigma'_{xz}\right)^2\right\}\end{aligned} \tag{6.14}$$

In terms of the effective stress, the von Mises yield condition is simply

$$\overline{\sigma} = Y. \tag{6.15}$$

Fortunately, for uniaxial strain conditions, as in planar one-dimensional wave propagation, these two yield conditions produce the same result.

During deformation in the plastic region, the increment of strain along each axis is the sum of elastic and plastic components:

$$d\varepsilon_{ij} = d\varepsilon_{ij}^{el} + d\varepsilon_{ij}^{pl}. \tag{6.16}$$

In metal plasticity, it is usual to assume no inelastic compression is possible, a constraint that takes the form

$$d\varepsilon_{xx}^{pl} + d\varepsilon_{yy}^{pl} + d\varepsilon_{zz}^{pl} = 0. \tag{6.17}$$

The yield strength of materials is usually determined in standard tension tests conducted under uniaxial stress conditions. Figure 6.3 shows the idealized stress–strain diagram of standard tests under uniaxial stress, the usual test condition for quasi-static loading. In this case,

$$-Y \le \sigma_{xx} \le Y, \qquad \sigma_{yy} = \sigma_{zz} = 0, \qquad \text{and} \qquad \varepsilon_{yy} = \varepsilon_{zz} \ne 0. \tag{6.18}$$

Until the yield strength Y is reached, the material responds elastically to the loading, and obeys Hooke's law, which can be expressed by the following relationship:

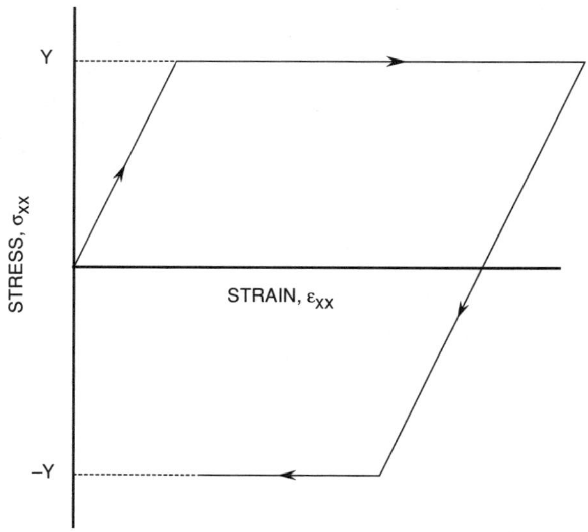

Figure 6.3. Idealized uniaxial stress–strain diagram in an elastic–plastic material under uniaxial stress loading conditions.

$$\sigma_{xx} = E\varepsilon_{xx}, \quad \text{where } E = \frac{3G}{1+\dfrac{G}{3K}}. \tag{6.19}$$

The coefficient E is called Young's modulus of elasticity. The stress–strain relationship is governed by Hooke's law until the yield strength of the material is reached (i.e., $\sigma_{xx} = Y$). Straining beyond the yield point does not cause an increase in stress. Upon unloading, the material again behaves elastically until reverse yielding occurs at the stress level $\sigma_{xx} = -Y$ After reverse yielding, the stress in the material remains constant with further straining. In subsequent loading cycles, each reversal in the loading direction causes an initial elastic response that persists until the stress is equal to the yield strength; thereafter, the stress remains constant.

In both compression and rarefaction waves, the boundary conditions are uniaxial strain where $\varepsilon_{yy} = \varepsilon_{zz} = 0$ and $\sigma_{yy} = \sigma_{zz} = \pi\ 0$. Figure 6.4(b) shows the stress–strain diagram for a solid body under one-dimensional compression during both loading and unloading. In the elastic region, the longitudinal modulus of the material is

$$\frac{d\sigma_{xx}}{d\varepsilon_{xx}} = V\frac{d\sigma_{xx}}{dV} = K + \frac{4}{3}G. \tag{6.20}$$

This uniaxial strain modulus is larger than the bulk stiffness

$$-V\frac{dp}{dV} = K, \tag{6.21}$$

which approximately represents the material stiffness during yielding.

The yield condition under uniaxial strain conditions is satisfied when

$$|\sigma_{xx} + p| = |\sigma'_{xx}| = \frac{2}{3}Y. \tag{6.22}$$

Thus, the longitudinal stress in an elastic-plastic body deviates from the hydrostatic curve $p(V)$ by not more than $2Y/3$. The longitudinal stiffness during plastic deformation is approximately the bulk stiffness. Combining the yield condition with the longitudinal stiffness relation, we obtain the longitudinal stress at initial yield:

$$\sigma_{xx} = -Y\left(\frac{K}{2G} + \frac{2}{3}\right). \tag{6.23}$$

This initial yield value for the longitudinal stress under shock wave loading is the Hugoniot Elastic Limit (HEL), which we discussed earlier, in Chapter 4, and is indicated by point A in Figure 6.2(b).

During unloading under planar wave propagation conditions the stress σ_{xx} initially decreases elastically, as from point B to point C in Figure 6.2(b). Reverse yielding at point C occurs after the deviatoric stress σ'_{xx} passes through zero and increases to yielding again (again meeting the criterion in Eq. (6.22). The decrease in the stress σ_{xx} to reach reverse yielding is approximately twice the HEL.

The pressure is computed from an equation of state. Since only three thermodynamic variables—the pressure, the specific volume or density, and the specific internal energy—are present in the conservation equations, a caloric equation of state is usually used in computations. The most frequently used equation of state to describe the mean stress is the Mie–Grüneisen relation:

$$P - P_{REF} = \frac{\Gamma(V)}{V}(E - E_{REF}), \tag{6.24}$$

where P and E are the pressure and energy at the current specific volume V, P_{REF} and E_{REF} refer to a point on a reference curve at the same specific volume V, and $\Gamma(V)$ is the Grüneisen ratio. Commonly used reference curves in Eq. (6-24) include the cold compression or potential curve and the Hugoniot.

Equation (6.24) provides a means for extending the information of a known reference P–V relation (such as the Hugoniot) to other values of internal energy. Three commonly used forms for the reference curve are the following:

(1) Series expansion: $$P = K_0\mu + K_1\mu^2 + K_2\mu^3, \tag{6.25}$$

where P is the pressure, K_0, K_1, and K_2 are the bulk modulus series with K_0 being the standard bulk modulus, μ is the compressive strain, $V_0 / V - 1$, and V_0 is the reference specific volume. Eq. (6.25) may be used to represent either the cold isothermal compression curve or the Hugoniot. In the latter case, Rice et al. [1958] provide values of the fitting parameters for a large number of metals, and Kohn [1969] provides values of the fitting parameters for metals, plastics and ceramics.

(2) Murnaghan: $$P = \frac{K_0}{b}\left[\left(\frac{V_0}{V}\right)^b - 1\right], \tag{6.26}$$

where b is a material constant with a value of about 5. The Murnaghan equation is often used in geomechanics to represent the cold isothermal compression curve of geologic media.

(3) Linear $U - u$: $$P_H = \frac{K_0(1 - V / V_0)}{[1 - s(1 - V / V_0)]^2}, \tag{6.27}$$

where U is the shock velocity, u is the particle velocity, and s is a dimensionless material constant representing the slope in the linear shock velocity–particle velocity relation:

$$U = c_0 + su. \tag{6.28}$$

This is the same as Eq. (2.12) in which c_o is the sound velocity corresponding to the initial equilibrium bulk compressibility of the medium, which is related to the bulk modulus, K_0, as follows:

$$K_0 = c_0^2 / V_o. \tag{6.29}$$

Equation (6.28) is one of the most widely used Hugoniot representations.[1] The Hugoniot pressure, P_H, in Eq. (6.27) is obtained by combining Eq. (6.28) with the jump condition equations for the conservation of mass and momentum (Eqs. (2.9) and (2.10)). In this derivation, strength effects are neglected and the material behavior is assumed to be hydrodynamic. A more detailed treatment may be needed when the material strength is not negligible in comparison with the Hugoniot pressure.

With this background on yielding in ideal plasticity (only a rough approximation for most materials) and the equation of state (pressure–volume–energy relation), let us now examine the wave processes illustrated in Figure 6.4 for a typical plate impact experiment. Figure 6.4(a) shows the configuration for the impact. An impactor on the left strikes a target plate with a velocity v, and we examine the resulting stress wave as it passes two points in the target. The stress wave histories at these points are shown in part (c) of the figure. At point 1, we see a rapid rise to a stress plateau, and then a gradual decrease in stress corresponding with the rarefaction fan proceeding from the rear of the impact plate (a stress history diagram appropriate for this impact is shown in Figure 3.23). The impactor in this case has responded as either a purely elastic material, or as a material with negligible yield strength so that no yield process is observed in the rarefaction fan. After the wave has passed through some of the elastic-plastic target material, considerable structure has developed. At point 2, the wave front shows the arrival of an elastic precursor with an amplitude corresponding to the HEL. The wave velocity of the precursor corresponds to the elastic wave velocity in the target (the equation for the longitudinal elastic wave velocity is given below in Eq. (6.30)). The remainder of the compressional wave arrives with the shock velocity U. This velocity corresponds with the slope of the Rayleigh line connecting points A and B in Figure 6.4(b). As indicated in this figure, the stress path during the shock loading is along the Rayleigh line, not along the locus of equilibrium states (i.e., the curved line between A and B). This curved line is parallel to the pressure–volume path with an offset of $2Y/3$ as noted above. Following the plateau on the second wave history in part (c) of the figure, the stress decreases by twice the HEL, at which point reverse yielding occurs. Following this reverse yielding the remainder of the wave travels at a slower range of velocities—a rarefaction fan corresponding with the bulk sound speed, which is a function of the stress level.

Figure 6.4(b) and (c) show the usual wave processes for the conditions under which the elastic longitudinal wave velocity is greater than the shock velocity,

1. Values of c_0 and s were presented in Table 2.1 for several metals.

that is, the slope of the stress path up to A in part (b) is steeper than the Rayleigh line. If the stress level is high, the shock wave may be overdriven—the slope of the Rayleigh line then becomes steeper than the elastic slope connecting the initial state to point A. In this latter case, loading proceeds along a Rayleigh line that connects the initial state to the final state at B and no precursor appears in the wave front.

Let us now examine approximate relations for the wave speeds illustrated in Figure 6.4(c). Because the longitudinal compressibility is different in the elastic and plastic regions, elastic precursors appear in both compression and rarefaction waves. Elastic precursors propagate with the velocity c_L of longitudinal elastic waves

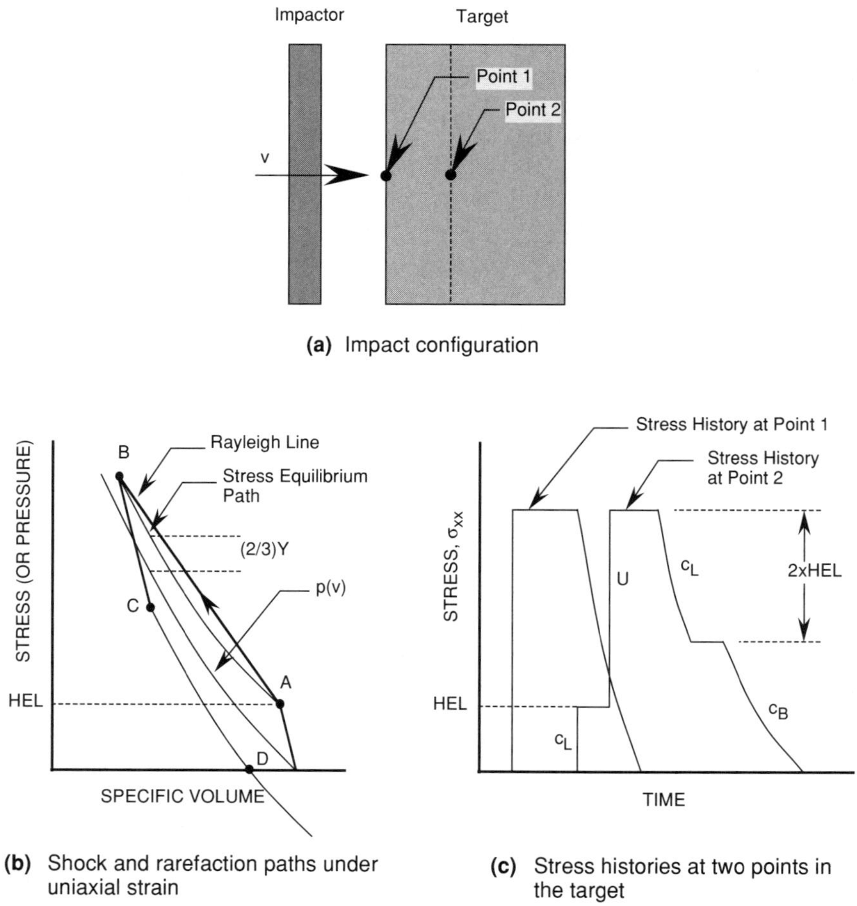

Figure 6.4. Evolution of a typical stress pulse in an elastic-plastic material subjected to uniaxial strain impact loading.

$$c_L = \sqrt{\frac{K + \frac{4}{3}G}{\rho}}. \tag{6.30}$$

Here K is, in general, the bulk stiffness at the current stress level. For the initial precursor, the appropriate value of K is the zero- or low-pressure bulk modulus. But for unloading from the peak stress at point B in Figure 6.4(b), we should use the bulk stiffness corresponding to that stress level. The curvature of the pressure–volume path indicates that this stiffness increases with increasing stress. The velocity of wave propagation in the plastic region is the bulk sound velocity, c_B,

$$c_B = \sqrt{\frac{K}{\rho}}. \tag{6.31}$$

The bulk sound velocity varies throughout a rarefaction fan as shown in Figure 6.4(c). When we use the zero-pressure bulk modulus, we obtain the velocity associated with the foot of the wave.

The longitudinal and bulk sound velocities are related through Poisson's ratio ν:

$$\frac{c_L}{c_B} = \sqrt{\frac{3(1-\nu)}{1+\nu}}. \tag{6.32}$$

It has been shown that Poisson's ratio is almost independent of pressure, so a constant ratio of the longitudinal and bulk sound velocities is a good approximation for treating most wave propagation problems.

The elastic–perfectly plastic idealization of material behavior is relatively simple, mathematically tractable, and serves the useful purpose of illustrating some of the important processes that take place in solid materials during stress wave propagation. It must however be emphasized that real materials rarely exhibit such simple response. In most instances, real materials exhibit a hardening behavior whereby the yield stress increases with increasing strain. Yielding in real materials is further complicated by the influence of several other variables including pressure, temperature, and strain rate. For example, Lassila, Leblanc, and Gray [1992] have reported on some of these effects in copper. Steinberg and Lund [1989] have provided a general model with some of these effects, including a Bauschinger effect (the variations of the apparent yield strength and shear modulus during unloading after yielding).

6.1.2. Temperature Computations

Temperature, as a fundamental state variable, is not contained in the usual conservation equations of continuum mechanics (conservation of mass momentum and energy). Consequently, most computations may be made without explicitly accounting for temperature changes during deformation. However, some of the phenomena that accompany dynamic deformation and fracture, such as phase transitions, shear banding, etc., may be temperature-dependent. This makes it necessary to account for temperature changes during the course of dynamic compression, rarefaction, plastic deformation and fracture processes. A variety of methods have been used to estimate the temperature in shock wave simulations. Here we introduce four methods, ranging from very simple to methods based on precise thermodynamic considerations.

In the first method, the temperature θ is simply derived from the total internal energy E:

$$\theta - \theta_0 = \frac{(E - E_0)}{C_v}, \tag{6.33}$$

where θ_0 and E_0 are reference values of temperature and internal energy, and C_v is the specific heat at constant volume, usually taken to be constant.

This method provides only a very rough estimate of temperature because it even neglects the fact that the internal energy is composed of elastic strain energy and other energy associated with thermal effects. Also, C_v is used although there is no accounting for changes in volume. The next method provides an initial accounting for these effects.

In the second method, if the Mie–Grüneisen equation of state of the matter is known, a reference cold compression curve is used. The temperature is then given by

$$\theta - \theta_0 = \frac{(E - E_c)}{C_v}, \tag{6.34}$$

where E_c is the energy on the cold compression curve, given by

$$E_c = E_0 - \int_{V_0}^{V} P_c dV, \tag{6.35}$$

where E_0 and V_0 are reference values of the internal energy and volume, and P_c is the pressure on the cold compression curve.

Steinberg and Lund [1989] give an equation for the cold compression curve and appropriate parameters for the curve for many materials. In this method, they are accounting approximately for the separation of the internal energy into elastic and thermal energies and for the change in specific volume.

The third method is outlined only briefly here to indicate the directions to be taken and to illustrate the shortcomings of the two previous methods. Here the temperature is treated as a basic thermodynamic variable, rather than an adjunct, which can be treated by an auxiliary equation when needed as in the first two methods. We recognize that for precise thermodynamic computations we must use a complete equation of state (Cowperthwaite [1969]). The usual energy-pressure–volume relation used in traditional hydrocodes is not complete. A complete equation of state provides all five basic thermodynamic variables: pressure P, entropy η, temperature T, internal energy E, and specific volume V (or density). For a solid we must also know the full stress and strain tensors. To explore the concept of a complete equation of state further, let us begin with the second law of thermodynamics, written as follows:

$$dE = \theta d\eta - PdV = \frac{\partial E}{\partial \eta} d\eta + \frac{\partial E}{\partial V} dV. \tag{6.36}$$

From this equation, we see that if we know E, η, and V, we can compute θ and P from the derivatives:

$$\theta = \frac{\partial E}{\partial \eta} \qquad \text{and} \qquad P = -\frac{\partial E}{\partial V} \tag{6.37}$$

Hence, one form of a complete equation of state is $E(\eta, V)$, with energy given in terms of entropy and specific volume. Another complete form is based on the quantities pressure, volume, and temperature.

The specific heat at constant volume is given by

$$C_v = \left(\frac{\partial E}{\partial \theta}\right)_V \tag{6.38}$$

and therefore cannot be independently specified nor treated as a constant: it must be derived from the equation of state and be consistent with the equation of state.

The foregoing method is thermodynamically rigorous, but neglects the facts that we are dealing with solids that may undergo additional rate-dependent (fracture) processes. A thermodynamically rigorous way to calculate temperature in materials that are undergoing rate processes was reported in 1967 in an important paper by Coleman and Gurtin [1967]. In their internal state variable formulation, all dissipation is described by evolution equations for internal state variables, α_n:

$$\dot{\alpha}_n = f_n\left(\theta,\ F_{ij},\ \theta_{,i},\ \alpha_n\right), \tag{6.39}$$

where θ is the temperature, $\theta_{,i}$ is the temperature gradient, and $F_{iJ} = \partial x_i / \partial X_J$ are the components of the deformation gradient tensor $\mathbf{F}$. The subscript n in Eq. (6.39) is used to represent the total number of internal state variables α. All other subscripts refer to coordinate directions. Capital letter indices like J in Eq. (6.42) (below) are referenced back to the original configuration, whereas small letter indices, like i in Eq. 6.39, are referenced to the current configuration.

Examples of processes that can be represented as internal state variables within the framework of the Coleman and Gurtin thermomechanical approach include microcrack nucleation and growth in an elastic material, which would result in inelastic behavior on the continuum level. In this case, one of the internal state variables (α_n) would be chosen to represent the crack deformation strain. A second example is the case of an elastic-plastic material where the internal state variable would be selected to represent some measure of the plastic deformation. Finally, for a reacting propellant, we could select internal state variables, α_k, to represent relative amounts of various chemical species. In general, the vector α_n refers to the complete set of internal state variables used in the specific application, and the set of Equations (6.39) completely describes their rate dependencies and associated dissipations in the constitutive model. Note that, depending on the process represented by any one of the internal state variables, the state variable can be a scalar, a vector, or a higher-order tensor.

In the subsequent developments, we choose two internal state variables: α_{ij}^{c}, which represents the cracking strain tensor; and α_{ij}^{p} which represents the plastic strain tensor. In this case then, the symbol α_n refers to the two internal state variables α_{ij}^{c} and α_{ij}^{p}.

The internal energy per unit mass, E, obeys the conservation of energy equation:

$$\rho \dot{E} = T_{ij} L_{ij} + \rho r - q_{i,i}, \tag{6.40}$$

where ρ is the mass density, T_{ij} are the components of the Cauchy stress tensor $\mathbf{T}$, L_{ij} are the components of the particle velocity gradient $\mathbf{L}$, r is the heat supply absorbed in the interior of the material, for example, by radiation from the external world, and q_i are the components of the heat flux vector $\mathbf{q}$. Sound, modern hydrocodes all use Eq. (6.40). Thus, increments in the internal energy E in the hydrocode should be calculated from the above formula. In many standard applications, both q_i and r are zero, but X-ray effects codes routinely account for r, and increasingly, they include heat flow as well. Note that the term $T_{ij}L_{ij}$ in Eq. (6.40) represents the total mechanical work done on a material element, and thus includes the work producing shear as well as the volumetric deformations, and includes plastic as well as elastic work.

Coleman and Gurtin developed a number of representations for constitutive relations that are thermodynamically consistent (obey the second law for all possible deformations). One such representation that may be useful for us is the following set of equations:

$$\theta = \hat{\theta}\left(F_{iJ}, E, \alpha_n\right), \tag{6.41}$$

$$S_{iJ} = \hat{S}_{iJ}\left(F_{iJ}, E, \alpha_n\right), \tag{6.42}$$

$$q_i = \hat{q}_i\left(F_{iJ}, E, \theta_{,i}, \alpha_n\right), \tag{6.43}$$

$$\alpha_n = \hat{\alpha}_n\left(F_{iJ}, E, \theta_{,i}, \alpha_n\right), \tag{6.44}$$

where a new stress tensor $\mathbf{S}$, with components S_{iJ}, has been introduced for mathematical convenience. This new stress tensor is closely related to the first

Piola-Kirkhoff stress tensor, $\widehat{\mathbf{S}}$ (i.e., $\widehat{\mathbf{S}} = \rho_0 \mathbf{S}$) and is given in terms of the Cauchy stress tensor as

$$\mathbf{S} = \frac{1}{\rho}\mathbf{T}\mathbf{F}^{-\mathrm{T}}. \tag{6.45}$$

In terms of the components of **S**, Eq. (6.40) becomes

$$\rho \dot{E} = \rho S_{iJ} \dot{F}_{iJ} + \rho r - q_{i,i}. \tag{6.46}$$

Coleman and Gurtin show that, within this framework, the entropy, η, can be expressed as

$$\eta = \hat{\eta}(F_{iJ}, E, \alpha_n). \tag{6.47}$$

Furthermore, the stress and temperature can be derived from Eq. (6.47) as

$$\theta = \left\{ \frac{\partial \hat{\eta}(F_{iJ}, E, \alpha_n)}{\partial E} \right\}^{-1}, \tag{6.48}$$

$$S_{iJ} = -\theta \frac{\partial \hat{\eta}(F_{iJ}, E, \alpha_n)}{\partial F_{iJ}}. \tag{6.49}$$

Thus, if we can define an entropy function $\hat{\eta}(F_{iJ}, E, \alpha_n)$ such that Eq. (6.49) produces realistic constitutive relations, we can calculate the temperature from Eq. (6.48).

To illustrate how the approach described above can be used to compute temperature in hydrocode calculations, we consider the equation of state

$$\hat{\eta}(F_{iJ}, E, \alpha_n) = C \ln\left[\left\{ \frac{E - H(F_{iJ}, \alpha_n)}{E_0} \right\} \left\{ 1 + g\left\{ \varepsilon_{ij} - \alpha_{ij}^{c} \right\} \delta_{ij} \right\} \right], \tag{6.50}$$

where $E_0 = C\theta_0$ is an arbitrary reference energy, θ_0 is a reference initial temperature, C is a constant with the dimensions of specific heat, g is a dimensionless constant, ε_{ij} is the strain tensor (see Section 6.1.1), and δ_{ij} is the kronecker delta and has a value of either 0 or 1 depending on whether $i = j$ or $i \neq j$. The parentheses ( ) are reserved to indicate functional dependence, and repeated indices denote summation.

We define $H(F_{iJ}, \alpha_n)$ as the elastic mechanical work done on a reference isotherm associated with the reference temperature θ_0:

$$H(F_{iJ}, \alpha_n, \theta_0) = \int S_{iJ}(\theta_0) d\varepsilon_{iJ}^{e} = \int S_{iJ}(\theta_0)\left[d\varepsilon_{iJ} - d\alpha_{iJ}^{c} - d\alpha_{iJ}^{c} \right], \tag{6.51}$$

where $S_{iJ}(\theta_0)$ is the stress on the isotherm. Note that we are using the Piola-Kirkhoff stresses, and corresponding measures of strains, in this expression for the work, consistent with Eq. (6.40). Note also that $H(F_{iJ},\alpha_n,\theta_0)$ depends on the complete elastic strain tensor.

Using Eqs. (6.48) and (6.50), we can now obtain the temperature as

$$\theta = \frac{E - H(F_{iJ},\alpha_n,\theta_0)}{C}. \tag{6.52}$$

Similarly, the stress can be computed from Eqs. (6.49) and (6.50) as follows:

$$S_{iJ} = S_{iJ}(\theta_0) - \frac{g\{E - H(F_{iJ},\alpha_n,\theta_0)\}}{1 + g\{\varepsilon_{kL} - \alpha_{kL}^c\}\delta_{kL}}\delta_{iJ}. \tag{6.53}$$

Note that θ_0 is a reference temperature for the unstrained case when $H = 0$. The temperature increase over θ_0 from Eq. (6.52) is intuitively obvious: it is the excess of internal energy over that on the reference isotherm divided by a specific heat.

Equations (6.53) are similar to the usual Mie–Grüneisen pressure–volume and elastic deviator formulations, except that the familiar thermal term "Γ" (the fractional term on the right-hand side) is a weak function of volumetric strain.

In a hydrocode application, the above method of calculating temperature is not without a computational burden, because significant computational storage is required to continually update the value of $H(F_{iJ},\alpha_n)$ in each computational cell. Furthermore, it is not trivial to fit experimental constitutive relation data to a $\hat{\eta}(F_{iJ},E,\alpha_n)$ function. (The example $\hat{\eta}(F_{iJ},E,\alpha_n)$ function of Eqs. (6.50) and (6.51) will fit available Hugoniot and shear moduli data in many cases, but not when the specific heat is a strong function of temperature.) Nevertheless, in cases where temperature effects are important, Coleman and Gurtin's approach provides a rigorous method for calculating the temperature in materials experiencing rate-dependent failure.

6.1.3. Thermodynamic Requirements

To provide physically reasonable and mathematically consistent constitutive relations for the fracture of materials, it is necessary to impose certain restrictions. Here we present some general physical requirements that any model must meet. Some thermodynamic requirements have already been presented in Section 6.1.2, during the discussions of temperature computations. These requirements are augmented later in Section 6.3.2 and 6.3.3 with special requirements for allowing the material model to be joined to a finite-element computer program.

The restrictions we need to impose stem from a variety of considerations including frame invariance, material symmetry, and thermodynamic requirements. Generalized thermomechanically consistent constitutive models must usually conform to the following criteria:

1. The material must not produce energy during cyclic loading. This requirement conforms to our accepted notions of energy conservation (first law of thermodynamics) and of entropy production (second law of thermodynamics). Brannon et al. [1995] have provided detailed thermodynamic requirements, as have Coleman and Gurtin [1967], Malvern [1969], and several others.
2. The constitutive relations must be invariant under a change of the frame of reference (principle of frame-indifference). For example, two observers, each representing a different frame of reference (possibly in relative motion with respect to one another) should observe the same state of stress in a continuous body.
3. Material symmetry restrictions (e.g., isotropy, transverse isotropy, orthotropy), place additional constraints on the form of the constitutive equations. A common example is that of an isotropic material, the properties of which are the same in all directions. In the case of elastic material behavior, material symmetry restrictions reduce the number of independent constitutive properties from 21 for a general anisotropic material to just 2 (Young's modulus and Poisson's ratio) for an isotropic material.

Models that conform to these restrictions tend to produce repeatable results and are not usually sensitive to small changes in input data or to strain states during a computation. Thus, they give results that meet our general expectations for reliability.

Processes that occur in shock or rarefaction waves (e.g., plastic flow, phase changes, chemical reactions, evolution of damage) are, in general, rate processes that depend on the temperature of the material. In many cases, such temperature dependence can be neglected, but in other cases, (notably chemical reactions) it is necessary in computational simulations to calculate the temperature. A specific and important example is the case in which burning occurs on microscopic crack surfaces in the propellant in a rocket motor, thereby leading to unstable burning rates and potential detonation. Although the initial nucleation and growth of the fractures may be to a first approximation temperature-independent, the burning rate depends strongly on the temperature. Furthermore, in fracturing material the fracture mode and kinetics are also typically temperature-dependent, so if the material is being heated by radiation, plastic flow, or exothermic reactions, the fracturing process may change significantly in time.

6.1.4. Numerical Problems in Models and Simulations

Numerical problems often arise in the treatment of fracture problems because the stress–strain path taken generally rises positively during initial loading until fracture begins and then falls with a negative slope during the final stages of fracture. In simple treatments of fracture this stress–strain path is taken as the constitutive relation: the negative slope leads to negative moduli, imaginary sound speeds, and runaway instabilities. In this section we attempt to face these problems and discuss approaches which have been taken to circumvent or minimize their effects. The following is only a background and some discussion of these problems—not a thorough solution. Here we separate the discussion into problems pertaining to a continuum treatment, nonlocal behavior, and large deformation.

6.1.4.1. Continuum Treatment

By *continuum*, we refer to a representation of material as a substance that is uniform and that remains uniform when examined at any scale. No real material qualifies as a continuum because no material is uniform down to the atomic level. However, it is useful to represent many materials as continua. Materials such as woven composites, concrete, and coarse-grained metals may be represented either as continua or as separate components depending on the level of detailed information required from the simulation. Within the context of the continuum treatment, we consider three kinds of problems:

1. *Instabilities:* An example of an instability is Euler column buckling. In this case, the material is stable, but the geometry of the structure is unstable. An immediate solution in simulating this behavior is to provide separate treatments for the material and for the geometry. Then we can correctly treat the material as stable and follow the dynamic processes in the change of geometry, which also represents well the actual behavior. The basic problem with unstable behaviors is that they are ill posed mathematically such that no analytical solution can be obtained. The general solution is to properly pose the problem. In section 6.1.5, we present several possible approaches toward recasting the problem so that it is well posed and solvable.
2. *Element-size dependence:* This problem is characterized by numerical solutions that depend on the size of the finite elements. For fracture problems (which are always dynamic), material rate dependence is required in the constitutive relation. Coleman and Gurtin [1967], among others, have provided a general internal state variable-based approach to developing material models, representing materials by a combination of elastic behavior plus a set of evolution equations for the internal state variables.[2] The state variables, α_i, can represent numbers and sizes of cracks, dislocations, plastic strain, etc. The evolution equations for the state variables, $\dot{\alpha}_i$, cannot have derivatives of time or space on the right-hand side of the equation.

2. The Coleman and Gurtin approach was discussed in Section 6.1.2 in connection with temperature calculations.

Our experience with the nucleation-and-growth fracture models tends to support the restrictions outlined by Coleman and Gurtin. Recently, Seaman and Curran [1998] examined the conditions necessary to make the computed results independent of the element sizes. In these models the rate dependence arises naturally from the observed damage processes, so the model does not provide the freedom to specify rate-dependent parameters for numerical or stability convenience. As noted in the 1998 paper, we discovered that mesh-independent results required a mesh size that was predictable in advance from the material properties. The element size is governed by the material rate dependence; that is, by the viscoelastic and viscoplastic response of the intact material. When this mesh size is used, the stresses in each element are reliably given, and the damage calculations can then also give results that are independent of the mesh size. Unfortunately, for many computations, this reliable mesh size is smaller than desirable for computational purposes.

3. *Localization:* Localization occurs when an object under approximately uniform strain and loading develops a region or band at a high strain surrounded by the remainder of the material at a lower strain. A typical example of this behavior is shear banding. A less common example is a plate undergoing spall. In spall usually a large central region of the plate has essentially the same strain, yet the region which first reaches this strain begins to fracture and thereby becomes softer. Subsequent straining magnifies the damage at this initial region, providing a marked band of damage.

   Localization includes both the ideas of instability (because of the softening that occurs in the stress–strain path) and element-size dependence. So the methods mentioned under these categories should be used. Recent work by Wright [2000] has shown that shear banding can be accurately treated numerically by providing the appropriate element size (small enough to define the details of the band) and a material model with the needed strain-rate and temperature dependence.

6.1.4.2. Nonlocal Behavior

In nonlocal behavior, the material response recognizes some spatial extent of the material. Concrete or other composites are examples of materials that may undergo nonlocal behavior. The stress in a finite-element of concrete is an average of the stresses in the separate particles and these stresses are based on strains that vary from place to place within the finite element. Several approaches are possible for treating nonlocality; among these are:

1. Imbricated Continuum method of Bazant (Bazant [1984] and Bazant et al. [1984]) in which the stress is computed from some weighted average of the strains in nearby elements. To represent concrete Bazant

has used elements of uniform properties, not some with properties of cement and some of aggregate. The element sizes are not related to the particle size being represented and the breadth of the weighting for determining the average strain is determined for numerical convenience.

2. The elements are identical and each represents only average properties of the composite, so the elements are of the continuum type. The size of the elements must be large enough to actually represent the average cross section of the material.

With these two treatments we recognize that we cannot gain information about details within an element, but only average behavior of the element. These approaches are inappropriate for treating the hypothetical problem of a knife cutting through a block of concrete (for normal concrete the element must be several centimeters across, but the knife blade is much smaller) or a shock front in concrete (because the shock front thickness is thinner than the finite elements).

6.1.4.3. Large Deformation

In representing dynamic loading of solids, we can expect large volume change and also large distortion of the finite elements used to depict the material response. The usual procedure for handling large distortions is through advection of material from one element to the next to maintain the elements in relatively undeformed shape while permitting the material to deform. Problems of advection are numerical and not part of the actual material response to be treated. Methods for treating some aspects of advection are discussed in Section 6.3.3.

6.1.5. Background on Stability

Our primary concern is to construct a mathematical model to describe the behavior of a material undergoing fracture. When the material exhibits an instability, we want the model to represent that instability, and when the material exhibits stable behavior, we want the model to represent that as well. Here we mention first some of the approaches that have been taken to address the stability problem.

Sandler and Wright [1983] discussed the stability problem in one dimension by analyzing wave propagation through a material exhibiting strain softening, that is, the stress–strain curve showed the usual positive slope for elastic loading to a peak, then the stress decreased with increasing strain. The usual elastic wave propagation in the rod of material occurred for stresses below the peak. But for loading past the peak, the first computational element absorbed all the additional strain and later elements experienced some unloading. Hence, the wave was essentially trapped in the first element, and the amount of material being loaded by the wave depended on the size of this first element. This landmark paper clearly showed that a simple, rate-independent model for fracture behavior is severely limited. Following this paper there were a number of studies to define the characteristics of a model that could satisfactorily represent fracture.

An extensive discussion of the stability problem occurred at a meeting in 1983 (published in 1984).[3] Sandler and Wright [1984] showed Sandler's result of the effects of using a rate-independent softening model in wave propagation. They also showed that for a rate-dependent model, the computation could proceed in the usual way and have a well-defined result; the problem appears to be well posed.

Wu and Freund [1984] considered wave propagation in a one-dimensional shearing situation, using a rate- and temperature-dependent material behavior. In this case, softening caused the loading to be trapped at the loaded edge of the material. The extent of the region in which the shearing deformation is trapped was determined from their numerical analysis to be dependent on the particular rate-dependence and loading conditions. Prevost and Loret [1990] have also studied localization in viscoplastic solids.

Read and Hegemier [1984] reviewed several aspects of softening. They emphasized the apparent softening in compression tests of rock and concrete, and they dealt with the question of whether softening is a material property, a boundary condition, a geometrical feature, or a structural problem. Their analysis makes clear that softening is not a material behavior. They developed a characteristic analysis showing that for rate-dependent behavior, one always gets stable computations. They considered Malvern's [1951] one-parameter viscoplastic model (hyperbolic behavior), a simple rate-dependent model in which the stress is proportional to the strain rate (parabolic behavior), and a rate-independent model (this model is not well posed; it gives imaginary characteristics). They also discussed the result of Wu and Freund [1984], pointing out that the rate-dependent material model gives a region of trapped deformation where the rate-independent model gives a discontinuity of unlimited strain.

Bazant [1984] has given several methods for producing stable computations with rate-independent models. In one of these, the softening band is given a fixed dimension; then the computation is stable, and the results can also be independent of the cell size. However, the size of the band must be prescribed in advance, so many of the phenomena of fracture are imposed, rather than arising from the computation. Bazant and Chang [1987], Bazant [1988a, 1988b], and Bazant and Pijaudier-Cabot [1988] have also used several "nonlocal " methods to provide stability. In these methods, the damage growth at one location depends on the stresses over a prescribed vicinity, rather than just the stress in one finite element.

3. Workshop on the Theoretical Foundation for Large-Scale Computations of Nonlinear Material Behavior, Evanston, Illinois, October 24 to 26, 1983.

Frantziskonis and Desai [1987a, b, c] constructed a model in which part of the volume is intact and part is fracturing. They show an evolution law for damage, in which the damage D is a scalar and its growth rate depends on the plastic strain rate.

$$\frac{\partial D}{\partial t} = g(\bar{\varepsilon}^P)\frac{\partial \bar{\varepsilon}^P}{\partial t} \qquad \text{or} \qquad D = F(\bar{\varepsilon}^P), \tag{6.54}$$

where $g(\bar{\varepsilon}^P)$ and $F(\bar{\varepsilon}^P)$ are functions of the plastic strain. Hence, in this formulation, fracture is not actually strain rate dependent, but only strain dependent. In their finite-element analyses, softening occurs, and the results are stable and relatively independent of the cell size. They indicate that a rate-dependent model would provide guaranteed stability, but their model has at least a qualified stability. Several other researchers have constructed fracture models by making a connection between damage and plastic flow—often making the damage develop as a function of plastic strain or a variable that describes an inelastic process very similar to plastic flow. Models like this are given in the works of Lubliner et al. [1989], Ju [1989], and Runesson et al. [1989].

Lade [1988] presented a useful comparison of the constitutive behavior and yielding of frictional materials and metals. He noted that for many frictional materials, the product of the void ratio, V_f, and the friction angle, μ is constant:

$$\mu V_f = \text{constant}. \tag{6.55}$$

The frictional materials do not exhibit normality, but respond more like a material with a nonassociated plastic potential. This potential generally causes plastic strain to have a smaller angle with the hydrostatic axis than a normality-law material would have. Lade showed that Drucker's postulate for stability is not obeyed in some of his samples. He also noted situations in which frictional materials can become unstable (as in liquefaction) although the stress state is within the yield surface. He presented alternative stability postulates to represent the behavior he observed in soils. Lade mentioned experimental data in which strain softening occurred uniformly in tests on dense sands; hence, in these tests strain softening was a constitutive property.

Our nucleation-and-growth rate-dependent models have been used to represent fracture under a wide variety of dynamic loading conditions. Many of these conditions are ripe for instability: negative slope of the stress–strain path and damage (volumetric strain) in a narrow band. However, instability has not occurred. We are in agreement with Wu and Freund [1984] and Read and Hegemier [1984] that softening is not a material property and that rate dependence is needed in the model to provide stability. Our models represent the ordinary material with standard properties plus rate laws for the fracture processes. The negative slope of the computed stress–strain path is associated with the fracture processes: the material constitutive relation always has a positive slope.

Thus, it appears that for stability in the simulations we need to treat separately the solid, intact material and the rate-dependent fracture processes. Also, appropriate rate dependence for the material behavior (viscoelastic and viscoplastic, for example) must be treated. We acknowledge here the fact that the required element sizes and time steps may be too costly for some practical problems.

6.2. Fracture Modeling Approaches

Fracture modeling approaches can generally be divided into two strategies for treating fracture: passive and active. In the passive approach, the stress–strain relations are not modified until a fracture criterion is reached. The fracture criterion is computed based on the computed stresses and strains although these stresses and strains are not modified by the developing damage; hence, the term "passive." The criterion may be based on a stress or strain or some more complex quantity. The fracture is usually assumed to occur instantly when the criterion is reached. The passive treatment may be used with a finite-element code, but it is a simple addition to the code and does not modify the usual stress computations that are undertaken for an intact material. When the fracture criterion is reached, the stresses are set to zero and remain unchanged for the remainder of the computation. Because the passive criterion does not modify the stresses until the time of fracture, the criterion is often computed after the simulation, based on the history of the stresses and strains.

The active approach, which is emphasized in this and later chapters, is exemplified by a fracture model in which the damage alters the material properties and therefore changes the stress–strain relations from those for an intact material. Severe damage can have a large effect on the constitutive relations. It is for these models of the active type that special problems arise during the process of joining the models to finite-element or finite-difference codes.

In the following sections we first provide some background on passive fracture models. Then we discuss several types of currently available active fracture models.

6.2.1. Passive Fracture Modeling Approaches

Many approaches have been used in past studies to model dynamic fracture. The simplest of these approaches is the threshold stress criterion whereby dynamic fracture is assumed to take place when the stress reaches some critical value. This approach is consistent with the Griffith criterion [1921] for brittle fracture under quasi-static conditions, but is not consistent with experimental observations of dynamic fracture where failure is caused by the propagation of many microcracks, not a single macrocrack. Experimental observations in many materials, both ductile and brittle, indicate that dynamic fracture is a time-dependent process. Thus, a time-dependent criterion is required to provide a realistic description of the time-dependent fracture under dynamic loading conditions.

Early attempts at correlating fracture stress under dynamic loadings with applied stress pulse parameters included a criterion based on the applied stress gradient (Breed et al. [1967]) and another criterion based on the stress rate at fracture (Skidmore [1967]). A well-known general criterion for time-dependent dynamic fracture that includes the stress rate at fracture criterion and the stress gradient criterion as special cases was developed by Tuler and Butcher [1968]. This criterion is

$$(\sigma_o - \sigma)^{\lambda} \Delta t = K, \tag{6.56}$$

where σ is the stress (taken to be negative under tension), t is time, σ_o is the stress below which fracture does not occur, and K and λ are material parameters. As noted by Barbee et al., [1970], when $\lambda = 1$, Eq. (6.56) is a simple impulse criterion, and when $\lambda = 2$, Eq. (6.50) can be shown to be equivalent to an energy criterion.

6.2.1.1. Grady's Energy Criteria for Fracture and Fragmentation

A more advanced energy-based set of criteria for spall fracture was proposed by Grady et al. [1981, 1982, 1985, 1988, 1990, 1995]. In this treatment, Grady developed criteria for the spall strength, P_s; time to fracture, t_s; and average fragment size, s, for brittle and ductile materials and for liquids. In each case, it is assumed that spall occurs when the sum of the strain energy and kinetic energy is at least as large as the fracture energy. For brittle solids, the fracture energy dissipated in the creation of new fracture surfaces is derived using a fracture mechanics approach and is characterized in terms of the fracture toughness of the solid, K_c. In this case, the expressions for spall strength, time to fracture, and fragment size are

$$P_s = \left(3\rho c_0 K_c^2 \dot{\varepsilon}\right)^{1/3}, \tag{6.57}$$

$$t_s = \frac{1}{c_0}\left(\frac{\sqrt{3}K_c}{\rho c_0 \dot{\varepsilon}}\right)^{2/3}, \tag{6.58}$$

$$s = 2\left(\frac{\sqrt{3}K_c}{\rho c_0 \dot{\varepsilon}}\right)^{2/3}, \tag{6.59}$$

where ρ is the density, c_0 is the bulk sound speed, and $\dot{\varepsilon}$ is the strain rate.

For ductile solids, spall is assumed to occur by the ductile growth of spherical voids. Accordingly, the fracture energy is derived based on the plastic work expended in growing the void. In this case, the expressions for spall strength, time to fracture, and fragment size are

$$P_s = \sqrt{2\rho c_0 Y \varepsilon_c}, \tag{6.60}$$

$$t_s = \sqrt{\frac{2Y\varepsilon_c}{\rho c_0 \dot{\varepsilon}^2}}, \tag{6.61}$$

$$s = \sqrt{\frac{8Y\varepsilon_c}{\rho \dot{\varepsilon}^2}}, \tag{6.62}$$

where Y is the yield strength and ε_c is the critical strain (or void volume fraction) for stable void growth.

In both the brittle and ductile cases, the inherent flaw structure is assumed to be "favorably disposed" such that spall occurs as soon as the energy criterion is satisfied. Single crystal structure is an example where the inherent flaw structure might not be "favorably disposed" and spall fracture occurs at considerably higher stresses than those predicted by the energy balance criteria above.

For liquids, Grady [1988] considered two cases, one based on the fracture energy being dominated by the surface energy (similar to brittle spall) and another based on the fracture energy being dominated by viscous dissipation (similar to ductile spall). In the former case, where the surface tension, γ, dominates the fracture energy term, the expressions for spall strength, time to fracture, and fragment size are

$$P_s = \left(6\rho^2 c_0^3 \gamma \dot{\varepsilon}\right)^{1/3}, \tag{6.63}$$

$$t_s = \frac{1}{c_0}\left(\frac{6\gamma}{\rho \dot{\varepsilon}^2}\right)^{1/3}, \tag{6.64}$$

$$s = \left(\frac{48\gamma}{\rho \dot{\varepsilon}^2}\right)^{1/3}. \tag{6.65}$$

In the case where viscous dissipation dominates the fracture energy term, the expressions for spall strength, time to fracture, and fragment size are

$$P_s = \sqrt{2\rho c_0^2 \eta \dot{\varepsilon}}, \tag{6.66}$$

$$t_s = \sqrt{\frac{2\eta}{\rho c_0^2 \dot{\varepsilon}}}, \tag{6.67}$$

$$s = \sqrt{\frac{8\eta}{\rho \dot{\varepsilon}}}, \tag{6.68}$$

where η is the viscosity of the liquid.

The dynamic fracture criteria developed by Tuler and Butcher [1968] and Grady [1988] consider spall to be an instantaneous event. As such, the usefulness of these criteria is limited to applications where damage evolution is not an important consideration. It is fairly well established from experimental observations that spall is an evolutionary process that involves nucleation, growth, and coalescence of microscopic flaws. The development of damage during spall modifies the stresses, and the changing stresses, in turn, modify the growth of damage. When the details of this damage evolution process are important, a different kind of fracture model is needed, a fracture model that accounts for the nucleation, growth, and coalescence of damage at every step of the deformation history. Several approaches for developing this kind of model are available; these are discussed in the following section. Later, in Chapter 7, we focus our attention on one of these approaches: the nucleation-and-growth approach, or NAG theory.

6.2.2. Active or Constitutive Models for Damage

Active models for damage are those in which the constitutive relations are modified by the developing damage. Two types of modifications are commonly made:

1. The imposed strain increments $\Delta\varepsilon_{ij}$ are decomposed into a portion $\Delta\varepsilon_{ijs}$ taken by the intact solid material around the voids or cracks and the portion $\Delta\varepsilon_{ijc}$ taken by the developing damage:

$$\Delta\varepsilon_{ij} = \Delta\varepsilon_{ijs} + \Delta\varepsilon_{ijc}, \tag{6.69}$$

as suggested by Herrmann [1971], for example. This relation may also be expressed as a product of deformation tensors.

2. The stresses σ_{ijs} computed in the solid material are adjusted to account for the fact that they act over only a portion of the cross section. Carroll and Holt [1972a and 1972b] gave the average stress on the cross section as

$$\sigma_{ij} = \frac{\rho}{\rho_s}\sigma_{ijs}, \tag{6.70}$$

where ρ and ρ_s are the gross density and the solid density (density of the intact material).

The first of these models allows the stress to increase initially under tensile loading, but later to decrease as the damage continues to increase. The second provides a small adjustment when the damage becomes large, allowing a somewhat faster decrease in the stresses.

A great variety of active or constitutive models have been proposed for fracture processes. Here we discuss a few of these models to provide a brief sampling of the major types. First we mention models that are continuum in nature because they deal mainly with continuum variables like average stresses and strains and relative void volume. Then we discuss so-called "microstatistical models," which specifically account for the number and sizes of the microvoids or microcracks.

We begin our discussion of continuum models with the Gurson [1977] model, a very successful model proposed for porous materials under compression. The model was developed based on the analysis of the deformation of a sphere or cylinder of material containing a single spherical void. But the resulting stress–strain relations are used to represent a block of material containing voids, without regard to their size or number. The model consists of a yield surface defined as a function of mean stress (or pressure P) and effective stress $\overline{\sigma}$. When loading causes the stress state to reach the yield surface, the void volume is allowed to change, but the stress state remains on the yield curve. For use in fracture studies, the Gurson model has been combined with some fracture initiation process or just an initial void volume. Then, under tension, the void volume expands according to plastic flow rules such as those of Rice and Tracey [1969]. The model is especially appropriate for slow-rate fracture, but it also captures the main response features up to strain rates on the order of $10^4\ s^{-1}$ or loading durations on the order of 10 μs (these are rough guidelines developed based on fracture in aluminum—see Curran et al. [1987]). The Gurson–Rice–Tracey model has the advantage that only an initial void volume and a threshold stress for void growth under hydrostatic tension are required to fully characterize the model—most fracture models require more parameters.

As noted above, the Gurson model treats a spherical void in the center of a homogenous material element. This element is assumed to be representative of the whole material at the point. The model does not keep track of the number of voids or of their individual radii or volumes, but handles only the single quantity of the void volume, treating it as a continuum variable. Many other models in the literature follow this strategy: the fracture processes may be derived with a particular crack or void process in mind, but only a relative volume of voids or cracks or fragments is actually treated. Models of this type include those of Cochran and Banner [1977] for spall damage in uranium; Davison et al. [1977] for a general development of a fracture model; Rajendran et al. [1988, 1989] for fracture of ceramics; and Johnson et al. [1990] for fracture in many materials with special focus on projectile penetration simulations.

The microstatistical or microfracture models, which deal specifically with the numbers and sizes of microvoids or microcracks, are models of the type pioneered by some of the present authors and described generally in Curran et al. [1987]. They treat the behavior of material containing an array of microdamage sites of various sizes, or a size distribution. They describe the nucleation (or initiation) and growth (or enlargement) of the microdamage and are therefore often referred to as nucleation-and-growth (NAG) models. Such models use nucleation-and-growth rate equations of the general form:

Nucleation rate $$\dot{N} = f\left(\sigma, \varepsilon, \sigma_{no}, \varepsilon_{no}, V_v, N, \ldots\right), \tag{6.71}$$

Growth rate $$\dot{R} = g\left(\sigma, \varepsilon, \sigma_{go}, \varepsilon_{go}, V_v, N, \ldots\right), \tag{6.72}$$

where $f()$ and $g()$ are functions of the current stress, strain, and damage states, N and R are the number and radii of the voids or cracks, σ and ε refer to the stress and strain state, σ_{no}, ε_{no} and σ_{go}, ε_{go} are the thresholds for nucleation and growth, and V_v is the void volume. A more detailed description of the NAG type of model is given in Chapter 7.

Other models developed using this approach include the models of Davison et al. [1977], Johnson [1981], Johnson and Addessio [1988], Johnson et al. [1996], Nemes and Eftis [1992a, 1992b], Espinosa et al. [1995, 1996], Dienes [1978, 1981, 1985, 1996], Dienes and Margolin [1979, 1980], Margolin [1981], Grady and Kipp [1979], Kipp and Grady [1980], and Horii and Nemat-Nasser [1986]. The models of Johnson [1981] and of Johnson and Addessio [1988] avoid the nucleation process by starting with a predefined void volume. Johnson and Addessio use the Gurson curve (instead of a simple stress value) as the threshold condition for void growth, thus making the model ready for treating quasi-static as well as high-rate fracture. Nemes and Eftis [1992a, 1992b] present a derivation of their ductile damage model, which carefully accounts for mechanics requirements in detail. Espinosa [1995] and Espinosa and Brar [1995] have derived a brittle fracture model that explicitly accounts for the size and orientation of cracks. The cracks may occur on any of nine planes and each crack may undergo both tensile and shearing fracture. The normals to these nine planes are oriented along x, y, and z directions and at 45° and 135° between pairs of directions. Espinosa has developed the model to treat high-rate fracture in ceramics. Earlier, Dienes [1978, 1981, 1985] developed a similar brittle fracture model for the fracture of oil shale and for rocket propellants. He used an analytical form for the crack size distribution capable of treating both nucleation and growth processes. Margolin [1981], Grady and Kipp [1979], and Kipp and Grady [1980] have also developed brittle fracture models similar to that of Dienes for treating fracture in rock.

Kanel et al. [1983, 1984] have proposed a simplified nucleation-and-growth model without referring to it by that name. The initial void size is given by

$$V_{co} = K_2 \sigma^4, \tag{6.73}$$

and growth is described by

$$\frac{dV_c}{dt} = K_1 \left[|\sigma| - \sigma_0 \frac{V_a}{V_a + V_c} \right] (V_c + V_{c0}), \tag{6.74}$$

where K_1, K_2, V_a, and σ_0 are constants of the model, V_c is the crack or void volume, and V_{a0} is an initial crack or void volume. The model uses the decomposition of volumetric strain according to Eq. (6.69) but not the minor adjustment of stresses in Eq. (6.70). The model has the advantage that it has few parameters and Kanel et al. [1984] have been able to fit the model to particle velocity data without requiring examination of the cross sections of recovered samples.

6.3. Fracture Model Implementation

It is necessary to connect the material or fracture model to a finite-element (FE) code to be able to simulate experiments or applied problems. With the model in an FE code, we can simulate the impact, explosion, radiant energy deposition or other processes that produce the conditions for fracture. Fracture then proceeds in a natural way along with the other mechanical and thermal processes in the event being simulated. With the FE simulation, we can follow the evolution of the fracture event and the locations in the objects being simulated where fracture occurs.

Ease of implementation of the constitutive model in a variety of finite-element codes is a critical factor that influences the extent to which a constitutive model is used to simulate fracture events. Recently, several colleagues in the field, notably, Brannon and Wong [1996], have been working to standardize the connection process so that material models can be easily moved from code to code. This is certainly a valuable trend and one that will grow in the future. We expect that codes with a large user community will all develop standard interfaces to facilitate the addition of new material models, and the material models will be equipped to match such standard interfaces.

We begin the discussion of model implementation by reviewing the major types of finite-element codes available for dynamic simulations, emphasizing those features that bear on how the connection is made to the material model. Next we outline general requirements for the model: these are in addition to the requirements listed in Section 6.1.3. Then we outline the major features required for a connection between the code and the material model. Finally, we present our thoughts on dealing with the problems that arise because of the advection processes in Eulerian and ALE (Arbitrary Lagrangian–Eulerian) codes.

6.3.1. Types of Simulation Codes

Several types of codes are used for the simulation of dynamic events in which fracture may occur. The ones considered here are called finite-element or finite-difference codes. In each case the material is discretized into small elements for analyzing the stress and motion of the objects. The common types are Lagrangian, Eulerian, Arbitrary Lagrangian–Eulerian (ALE), and Smooth Particle Hydrodynamics (SPH) codes. A recent innovation is a meshless or grid-free finite-element code, which has characteristics of both the SPH and ALE codes. Here we wish to describe these different types of codes only to the extent needed to facilitate the discussion of how to best connect material models to them.

In the Lagrangian method (Wilkins [1964]), the elements and nodes bounding the elements are joined in a fixed grid arrangement, and the grid distorts with the material so that a mass of material contained in an element always remains in the element. The Eulerian method (e.g., McGlaun [1990]) also uses a fixed grid of nodes and elements, but the grid is fixed in space, and the material flows through the grid. The ALE codes (Dube et al. [2000] and Couch et al. [1993]) may be used as pure Lagrangian, pure Eulerian, or with some intermediate procedure for allowing the material to flow through the grid. Usually, in ALE codes, objects in the problem which do not distort much are treated in a Lagrangian manner, but objects, or regions expected to undergo large distortion, the element boundaries are shifted to minimize distortion of the grid. The SPH codes (e.g., Stellingwerf and Wingate [1993]) have nodes that are fixed in the material and move with the material; the elements are masses connected to each node. The SPH method is especially appropriate for fragmentation problems and for problems involving large distortion. For large distortion problems one can also use a *meshless* method. The so-called "meshless" methods are characterized by nodes fixed in the material and elements that contain the material between each set of nodes. One of the defining features of meshless methods is that the element boundaries are redefined at each step, allowing one to account for very large relative motion of the nodes. The method has been used for crack propagation in which the location of the advancing crack was followed explicitly throughout the simulation Belytschko [1994].

To aid in the discussion of the characteristics of the different types of codes, let us consider two operations in the codes: the process of conservation of momentum and the provision for calculating motion of the material. The conservation-of-momentum computation occurs by gathering the forces acting on a node (with a mass attached) and computing the change in velocity caused by these forces (i.e., using F = Ma). In the Lagrangian, Eulerian, and ALE methods the gathering of these forces is straightforward because the nodes, and elements have a fixed-neighbor relationship throughout the calculation. An exception to this fixity occurs in ALE codes when elements are lost or added to provide for local distortions. By contrast, in the SPH codes the forces are accumulated by searching through the node array to find which nodes are near enough to contribute forces to the node under consideration.

The second operation is the provision for motion of the material with respect to the grid of nodes and elements. In the Lagrangian method, no flow through the element is permitted: the material remains fixed in each element. With the Eulerian technique, the material flows freely through the grid; hence, computations must be made at each computational time step to determine what material is currently in the element. In fact, because of the material velocities associated with each node, some material is regularly being transferred from element to element by the "advection" procedure. The ALE codes also provide for advection of the material, but they also permit the grid to move, so there is a motion of material through a moving grid. In the SPH method, the nodes remain attached to their points in the material, so there is no advection process.

With this background in mind, we now consider the process of connecting a material model to one of these finite-element codes. The next section treats major considerations for connecting to any of the codes. The following section then deals with the special problems associated with advection.

6.3.2. Recommended Model Features for Connecting to Finite-Element Codes

In addition to the physical and thermodynamic requirements specified in Sections 6.1.2 and 6.1.3, there are many features needed to facilitate the interaction of a model with the host finite-element code. We list here features that we feel are important: they are not absolute requirements, but features that we suggest to simplify the connection. Our list agrees (and is a subset of) the Model Interface Guidelines (MIG) developed by Brannon and Wong [1996]. MIG refers to a convention and also to a code for facilitating the connection between a model and a code. This new convention is largely concerned with modularity, well-defined work of code and model, and the transfer of variables between the two. MIG describes requirements for the material model, the host code, and a routine (or routines) to provide the interface between these two.

We recommend that the model be constructed in a modular way and we describe our definition of "modular." The model should be so constructed that it can accept any units for time, distance, stress, etc. that are appropriate for the main code. We describe the types of data that may be needed in the model and that are available from the host finite-element code, and the procedure for transferring data between the model and the host code. Finally, we mention here the decisions on whether the model treats "local" or "nonlocal" phenomena.

6.3.2.1. Modularity

Modularity is a positive requirement for a material model. We often hear of material models that are so deeply embedded into a finite-element code that it is nearly impossible to separate the model from the code. This approach to model implementation may lead to a faster running code, but it has disadvantages. For example, how can one verify that the model does what it is expected to do? Or how can we borrow it to put into another code? Here we wish to give a definition of and requirements for modularity to circumvent disadvantages. We see the material model as permanent and complete; it has a life of its own outside the finite-element code. Hence, it can be fully tested and verified separate from the code and it can be readily transferred to other codes. Toward producing such a modular material model and connection, we propose the following features:

1. The model is a full constitutive relation (it completely generates the stresses, i.e., it has a well-defined task).
2. The model makes a clean and well-defined connection to the host code. This means that it is clear as to which variables are being transferred and whether they are updated in the model, or in the host code, or both.
3. The material model can be tested with a driver code to represent a single element under a variety of loadings. The driver should present a standard connection to the model, just like the standard connection in

the finite-element code. The driver may be considered to represent the behavior of one element of the material.

4. The model is prepared to interface with 1-, 2-, or 3-D codes. Thus, the geometric constraints of the problem or code are communicated to the model only through the strain increments, but the model is prepared for general 3-D conditions.
5. The model is internally modular and therefore can be used with other material models that are modular.

An internally modular fracture model is actually an assemblage of at least four elements: (1) an equation of state to provide the pressure for a given density and internal energy, (2) a deviator stress model, (3) a set of relations providing the rate of change of the damage, and (4) the solution procedure for computing the stress and damage for a given strain increment and time step. A modular fracture model has these four components well separated so that any of the four can be exchanged without a major reconstruction of the whole model and redesign of the interaction of the components. Thus, the material model is internally modular as well as in its connection to the host code.

6.3.2.2. *Units*

The units used in the host code and the model should be at least clearly specified. The best alternative for both the code and the model is to use a consistent set of units. "Consistent" units are units that give

$$\text{Force} = \text{Mass} \times \text{Acceleration} \tag{6.75}$$

without a calibration factor. Samples of such sets of units are provided in Table 6.1.

Table 6.1. Basic units and some derived units in four systems of units commonly used in shock wave studies.

		Unit systems			
		SI	**cgs**	**English**	**Shock wave**
Basic units	**Length**	m	cm	in	cm
	Mass	kg	g	snail*	g
	Time	s	s	s	μs
Derived units	**Velocity**	m/s	cm/s	in/s	cm/μs
	Density	g/m^3	g/cm^3	snail/ in^3	g/cm^3
	force	N	dyne	lb.	$g.cm/\mu s^2$
	Pressure	Pa	$dyne/m^2$	psi	Mbar
	specific energy	m^2/s^2	cm^2/s^2	in^2/s^2	$cm^2/\mu s^2$

* 1 snail = 12 slugs.

To verify the self-consistency of a given set of units, we define a series of conversion factors for going from a known self-consistent system, such as cgs, to the new system, and then check to ensure that the conversion factors satisfy Eq. 6.75. To do this, let us define the following conversion factors for force, mass, distance, time, stress, and density (here we are converting to a set of English units):

$$F_{cgs} = f_{unit} \times F_{engl}, \qquad M_{cgs} = m_{unit} \times M_{engl}, \tag{6.76a}$$

$$X_{cgs} = x_{unit} \times X_{engl}, \qquad T_{cgs} = t_{unit} \times T_{engl}, \tag{6.76b}$$

$$S_{cgs} = s_{unit} \times S_{engl}, \qquad D_{cgs} = d_{unit} \times D_{engl}. \tag{6.76c}$$

In this equation, F is force, M is mass, X is distance, S is stress, T is time, and D is density. Quantities in the cgs system are designated with the subscript cgs Likewise, quantities in the English system are designated with the subscript engl. Quantities with the subscript unit are conversion factors for converting the various quantities that appear in the equation from the cgs system to the English system. With these conversion factors defined, we can now substitute into Eq. 6.75 above and obtain

$$f_{unit} = m_{unit} \times \frac{x_{unit}}{t_{unit}^2}. \tag{6.77}$$

Alternatively, we can substitute the conversion factors into the differential form of the conservation of momentum equation,

$$\frac{\partial \sigma}{\partial x} = \rho \frac{\partial u}{\partial t} \tag{6.78}$$

to obtain

$$\frac{s_{unit}}{x_{unit}} = d_{unit} \times \frac{x_{unit}}{t_{unit}^2}. \tag{6.79}$$

Units in the cgs, SI, English, and shock systems are shown in Table 6.2, with conversion factors to change from cgs units to the given system.

Table 6.2. Conversion factors for relating cgs units to the other systems of units.

Quantity	cgs	SI	English	Shock
Force	dyne	10^{-5} N	2.24809×10^{-6} lb	No unit
Mass	g	10^{-3} kg	5.710×10^{-6} snail	1 g
Distance	cm	10^{-2} m	0.3937 in	1 cm
Time	s	1 s	1 s	10^6 μs
Stress	dyne/cm^2	10^{-1} Pa	$1.4504 \times 10m^{-5}$ psi	10^{-12} Mbar
Density	g/cm^3	10^3 kg/m^3	10686.9 snail/in^3	1 g/cm^3
Temperature	°C or °K	°C or °K	(32 + 9/5 °C) °F	°C or °K

The standard sign convention for stresses and strains is + for tension or expansion. But pressure is usually + for compression. Temperature may also be a factor: the standard units are degrees Kelvin, Celsius, Rankine, Fahrenheit, or electron volts (11,604.5 °C / electron volt).

6.3.2.3. Data Types

The primary data for a material model are mid-element quantities such as stress, internal energy, strain rate, and history variables describing the state of damage. Other information that may be required by the model are element location (coordinates of the element boundaries), energy flux at the element nodes, and quantities providing the state of damage in neighboring elements. Most material models require only mid-element quantities and such a requirement is the easiest for the code to meet.

Among the mid-element quantities to be passed from the code to the model there are several classes of variables that may be stored and generated differently in the code:

1. *Material properties:* The material properties may be in a single array, or they may be individually named variables. They are generally permanent, not varying during the simulation.
2. *Current variables* (strain increments, etc.): These quantities are probably generated just before the call to the material model. The strain may be provided as 6 components (rates or increments) or by the velocity gradient (9 components).
3. *Element arrays* (stresses, density, energy, yield strength, etc.): These are permanent arrays during the computation, but during a simulation they are constantly updated either by the code or the model. Density and internal energy may be provided by the code with quantities both before and after the current time step, or just before and allow the model to update using the velocity gradient and stress tensor.
4. *History variables:* These are special variables intrinsic to the model, like plastic strain, damage, and orientation of cracks. These variables are element-specific quantities, and they are updated by the model during every computational cycle. They are stored in an array the size of which depends on the model.

6.3.2.4. Data Transfer

During computations, data of all the types mentioned above must be transferred from the finite-element code into the material model and the model results must be transferred back to the code. In making connections between a code and a material model, the major problem is treating the data transfer: verifying that all data are transferred in both directions with the correct form and that the variables are updated in the code or in the model and not in both.

Both the finite-element code and the material model are large, independent components that can be separately verified. Hence, the material model may be tested in a one-element code, and also in one-dimensional or multidimensional codes. Also, the code may be Eulerian, Lagrangian, ALE, or of another type. Furthermore, the material model can be used simultaneously in several ways by the code. For example, an elastic-plastic model may be called to treat an aluminum object, one of the components of a composite material, or the solid aspect of a porous material. It may also be called to treat the intact material of a fracturing object, or the unreacted portion of an explosive. Only in the case of the aluminum object is this model provided the total stress quantities for the element: in all other cases this elastic-plastic model is called by other models that treat the composite, the porous behavior, the fracture processes, or the explosion. In planning the code-model connection, it is therefore desirable to account for as many of these considerations as possible.

Several means are available for the data transfer between the finite-element code and the material model:

1. Allow the data to be available throughout the routines of the finite-element code and the routines of the material model. Such a transfer provision is given by COMMON statements, or common blocks, in Fortran and by header files in C. This method may provide the fastest data transfer, but it also makes independent verification almost impossible. A one-element testing code would have to contain the full set of common data to be able to test the material model. The model has access to the total energy, density, and stresses for the element and therefore cannot treat these variables for a component of the material as required by a composite, fracturing, porous, or explosive representation.
2. Transfer information to the material model only through the call statement to the material model. This is the preferred method in most cases because it does not suffer from the disadvantages associated with the previously mentioned method of data transfer.
3. Transfer some data through the call statement and some through COMMONs or header files. For example, the material property data might be included in a common block, but the data specific to an element might be transferred to the material model through the call statement. This is an easy method and provides for some independence of the model, but does not allow the model to represent a portion of the material in the element.

The call statement is the easiest for the programmer to implement correctly in connecting the model to the code and to verify that the information transfer is occurring correctly because there are no alternate means for modifying or transferring the data. An added advantage is that there are no changes needed in the model to permit connection to a new code. This permanent nature of the model is important so that the model can be verified in one code and used in any other code with the expectation of identical results.

6.3.2.5. Locality

Since both local and nonlocal behaviors may be treated by the model, a decision must be made as to the nature of the fracture process (local or nonlocal) at the outset of model construction. The local approach takes the viewpoint of the material particles and requires that the constitutive relations involve only quantities local to those particles. For example, the particle knows mainly its current stress state, strain increments and strain rates; but sees no overall wave shapes, has a limited memory of its own past, and has no knowledge of the future. The particle cannot tell if the loading is sinusoidal nor can it discern the period or frequency of loading. The knowledge of the past is limited to the prior stress state, plastic strain, damage, and similar current quantities that may be stored, but there is no history of stress, for example. In fracture problems the particle does not have an overall knowledge of the damage, such as the crack length and orientation, but knows only the motion of its own boundaries and the stress within. Hence, the particle cannot know the stress intensity factor, for example, which would require knowledge of the crack length, orientation, and an appropriate far-field stress normal to the crack plane. As an example, Needleman and Tvergaard [1991] have been able to simulate a propagating crack using only a ductile fracture model based on local stress and strain states, but they have been able to produce crack growth which would normally require nonlocal knowledge such as the stress intensity factor and fracture toughness.

The nonlocal approach allows the computation to proceed with a broad knowledge of the past and present, but not of the future. In this procedure, the program can compute global quantities such as the stress intensity factor and compare this factor with the fracture toughness; then decide whether an element does or does not fracture. This second approach allows computational simulations to represent some analytical approaches taken in fracture.

Only local behavior is readily handled by a general-purpose code, because then, the only variables needed by the model are those mid-element quantities stored by the code specifically for each element. Nonlocal variables are processed outside the normal computational cycle. Then they must be provided to the elements using some special method for determining which elements are related to which nonlocal variables.

6.3.3. Advection Requirements for ALE and Eulerian Codes

Advection describes the process of moving some material and its associated properties and state variables from one element to another and determining new properties for the combined material in the second element. Such advection processes occur in Eulerian codes and in ALE codes running in a non-Lagrangian mode.

To provide for advection, it is necessary that all the historical variables be advectable. That is, all the quantities must retain their meaning when they are mass- or volume-weighted and summed with corresponding quantities from adjacent elements. Integer indicators (for example, with a value of 0 for solid and 1 for porous) cannot be used because a fractional state of the combined material would not have meaning. A second-order accurate advection process may produce results that are outside the normal range of definition of the quantities; such as, a negative number of cracks, or a level of damage greater than one (full damage). The constitutive relation must be equipped to handle history variables that arrive from the main code with unexpected and unusable values, adjust the variables to the acceptable range, and maintain reasonable accuracy throughout the process.

Advection presents special problems for microfracture models because there are many history variables, all of which must make sense when they are averaged with the corresponding variables from adjacent elements. One can readily appreciate the difficulty by imagining combining a small crack from one element with another crack with a different orientation and size from another element, and obtaining a result that has physical sense.

Historical variables that are isotropic (damage or temperature) or specific (number per unit volume or mass) are readily advected. Quantities such as crack size and orientation require special treatment. Here, to get into the details of the advection problem, we discuss specific problems and solutions for the SRI NAG microfracture models. These models do not use specific cracks, but numbers of cracks per unit volume; therefore, many of the special fracture variables advect readily. But crack sizes, orientations, and some other quantities need special representation. Here we outline some requirements for advection and how we have approached these requirements.

1. There can be no redundancy in the advected variables describing the material state. For example, damage can be treated as a separate advectable variable or derived from the numbers and sizes of voids or cracks. To avoid redundancy, we treat the numbers and sizes as primary variables and recompute the damage at each step.
2. All advected quantities must be examined to verify that they remain within their defined limits. Crack volume and numbers of cracks, for example, must remain nonnegative.
3. Crack orientation angles require special treatment so that, for example, cracks normal to $\theta = 0^\circ$ and those from another element normal to $\theta =$ 180 (hence, having the same actual orientation in space) are not combined and averaged to form cracks at $\theta = 90^\circ$. In addition, θ (the angle in the x–y plane) is required to remain within the range $0^\circ \leq \theta \leq 180^\circ$, and ϕ (the angle measured from the positive z-direction) is kept in the range $0^\circ \leq \phi \leq 90^\circ$. We chose to represent crack orientation by storing $\cos\theta$, $\sin\theta$, and ϕ and to require that the normal be directed such that its dot product with a line in the x = y = z = 1 direction is positive. This approach has given reasonable results, although it is not an exact or universal solution.

4. Some damage variables are weighted by the amount of damage to make them advectable. For example, the crack orientation angles are weighted by the amount of damage: we actually store $\tau \cos\theta$, $\tau \sin\theta$, and $\tau\phi$, where τ represents the damage (varying from 0 to 1). With this method, when the orientation variables are summed for partial elements during advection, the more heavily damaged material controls the new orientation.
5. The crack size distribution consists of a table of numbers and crack radii representing a range of sizes from submicron cracks to the largest present. To make this table advectable, the crack radii are remapped back to a reference set at the end of each growth operation. Then, only the numbers of cracks must be advected. The maximum crack radius is given a special treatment that allows it to be advected: $R_{\max}$ is stored as $\tau R^n_{\max}$, where $n = 3$. This special storage is not exact but it allows the more heavily damaged element and the one with the largest $R_{\max}$ value to dominate in the averaging process associated with advection.
6. When nucleation of new cracks occurs, their orientation is computed as normal to the current maximum tension stress. This new group is then added to the existing cracks by combining the orientations of the new and existing cracks. This orientation treatment allows for a changing stress orientation and also provides that advected cracks do not absolutely govern the orientation of cracks in a hitherto undamaged element.
7. The differences between advected quantities cannot be relied upon to produce accurate results. For example, the crack size distribution can be stored either as cumulative numbers of cracks, N_i, greater than a given radius, R_i, or as numbers of cracks, ΔN_{ij}, within a certain size range (from R_i to R_{ij}). The ΔN values are actually used in the computations associated with nucleation and growth of damage, whereas the N values are useful for display. We chose to use the numbers within a range of radii and then demand that such numbers be always positive when they return from the main code, rather than storing the cumulative numbers. Our attempts to use cumulative numbers showed that the ΔN_{ij} derived from the cumulative numbers were more likely to become negative during advection. A second example occurs with a multiple-variable viscoelastic model in which the deviator stresses S_{ij} are obtained by summing components N_{ijm} from each of M viscoelastic elements:

$$s_{ij} = \sum_{m=1}^{M} s_{ijm} \,. \tag{6.80}$$

8. Components from each viscoelastic element should be stored: we cannot rely on advection to allow us to obtain the Mth component from

$$s_{ijM} = s_{ij} - \sum_{m=1}^{M-1} s_{ijm} . \tag{6.81}$$

9. It may be necessary to store deviator stresses in the historical array, rather than depending on the code to provide them. Problems arise, for example, in treating mixed cells (i.e., elements with more than one material). Then the code may provide to one material model deviator stresses that reflect properties of the other material. For a simple plastic material, an inappropriate set of deviator stresses is quickly corrected by the material model; but for a viscoplastic or viscoelastic material, the deviator stresses persist for some time, so a large overstress provided by the code can cause seriously inaccurate stress states for some time.
10. Integer indicator variables can be replaced with more physical quantities. For example, instead of indicating a state of porosity with an integer indicator, one can store a measure of the porosity. To distinguish solid, liquid, and vapor states, one can use the energy or temperature, density, and pressure, rather than an indicator. This change in the material model to avoid the use of the indicator requires extra computations to assess the material state, but makes the current state free of reliance on a nonphysical indicator.

These items are listed with our provisional solutions to indicate the kinds of problems that are encountered and that must be solved in some manner to make the connection of the material model to a code using advection. We have adopted the attitude that advection forces us to choose for storage only variables with a clear physical meaning. Providing for the use of such variables has required us to change how we undertake some processes, but in each case we have found that the resulting code is more strongly related to the physical processes we wish to represent.

6.3.4. Calibration of the Simulation Tool

Numerical simulations of dynamic fracture events often involve complex interactions. Due to this complexity, calibration of the simulation tool is often necessary to gain confidence in the simulations. This process involves calibration of the various components of the simulation tool, and it usually includes the use of laboratory data, preferably obtained under conditions similar to those prevalent during the events of interest. For convenience, we consider the simulation tool to consist of the following four parts:

1. the computer program (hydrocode) discussed in the preceding sections,
2. the layout of the finite elements and boundary conditions to describe the problem of interest,

3. the constitutive or material model for the material or materials involved, and
4. the material parameters for the material model or models.

The calibration of each of these features of the simulation tool is discussed separately in the subsequent paragraphs.

The *computer program* must operate in a manner consistent with the basic conservation relations for momentum, energy, and mass at all times, even during special operations such as rezoning and advection. These requirements should be met also throughout all the material, even along boundaries, slide lines between materials, and within multimaterial elements. Such verification can occur independent of any experimental data because these requirements are independent of the material and also because the simulation tool can be interrogated to provide values for the conserved quantities at any time. Generally, the code developer will have conducted extensive verifications and be able to describe how well the conservation relations are satisfied under most circumstances.

The *layout of the elements* for the problem is of interest even for a material model that is known to be "element size independent." Even an elastic material requires enough elements to represent the stress (or other) gradients appearing in the problem. The sizes, shapes, and types of elements can be verified in some cases by comparing with analytical solutions for simple material models. For more complex models, several simulations at a range of element sizes are usually required to determine an appropriate element size for an accurate treatment. Alternatively, when the material model represents a composite material, for example, it may be necessary that the elements have a size that represents the actual coarseness of the material.

Verification of the constitutive relations requires matching the simulation results to experimental data from the simulated experiments. Here we distinguish two aspects of the constitutive relations:

1. the *material model*, hence, the type of behavior to be represented, and
2. the *material parameters* to specialize the model to represent a particular material of interest.

The material model contains features such as viscoplasticity or ductile fracture that best represent the behavior of the material. The user selects a model that contains the features important in the application under study. This selection is a very important step in the calibration because real materials often behave in a more complex manner than a model can accurately describe. Therefore, we are usually in the position of selecting (or designing) a model for the most critical features of the behavior while sometimes neglecting other less important features. Our selection may be somewhat verified by attempting to match simulation results with experimental data. Some aspects of the data may not be well represented in which case we either modify the model, or select a different model to improve the agreement.

The material parameters (bulk modulus, density, yield strength, etc.) are chosen to allow the material model to match the behavior of a particular material. The parameters affect the amplitudes of the response, whereas the material model determines the type of response. Conceptually the roles of parameters and model type are distinct, but in practice it may be very difficult to determine which to adjust to improve the match for a given set of experimental data.

Often, numerical simulations are relied upon to predict the material behavior in a complex event for which there's a lack of experimental data. In this case, the simulation tool is calibrated using data from simpler (and usually smaller) experiments. These simpler experiments must be carefully selected to test all aspects of the material model (and the simulation tool) which will be exercised in simulating the event of interest. Below is a list of desirable features that should be represented in the calibration experiments:

1. Stress and strain levels should be similar in the event of interest and the calibrating experiments.
2. Temperatures and temperature ranges should be similar.
3. Strain rates should be similar. The event of interest often has a range of strain rates, whereas the calibrating experiments are usually conducted at a single strain rate or a narrow range of rates. Therefore, it is necessary to conduct experiments at various strain rates to span the range of the rates in the event.
4. Stress and strain states. Most laboratory experiments are conducted under relatively simple loading conditions, usually uniaxial stress or strain. The loading paths experienced by the material under these conditions may differ greatly from the three-dimensional stress and strain states of the events we may wish to simulate. To the extent that the model cannot be calibrated under the same loading conditions as the events of interest, it must be recognized that model predictions may become less accurate as the difference increases between the loading conditions of the calibration experiments and those of the events of interest. Confidence in the model can be improved by calibrating against as many simple experiments as practically possible; and whenever possible, by including calibration experiments performed under loading conditions that differ from the usual uniaxial stress and uniaxial strain loading paths. Experiments of this type were performed by Erlich and Gran [1988], who reported a study of dynamic ductile fracture with applied lateral loading so that the stress and strain states differed from the usual ones experienced during plate impacts.
5. Scale should be similar. Often, we wish to use small-scale experiments to calibrate a model for representing a large event. But this may not be altogether satisfactory because we may be overlooking features like grain size and flaw size that do not scale geometrically (Holzapple and Schmidt [1982]).

When it is not possible to meet all the requirements for validating the simulation tool, it becomes necessary to provide some auxiliary tests and verifications that improve our confidence in the calibration. For example, when the scales are different, we may provide experiments and corresponding simulations over a range of scales to demonstrate the scalability in the event under study. For different stress states, we may also provide some experiments with two- or three-dimensional stress states that span the types of stress states expected in the event of interest. Finally, if the actual event exhibits strain rates over a large range, we probably need to conduct experiments at specific rates over this entire range of strain rates.

From the forgoing discussion, it is clear that in general we must make a considerable effort to demonstrate that our simulation tool is calibrated, and generally, we will conclude that our simulation tool is only approximately calibrated. Probably our conclusions regarding the calibration will be that the hydrocode conserves mass and momentum except under certain (known) circumstances, and that energy is only roughly conserved during rezoning (remeshing) and advection. The layout provides results that are accurate to a specified percentage under known conditions that are somewhat like those in our event of interest. The material model captures what we feel is the dominant behavior in most of our laboratory experiments, but neglects some features that are judged not to be very important. And the material parameters are determined from experiments that only approximately meet the requirements of the preceding paragraph.

7 Nucleation-and-Growth-Based Constitutive Models for Dynamic Fracture

One method for constructing micromechanical models for representing the fracture processes in materials is the Nucleation-and-Growth (NAG) approach. Early examples of this type of development are given by Davison et al. [1977], Curran et al. [1987], Margolin [1981], Dienes [1978, 1979, 1980, 1981, 1985, 1996], and Johnson [1981]. More recent work in this area has been done by Rajendran et al. [1988, 1989], Nemes and Eftis [1992a, 1992b], Johnson and Addessio [1988], Tong and Ravichandran [1995], Johnson et al. [1999], and by Wang [1995, 1997]. Here we follow the approach used in the SRI work, but the approach is similar in the other references. The SRI Nucleation-and-Growth (NAG) approach to fracture arose naturally from observations of the damage in cross sections of metal targets impacted in gas gun experiments (Barbee et al. [1972]). The experimental data showed increasing numbers of voids or cracks as a function of the stress level and of the load duration. From these observations we were led to postulate the existence of nucleation processes for the initiation of the voids or cracks and growth processes for their gradual development under continued loading. We postulated that the damage occurred in response to the tensile stresses experienced by the material and that the damage was indifferent to the method of load application (explosive, impact, or thermal radiation) that had caused the tensile stresses. Therefore, it became our intention to develop an understanding of the fracture processes at a level (i.e., scale) where the damage evolution in the material can be treated directly. We hoped that such an understanding would lead to the development of a generalized approach that can be used to treat fracture under many different types of loading conditions.

The intended approach is one in which the model can predict many aspects of the fracture phenomena: modifications of the stress or particle velocity histories by the developing damage and location, number, and size of cracks or voids throughout the specimen. In this way the model (and the whole approach), with its foundation in mechanics and micromechanics, is strongly constrained by the available data. The processes and parameters in the model should be uniquely defined by the data. We expect the processes and parameters thus identified based on fracture experiments to be basic to the material and not merely fitting functions that describe certain kinds of fracture under a narrow range of loading conditions and deformation rates. Rather we expect that the processes we study in fracture are the same processes that also control the shock front thickness, creep, fatigue fracture, and other material phenomena.

For our micromechanical approach, we have chosen to work at a particular level of detail, and this level has been a characteristic of the NAG approach:

1. The damage is examined microscopically (nominally at about 100X) and each microvoid or crack is measured. Thus damage in the model has characteristic dimensions of 1 μm up to about 100 μm or somewhat larger.
2. The microdamage is considered statistically. That is, we count microvoids or cracks, but we mainly work with crack or void size distributions, not individual cracks or voids.

Because we always start with observations of the damage, it is clear that NAG has a strong experimental basis. The theoretical models are always developed based on observed types of damage and measured processes. Yet the models are also firmly based on established mechanics principles and known constitutive properties and processes for intact material. Hence, the models represent a unique combination of observed processes for the development of damage and standard mechanics principles for the behavior of intact material.

This chapter explores the development of both the ductile and the brittle fracture NAG models. In each case we describe first the experiments on which the model is based, next the rate processes derived from the data, and then the development of the model.

7.1. Nucleation and Growth of Voids and the Model for Ductile Fracture

"Ductile" fracture here means damage processes that occur by the development of voids in the material; hence, large amounts of plastic flow accompany the fracture (the plastic flow fits with the usual meaning of "ductile," which refers to large plastic deformation). As described in Chapter 3, plate impacts are used for the fracture tests. In these tests, full damage usually occurs with tensile stress pulse durations on the order of a microsecond. In addition to the usual measurements in these tests, detailed post-test metallographic studies are made of cross sections through the specimens. The observed voids are counted and organized into size distributions. These observations of damage are the basis for a proposed fracture model that accounts for the nucleation (initiation) and growth (enlargement) of voids in ductile fracture and comprise the topic of this section.

Here we consider in more detail the damage seen in cross sections of impacted plate specimens such as those shown earlier in Figure 3.17. The voids tend to be circular in plane sections and are therefore in fact spherical. They appear in a range of sizes and are distributed through the plate thickness — they are not of a single size nor do they occur in a single plane. These plate specimens were impacted by a flyer plate in a gas gun and were therefore subjected to plane, one-dimensional loading. Therefore, the tensile stress history and hence the damage per unit volume should vary only with distance through the specimen in the direction of shock wave propagation. In this section we shall see that tensile stresses occurred in the locations where voids were observed, and that longer durations of tensile stress loading coincided with the locations of the highest void densities and of the largest size voids. Hence, we have concluded that number and sizes of voids are closely related to the transient tensile stresses in the specimens. Our approach is to count the voids and characterize their size distribution so that they can be quantitatively related to the stress histories. Evolution equations for the nucleation and growth rates of the voids are then derived from the data.

A model termed DFRACT (Ductile FRACTure) was developed at SRI (Barbee et al. [1970, 1972], Seaman et al. [1976], and Curran et al. [1987]) to describe the fracture processes that occur by the nucleation and growth of voids in very ductile materials. The model represents fracture under conditions of projectile impact, air shock loading, deposition of intense thermal radiation (from X-ray, electron beam, or laser sources), or explosion. This model is appropriate for loadings with tensile durations of a few microseconds or less. The model has been implemented in a computer subroutine for use with wave propagation or structural analysis programs.

The model has a micromechanical basis: it contains algorithms for describing nucleation and growth of statistical distributions of voids, and stress–strain relations that account for the developing damage. Thus, the DFRACT model is a Nucleation-and-Growth, or NAG, model. The model is based both on metallographical observations of damaged specimens and on the customary continuum mechanics treatment of constitutive laws. The model derivation given here includes the nucleation and growth processes, the damage-affected stress–strain relations, and the solution procedure to determine the current stress and damage state under imposed strains.

7.1.1. Experimental Aspects of Ductile Fracture

Dynamic ductile fracture experiments have been conducted by a large number of experimenters and with a variety of instrumentation. Loading to the specimen is generally accomplished either by plate impact (e.g., Curran et al. [1987], Cochran and Banner [1977], Christman et al. [1971, 1972], and Zhao et al. [1992]) or by using an explosive in contact with the specimen plate (e.g., Novikov et al. [1966], Kanel et al. [1984a, 1984b], Kanel and Glusman [1983], Tarver and Maiden [1988], and Christy et al. [1986]). Recent studies have also included spall induced by laser loading (e.g., Boustie et al. [1992a, 1992b], Tollier et al. [1994, 1996], Arad et al. [1998], and Auroux [1999]). Often the sample is instrumented to obtain the stress or velocity history at the rear of the specimen plate. The recorded data provides an indication of the damage induced in the plate and may serve as a test for attempted simulations of the event with a fracture model (Romanchenko and Stepanov [1980], Kanel et al. [1994a, 1994b], and Kanel and Utkin [1995] provide methods for deriving fracture strength from the experimental records). In other studies the sample is merely captured and sectioned to determine the extent of damage induced. This method based on sectioned samples is described here.

One-dimensional plate impact experiments have been used extensively in our model development efforts for determining the fracture processes and their associated parameters. To determine the fracture processes for a material it is necessary to conduct a series of experiments at a range of impact velocities (i.e., peak stress) and with several specimen thicknesses (i.e., load duration). The minimum impact velocity should be chosen to cause no observable damage in the specimen, while the maximum velocity should produce the maximum damage of interest. The range of specimen thickness, which governs the duration of loading, should be chosen to produce loading pulses with the durations of interest.

In analyzing the experimental data, we assume that the sample is isotropic and homogeneous. We further assume that spherical voids are nucleated at random with a concentration and total growth that are dependent only upon the tensile stress history. The voids may actually be nucleated at inclusions or at grain boundaries, but if the inhomogeneities are at a sufficiently small scale, the assumptions are still appropriate. Under these assumptions the compressive stresses which precede damage do not contribute directly to the growth of damage, although they may precondition the material through work hardening or heating.

Our quantitative analysis of the observed void damage in the specimen consists of the following steps:

1. Sectioning, polishing, and photographing the fractured specimen so that the damaged region can be studied.
2. Determining from the photographs of the cross section the areal density $n(r, x)$ of voids that intersect the surface with a radius r at position x.
3. Using a statistical transformation of $n(r, x)$ to determine the volume density $\rho(R, x)$ of voids of radius R at position x. Then the volume density is summed to provide the cumulative density function

$$N(R,x) = \int_R^\infty \rho(R',x)\,dR'. \tag{7.1}$$

4. Determining the nucleation and growth rates by correlating the function $N(R, x)$ with the computed stress history $\sigma(x, t)$.

These steps are discussed in more detail in the remainder of this section.

7.1.2. Determination of the Void Size Distribution

The photomicrographs of the damaged specimens show a range of void sizes, with larger ones concentrated near the expected spall plane and many small voids throughout the region near this plane. In this section we discuss quantifying the data and organizing it into patterns that can be used to derive equations for the nucleation and growth processes.

For the metallographic examination, the specimens are sectioned perpendicular to the spall plane (several sections can be made to increase the number of observations and therefore improve the statistics of the damage counts). After sectioning, the surfaces are polished—this polishing must be performed carefully because with soft materials, the removed material may be moved into the voids and thus change the observable dimensions of the voids. In some cases it may be necessary to etch the surfaces to make identification of the voids easier. Following these preparations, the fracture surfaces are photographed through a microscope for the quantitative study.

With the photomicrographs available, the voids are counted. A cross section of an aluminum target was shown in Figure 3.17 of Chapter 3 in a form ready for counting. Through-thickness intervals were marked on the photomicrograph and the voids within each interval were measured and counted. More recently we have simply measured the radius and position of each void. In either case we attempt to choose distance intervals such that within each interval the void damage level is fairly uniform. After counting we often group the voids within radius intervals, as shown in the sample count in Table 7.1 for a laser radiation in an 1100 grade aluminum. The radius and location of each void on the examined cross section was recorded and then the void counts were collected into intervals as shown in the table. The next step in processing these data is to divide by the area counted (50 μm wide and 2000 μm long) to determine the number per unit area, $n(r, x)$. The distribution $n(r, x)$ is the areal density of voids at position x (or, in the vicinity of x) that intersect the surface with a radius r (or in an interval from r_i to r_{i+1}). Next, we form cumulative numbers $\sum n(r,x)$ by summing the number densities from the largest radii to the smallest ones. The division by the area over which the count was made makes the result independent of the x-interval size. The summation over the whole distribution makes the result independent of the size of the radius interval. The cumulative void counts for the values in this table were shown earlier in Figure 3.20 (these are cumulative counts, not void densities per unit area).

Using the standard technique of Scheil [1931, 1935], the volume density $\rho(R,x)$ of voids of radius R is computed from $n(r,x)\Delta r\Delta x$. This transformation is an inverse transformation and therefore may be unstable, yielding a noisy volume density distribution. For this reason, it is often necessary to perform a smoothing operation on the original surface density data. To make the volume

Table 7.1. Void counts on a cross section of 110 grade aluminum.

Range of radii (μm)	Distance from the rear surface (μm)				
	0–50	50–100	100–150	150–200	200–250
0.00–2.40	2	7	2	1	0
2.40–2.95	6	12	6	2	2
2.95–3.57	4	18	17	1	1
3.57–4.30	7	11	7	8	0
4.30–5.17	3	5	7	3	0
5.17–6.20	2	11	9	5	1
6.20–7.64	3	10	2	0	0
7.64–9.45	0	7	2	1	0
9.45–11.9	0	4	3	0	0
11.9–15.3	0	3	1	0	0
15.3–20.2	0	3	0	0	0
20.2–27.5	0	1	0	0	0

density distribution independent of the radius intervals chosen, it is convenient to work with the cumulative density function:

$$N(R,x) = \int_R^\infty \rho(R',x)\,dR', \tag{7.2}$$

which is the density of voids at position x with a radius greater than R. A group of researchers at Los Alamos National Laboratory (e.g., Zurek et al. [1998], Hixson et al. [1998], Tonks et al. [1998], and Thissell et al. [1998]) has developed a method for looking into the voids on a sectioned surface with instruments that can measure the depth of the void so that the true void radius and connectivity with other voids can be determined. With this new method there is still the need to derive the void size distribution in the volume of material, but the uncertainty about the void sizes observed has been greatly reduced.

An example of the void distributions resulting from the volume transformations of surface data is shown in Figure 7.1 for an impact that caused a tensile stress of 9.2 kbar for about 0.5 μs in a plate of 1145 aluminum. Shown are the cumulative numbers of voids per unit volume as a function of the void radius. Each curve is from a different 80-micron-wide zone through the thickness of the specimen, numbering from the region of high damage toward the free-surface (counts from zones on the side of the high damage region toward the impact plane are not shown). The ordinate is the volume-related void density. The intercept at $R = 0$ is the total number of voids per cubic centimeter. We note here that more and larger voids were present in zones labeled F01 and F03 than in the adjacent zones.

Based on the observed patterns of the void densities, we fit the void size distributions to the form

$$N_g = N_0 e^{-\left(R/R_1\right)}, \tag{7.3}$$

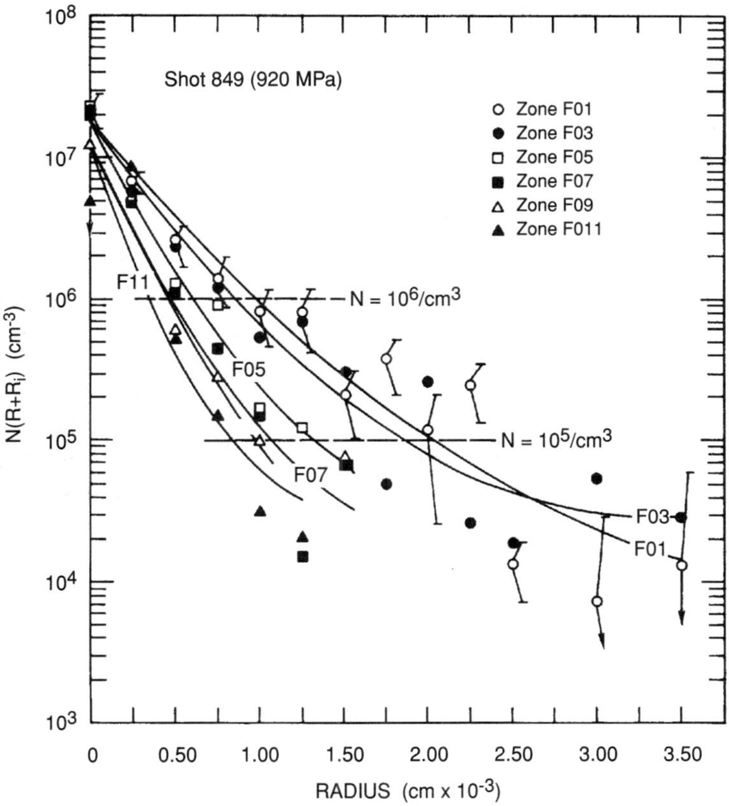

Figure 7.1. Distribution of voids (error lines are slanted to avoid overlap).

where N_g is the number of voids per unit volume with radii greater than R, N_0 is the total number of voids per unit volume, and R_1 is a size parameter. These two parameters N_0 and R_1 are used extensively in deriving the nucleation and growth processes and in comparing simulation results with the experimentally observed damage.

With the damage quantified, we now turn to the stress and strain histories experienced by the material and attempt to generate relations between the damage and the stresses (and/or strains). To obtain the stress histories throughout the sample, we can perform a wave propagation analysis, neglecting the presence of the damage. For an impact experiment, an x–t plot of the stress waves can be constructed as shown in Figure 7.2. A projectile, or flyer F moving from left to right impacts a sample S at time $t = 0$, inducing compressive shock waves (1 and 2) that propagate into the sample and projectile away from the impact plane. When these impinge on the free-surfaces they are reflected as rarefaction waves. These waves interact in A_1, a region in which the stress varies from compressive at the bottom to tensile at the top. The tensile waves (3 and 4) then propagate to the free-surface of the sample and are reflected as recompression waves. Tensile waves labeled 4 are reflected at the interface between the projectile and the sample because this interface cannot sustain a tensile wave. The time at stress (for this first tensile pulse) at any position x is then given by a vertical section of the tensile region T (Figure 7.2(a)). This time-at-stress was plotted as shown in Figure 7.2(b). This simple analysis defines the time at peak tensile stress for any plane x parallel to the impact surface.

By comparing void distributions, such as those of Figure 7.1, with computed stress histories such as those in Figure 7.2, we see that larger voids and larger numbers of voids are associated with longer durations of the peak stress. When we examine void distributions from a number of impacts at several stress levels and a range of target thicknesses (and therefore, different stress durations), we also notice that the numbers and sizes of voids increase with stress level as well as duration. Therefore, a combination of nucleation and growth mechanisms would explain the presence of the voids. In the following paragraphs we outline methods for relating the void distributions with the computed stress histories to derive nucleation and growth relations.

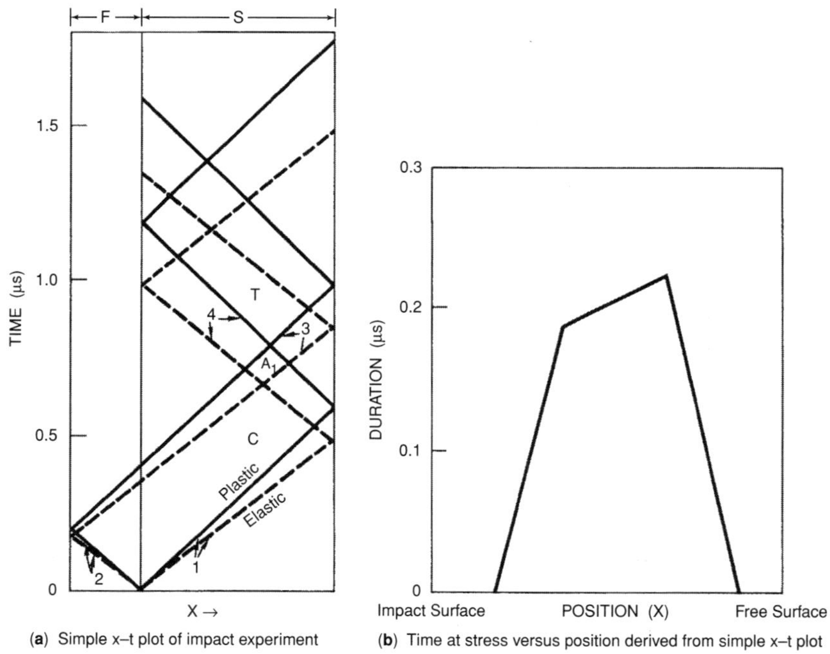

Figure 7.2. Computed stress waves in an impact specimen and the derived duration of peak tensile stress for a simple square wave with a precursor.

7.1.2.1. Void Nucleation Rate

The void nucleation rate is determined from the void densities from several impact experiments at a range of stress levels and stress durations. An initial estimate of this nucleation rate is obtained using the stress histories from no-damage computations, as indicated above. This estimate is made by plotting $N_0/\Delta t$ versus the mean stress (pressure), σ_m. For aluminum data, we found that the nucleation rate fit the following functional form:

$$\dot{N} = \dot{N}_0 \left[\exp\left(\frac{\sigma_m - \sigma_{n0}}{\sigma_1} \right) - 1 \right], \tag{7.4}$$

where $\dot{N}_0$ is the nucleation rate constant, σ_{n0} is the threshold stress for nucleation, and σ_1 is the stress sensitivity factor. All three of these parameters are new material properties that are obtained from the fracture experiments.

7.1.2.2. Void Growth Rate

An initial estimate of the void growth rate $\dot{R} = (\partial R/\partial t)_N$ is calculated by comparing the function $N(R, x)$ from several tests with computed stress histories at the same x-locations as where the data were taken. The approach to this calculation is explained with the aid of Figure 7.1. The distributions in the figure are taken from different zones in the material. Associated with each zone is a load duration that varies from zone to zone. The amount of damage also varies from zone to zone, increasing with load duration. Therefore the distributions can be visualized as a historical sequence of the damage at one location. Considered in this way, the curves provide a velocity when we divide the change in radius of a particular void by the difference in the stress duration. We note that this examination of growth neglects the fact that damage evolution involves both nucleation and growth, so that what we are interpreting as $(\partial R/\partial t)_N$ is partially associated with nucleation. Nevertheless, the simpler interpretation is useful for discussion of the underlying physics and provides a good initial estimate of the growth rate parameters.

An initial indication of the nature of the growth process can be seen in Figure 7.1. The size distributions are approximately straight lines pivoting about 2×10^7 cm^{-3} on the ordinate, with curves of shallower slope corresponding to longer stress durations. An expression that represents these ideas is

$$\dot{R} = AR, \tag{7.5}$$

where A is a constant for one test, but more generally is a function of the nominal mean stress in the test. By multiplying Eq. (7.5) by $4\pi R^2$, we can express the growth law in terms of void volume V_v as follows:

$$4\pi R^2 \dot{R} = \frac{4}{3}\pi R^3 (3A) \qquad \text{or} \qquad \frac{dV_v}{dt} = 3AV_v. \tag{7.6}$$

This volume expression is useful because we may consider V_v to be the relative volume of the entire void size distribution instead of the volume of a single void.

The growth rate process in Eq. (7.5) is consistent with the following equation for growth derived by Poritsky [1952] for expansion of a void under tension in a viscous fluid:

$$\dot{R} = \frac{\sigma_m - \sigma_{go}}{4\eta} R, \tag{7.7}$$

where η is the material viscosity and σ_{go} is a mean threshold stress. In Poritsky's equation, σ_{go} is the internal tensile pressure in the void, but here we assume it corresponds to a strength of the solid material. The derivation of Carroll and Holt [1972a] indicates that σ_{go} is several times the yield strength of the material. To obtain initial estimates of η, plots can be made of the void distribution parameter R_1 (in Eq. 7.3) versus $(\sigma_m - \sigma_{go})\Delta t$. The slope gives η and the intercept on the ordinate is the value of R_1 at nucleation, or R_n

Thus, we have a void growth rate that has been derived directly from the experimental observations of voids, yet it is also consistent with theoretical derivations. This is one of the processes to be inserted into any model for describing ductile fracture. A more complete growth law for voids, including inertia effects, is outlined in Section 7.1.6.2.

7.1.2.3. Nucleation Void Size

Nucleation may occur physically through the debonding of the matrix material from an inclusion or through a vacancy development at a triple point among metal crystals. We can determine the size distribution at nucleation either by counting the inclusions or other defects that lead to void development or by extrapolating from the counted void size distribution (as in the determination of R_n above). It is desirable to follow both procedures, if possible, to verify the physical appropriateness of the modeling approach to the material under study.

7.1.2.4. Damage Characterization Parameters

Thus far, we have characterized the damage with the parameters N_0 and R_1 (or R_n at nucleation). However, the void volume is more important in determining both the strength degradation and the effect on the stress waves. The relative void volume at any time is

$$V_v = \frac{4\pi}{3}\int_o^{N_0} R^3 dN = \frac{4\pi}{3}\int_o^{\infty} R^3 dN\left(-\frac{dN}{dR}\right) dR = 8\pi N_0 R_1^3. \tag{7.8}$$

Thus, in addition to the total number of voids per unit volume N_0, either the characteristic radius R_1 or volume V_v may be specified.

7.1.3. Damage Processes in the Ductile Fracture Model

Damage in the model is represented by the total number of voids per unit volume and the relative void volume (V_v). The void density quantity N_0 is actually the total number of all voids per unit of initial volume. With this definition of N_0 the number does not change with expansion of the material, but only through nucleation of new voids. But, the relative void volume is the amount of void volume relative to the current total volume; hence, it changes because of nucleation and growth, and from changes in the overall volume of the material. We assume that the void size distributions retain the form of Eq. (7.3) throughout their history. Here we consider the change in void volume caused by nucleation separately from that caused by growth. The total change in void volume is obtained as the sum of these two contributions.

7.1.3.1. Nucleation

Nucleation in the model occurs as the addition of new voids to the existing set. These new voids are presumed to occur in a range of sizes with a size distribution given by Eq. (7.3). That is, some voids are initiated at a large size, others at a small size. This may seem peculiar but it fits with the assumption that voids are nucleated at inclusions and the inclusions have a range of sizes. It thus appears that the larger voids are not necessarily nucleated first. At nucleation in the model, the parameter R_1 equals R_n, the nucleation size parameter (a material constant). The number of voids nucleated is governed by the nucleation rate function in Eq. (7.4). The void volume nucleated in a time interval Δt is found by integrating as in Eq. (7.8) for the ΔN new voids obtained from Eq. (7.4):

$$\Delta V_{vn} = 8\pi \Delta N R_n^3 . \tag{7.9}$$

7.1.3.2. Growth

Growth in the model is represented by Eq. (7.7), altered as in Eq. (7.6) to represent the increase in void volume due to growth:

$$\left(\frac{\partial V_v}{\partial t}\right)_N = \frac{3}{4}\frac{\sigma_m - \sigma_{g0}}{\eta} V_v . \tag{7.10}$$

Because this expression is independent of the size and number of voids, we can use it also for the growth of the volume of the entire void size distribution. The growth of the void volume during a time interval Δt is obtained by integrating Eq. (7.10) to obtain the change in volume ΔV_{vg} caused by void growth. Thus,

$$\Delta V_{vg} = V_{v0}\left[\exp\left(\frac{3}{4}\frac{\sigma_m - \sigma_{g0}}{\eta}\right)\Delta t - 1\right], \tag{7.11}$$

where V_{v0} is the void volume at the beginning of the time interval. Because every void in the distribution grows by the same exponential factor, even the size parameter R_1 grows according to Eq. (7.7)

$$R_1 = R_{10} \exp\left(\frac{\sigma_m - \sigma_{go}}{4\eta} \Delta t \right), \tag{7.12}$$

where R_{10} is the size parameter at the beginning of the time interval. Then the new void volume can also be found from Eqs. (7.9) and (7.11):

$$V_v = V_{v0} + \Delta V_{vn} + \Delta V_{vg}, \tag{7.13}$$

where $V_{v0} = 8\pi N_0 R_{10}^3$, the void volume at the beginning of the time internal.

7.1.4. Stress–Strain Relations and Solution Procedure

In the NAG approach to fracture, the material is treated as consisting of two components: solid (intact) material, and voids. The void processes are treated by the foregoing equations for nucleation and growth, and the intact material is treated by the usual constitutive relations. The moduli of the material are not modified for the presence of the voids; rather, the strains are adjusted to account for the change in void volume and the stresses in the solid are adjusted to account for the presence of the voids. The resulting stresses are the same as those that would be obtained by altering the moduli.

7.1.4.1. Stress–Strain Relations with Damage

In the DFRACT model, the stress–strain relations for material undergoing fracture are influenced by the presence of the voids. These stress–strain relations are constructed by requiring that the sum of the volumetric strains in the solid material and in the voids must equal the imposed volumetric strain. Thus, to conserve volume we require that

$$V_s + V_v = 1, \tag{7.14}$$

where V_s and V_v are the volume fractions of solid and void. This consistency in the model is essential to allow the damage to have an appropriate effect on the stress–strain relations, and thus on the moduli and effective wave velocities. This damage effect is important in defining the breadth of damage near a spall plane, and in providing for multiple spalls.

For the derivation of the stress–strain relations, the stress in the solid material is separated into mean stress and deviatoric components. The mean stress is related to the specific volume and internal energy through the Mie–Grüneisen equation of state

$$-\sigma_m = K_1\left(\frac{\rho_s}{\rho_{s0}} - 1\right) + K_2\left(\frac{\rho_s}{\rho_{s0}} - 1\right)^2 + K_3\left(\frac{\rho_s}{\rho_{s0}} - 1\right)^3 + \Gamma\rho_s E, \tag{7.15}$$

where K_1 is the bulk modulus, K_2 and K_3 are higher order terms in the series for bulk modulus, Γ is the Grüneisen ratio, E is internal energy, ρ_s is the solid density, and ρ_{s0} is the initial density of the solid. The mean stress has a negative sign here because we are treating stress as positive in tension, but the usual sign convention of Eq. (7.15) is for pressure, which is positive in compression. The mean stress computed from Eq. (7.15) is necessarily an average, because the actual stress states vary greatly through partially fractured material.

The mean stress on the gross section of the fractured material can now be related to the mean stress in the solid components according to a relation derived by Carroll and Holt [1972b] for porous material:

$$\sigma_{mg} = \frac{\rho}{\rho_s}\sigma_m, \tag{7.16}$$

where σ_{mg} is the mean stress on a section and ρ is the average density of the porous material. A combination of Eqs. (7.15) and (7.16) relates the average pressure P to the energy E and density ρ. The deviator stresses are computed by the usual elastic and plastic relations. These stresses are also modified as in Eq. (7.16) to account for the presence of the voids.

7.1.4.2. Solution Procedure

During its use in a wave propagation code, the ductile fracture model is provided with a strain increment (or rate) tensor and the model must provide the new stress tensor, in addition to updating the number and size of voids. The determination of the stresses requires the simultaneous solution of the nucleation rate equation, Eq. (7.4), growth rate equation, Eq. (7.10), pressure–volume relation, Eq. (7.15), and the consistency equations, Eqs. (7.14) and (7.16). We have successfully used two solution procedures:

1. First-order predictor–corrector technique, treating the growth equation as primary, and all other equations as secondary. The predictions are made by a Newton-Raphson method, followed by *regula falsi* for the subsequent iterations.
2. A Runge-Kutta method, treating the differential equations for growth and nucleation simultaneously and handling the equations for mean stress and deviator stress as subsidiary.

The results are equivalent, although the first is faster at later times when the damage is large and the second is better for small void volumes and very rapidly changing stresses. Simulations that have been made with the DFRACT model in a wave propagation computer program are shown later in Chapter 8.

7.1.5. Interaction of Stress and Damage

Even fairly small levels of damage significantly modify the stress history in the sample from that which would be obtained without damage. When damage occurs (that is, void volume increases, thereby reducing the imposed tensile stresses), recompression waves emanate from the damaged zone and alter stress histories in adjacent material. A sample of a computed stress history that was altered by a relative void volume of about 0.5% at the spall plane is shown in Figure 7.3. These recompression waves appear to overtake the tensile waves and attenuate them. Therefore, the peak tension is not usually reached throughout the tensile region of the target, and stress histories at points adjacent to the spall plane are modified even more than those at the spall plane. From these computed results we can conclude that for high damage, the peak stress and the stress duration from no-damage calculations are only rough approximations to the actual values. Hence, it is necessary to use a fairly accurate damage model for the wave propagation simulations to obtain reliable estimates of the stress histories.

Because the stress histories are strongly modified by the presence of the damage, initial estimates of the fracture parameters based on no-damage wave propagation calculations may be fairly inaccurate. Therefore, after the fracture model is constructed, the impacts in the experimental database are simulated with the damage model. It is often necessary to adjust the fracture parameters to

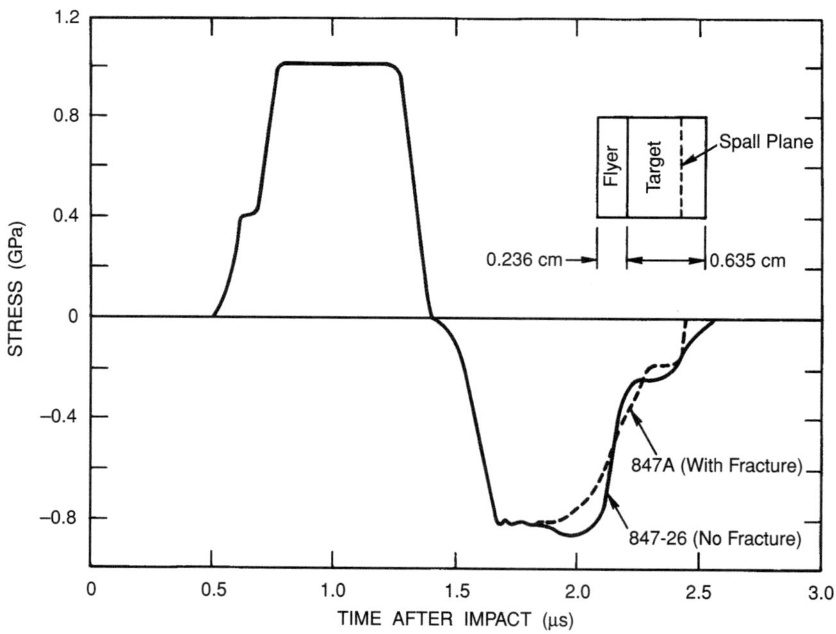

Figure 7.3. Effect of damage on stress history at spall plane of shot 847 (0.82 GPa).

achieve good correlation with the observed fracture data. This usually requires an iterative procedure that involves varying the model parameters one by one while repeating the simulations until a satisfactory agreement between the simulation results and the experimental data is obtained.

7.1.6. Current Studies for Amplifying the Ductile Fracture Treatment

The foregoing discussion of the ductile fracture model is based on many simplifying assumptions about the fracture process. Considerable work is currently underway to improve our understanding of ductile fracture. In this section we note some of these current efforts, which are mainly being undertaken by other authors.

7.1.6.1. Gurson Model as Threshold for Growth

In the foregoing analyses, the threshold stress was given as simply a stress quantity, defined for each material. However, we can be sure that the actual threshold depends on the state of stress, as suggested by the Gurson model (Gurson [1977]) for quasi-static fracture. Gurson developed a yield-like, continuum model for treating void growth based on the analysis of a spherical void in a finite-sized sphere of perfectly plastic material. The yield curve equation is

$$\frac{\bar{\sigma}^2}{Y^2} + 2q_1 v \cosh\left(\frac{3q_2\sigma_m}{2Y}\right) - 1 - q_1^2 v^2 = 0, \tag{7.17}$$

where $\bar{\sigma}$ is the Mises stress, Y is the yield strength, v is the relative void volume, and $q_1 = 1.5$ and $q_2 = 1.0$ are dimensionless parameters introduced by Tvergaard [1981, 1982] based on his analysis of the response of a void under general loading conditions.

In this model for spherical voids, the apparent threshold stress for void growth is equal to the yield strength when the stress state is pure shear (along the ordinate in Figure 7.4), but it is several times greater than the yield strength for hydrostatic tension (as seen along the abscissa). This value on the abscissa is given by the analysis of Carroll and Holt [1972a] for the collapse or expansion of a spherical void in a spherical shell of elastic-perfectly-plastic material, and has the following equation:

$$\sigma_g = \frac{2Y}{3} \ln v, \tag{7.18}$$

where σ_g is the mean stress at which yielding, which begins at the void boundary, reaches the outer edge of the shell so that with continuing applied stress the void can expand without elastic resistance.

In the standard plate impact experiments, the stress state initially moves along the elastic loading line shown in Figure 7.4; then as yielding occurs, the

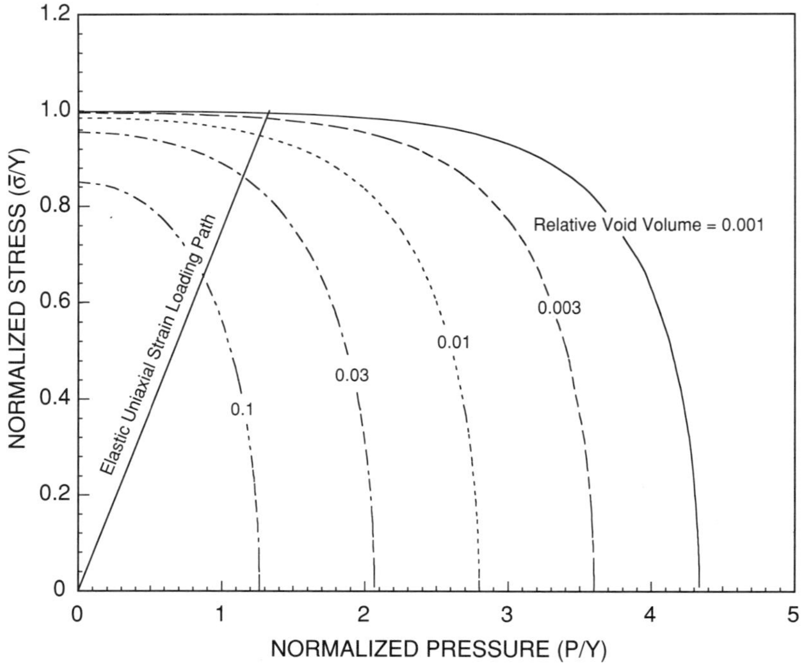

Figure 7.4. Gurson's yield function for several different void volume fractions. A uniaxial strain loading path is also shown in the figure.

point moves around to the right and down along the appropriate Gurson curve. In most cases, the point on the Gurson curve at which fracture occurs is approximately equal to the intersection along the abscissa, so for plate impacts the threshold is not seriously modified by adding this refinement; however, the yield value is considerably reduced. Also, we may wish to simulate other stress states.

7.1.6.2. Inertia Effect on Void Growth

In the preceding analyses of void growth, we used Eq. (7.7) in which the growth rate is a function of the current radius, applied mean tension, and material viscosity. With this relation all voids grow proportionally at the same rate, independent of their initial size. A more complete representation of the void growth rate can be obtained from the analysis of Poritsky for the growth (or collapse) of a spherical void in an infinite, linear viscous medium under uniform hydrostatic loading:

$$\frac{d^2R}{dt^2} + \frac{3}{2R}\left(\frac{dR}{dt}\right)^2 + \frac{4\eta}{\rho R^2}\frac{dR}{dt} + \frac{2\gamma}{\rho R^2} = \frac{\sigma_m - \sigma_{g0}}{\rho R}, \tag{7.19}$$

where η is the material viscosity and γ is the surface tension. Here σ_m is the mean applied stress and σ_{g0} is a threshold stress, representing the resistance due to the yield strength of the solid material (not present in Poritsky's analysis). This equation pertains to a void in an infinite medium, so the porosity is negligible. Equation (7.19) can be solved for assumed viscosities, surface tensions, and densities of materials of interest. Typical results, expressed in terms of growth rate as a function of void radius are shown in Figure 7.5. Surface tension may have an effect only for very small voids (less than 0.1 μm in the figure). Belak [1999] noted that, based on his molecular dynamics computations of the growth of voids in metals, there is no surface tension effect. For aluminum the ratio (dR/dt)/R is nearly constant for radii between 0.1 μm and 10 μm. In this range, growth is governed mainly by material viscosity. However, for larger voids there is an important effect of inertia and hence, material density. Increasing the density of the material while holding all other parameters constant causes the growth curve to shift to the left by an amount equal to the ratio of densities. Then there may be no void size range in which viscosity dominates sufficiently that Eq. (7.7) is appropriate.

When Eq. (7.19) is used to determine the growth rate, larger voids no longer grow in proportion to their size. Hence, the exponential size distribution of Eq. (7.3) will no longer represent the larger voids. But this is also the range of voids most likely to be affected by coalescence, so other phenomena may also be important for these voids.

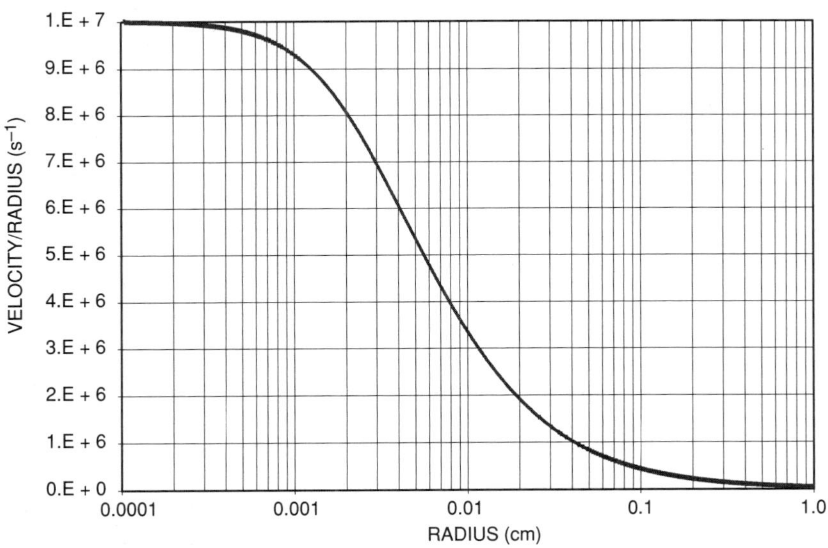

Figure 7.5. Void growth rate as a function of radius for a spherical void in an infinite, linear viscous medium under uniform hydrostatic loading. The parameters used (ρ = 2.7 g/cm3, η = 250 dyne•s/cm3, σ_{go} = 2 kbar) represent the behavior of aluminum.

Many other analyses have also been made of a void in a finite volume and considering inertia effects and a single representative void. Unfortunately, in general we need the growth of a distribution of void sizes in a finite volume, so that the relative void volume and the void radius R can vary independently.

7.1.6.3. Thermal Effects at the Void Surface

Wang [1995] and Tong and Ravichandran [1995] have reported from their analyses of a void in a solid under dynamic tensile loading that the material at the edge of the void becomes very hot, approaching melting. Wang included a thermal strength reduction effect that reduced the material strength and viscosity to zero at melting. From this study he concluded that the resistance of the material to rapid void growth is much less than that considered by other authors, especially Johnson et al. [1981, 1996] and Curran et al. [1987]. His results suggest that the actual material viscosity governing void growth is much higher than had been thought earlier. For example, he determined a standard-state viscosity of 500 Pa•s instead of 20 Pa•s from the data of Johnson [1981]. The larger viscosity may be more in line with values from dislocation analyses. It is important to get the correct mechanism so that a reliable treatment can be made over a large range of strain rates and also so that the viscosity value fits other data from high–rate loading. Perzyna [1986] had reported earlier that the temperature change from plastic strain at the void surface "is a substantial fraction of the melting temperature of the material."

Seaman and Curran [2001] have recently conducted simulations of a spherical void in a spherical volume under a suddenly applied tensile field, and continued the computations to significant amounts of void growth. It was found that the rim of the void approaches melting for fairly small void expansions, so that most of the void growth occurs while the void boundary is near the melting temperature. A temperature-dependent viscosity that approached zero as the material passed through incipient melting was used in the simulations. The simulation results indicate that there was little further increase in the internal energy at the void boundary. The softening of the material allowed at most a 35% increase in the void growth rate over that expected without a thermal effect. Hence, some adjustments are needed to account for thermal effects near the rim of the void, but the effect in most cases is not very significant.

7.1.6.4. Coalescence of Voids to Form Damage Planes or Fragments

Recent studies by Zurek et al. [1998], Tonks et al. [1995, 1998], and Thissell et al. [1998] have been addressing the problem of coalescence, building a data base of experimental observations, and developing plans and a theoretical basis for a model. A major problem for the model is to move beyond the isotropic modeling of Gurson and of the SRI model DFRACT. Usually, at some point as damage develops, voids coalesce to form a rough plane so that separation occurs in one direction. Their model plan has the effect of producing anisotropy within a statistical framework, that is, without requiring a layout of specific voids in specific locations and then following their detailed interactions. So it has the needed generality for handling 3-D damage-induced anisotropy without the gigantic memory requirements of treating individual voids.

Some simpler approaches to describing coalescence have been derived by other authors. A coalescence process was added to the Johnson [1981] model by Giles and Maw [1998] and verified by simulations of laser radiation experiments in aluminum. Curran and Seaman [2001] have developed a simple approach to coalescence, like the treatment of coalescence in brittle fracture (Guéry and Seaman [1996]). In this coalescence model the damage remains isotropic, but the void size distribution changes to have a few very large voids formed by coalescence. Coalescence has the effect of accelerating the void growth for large voids to produce full damage earlier than would otherwise occur. This effect is similar to that of the extension of Needleman and Tvergaard [1991] to the Gurson model to allow full damage to occur at a small void volume.

7.1.6.5. Effect of Void Shape

Generally, the models in use treat all the voids as spherical and experimental observations show that most voids are approximately spherical. However, under conditions of high shear (as noted by Giovanola, in Shockey et al. [1985]) or coalescence, the voids can become elongated or have other more complex shapes. Thissell et al. [1998] show evidence of odd-shaped voids formed by a combination of coalescence and high shear deformation.

Budiansky, Hutchinson, and Slutsky [1982] presented an analysis for ellipsoidal voids in a linear viscous material. Their results show growth rates for the axes of the void in a stress field whose directions coincide with the orientation of the ellipsoid. Nemat-Nasser and Hori [1987] have studied the growth and collapse of elliptical (two-dimensional) voids.

Gologanu et al. [1993] treat the behavior of an axisymmetric prolate or oblate ellipsoid in a finite volume of plastic material. They condense their results into a suggested modification of the Gurson yield-surface model to account for the ellipsoidal nature of their voids. Castañeda and Zaidman [1994] presented an alternate method of deriving a constitutive relation for a material with ellipsoidal voids. They discuss the effects of the changing shape and volume of voids under loading, noting differences between their model and that of Gurson, who considered only the effects of the changing volume of voids.

7.1.6.6. Transition from Voids to Cracks and from Cracks to Voids

Chang and Nemat-Nasser [1992] have studied situations in which voids were collapsed dynamically to form crack-like features in copper, steel, and iron. The surfaces of these cracks resembled those associated with brittle fracture, although they were present in very ductile material. Grady [1987] has proposed that in certain metals a transition from brittle spall (dominated by fracture toughness) to ductile spall (dominated by yield) can occur with a variation in the rate of dilatant strain.

7.1.6.7. Microvoid Interaction

Generally, in ductile fracture the voids are treated as if they exist in isolation, although they may be very close together, especially when the damage is high. Kachanov [1993] and Tsukrov and Kachanov [1993] have provided an analysis of the interaction of voids in elastic media so that one can estimate the importance of this interaction effect. A specific case in which interaction is definitely important is that in which a large void interacts with several nearby small voids. The small voids then grow in the enhanced strain field of the larger void and may eventually coalesce with the larger void. Needleman and Tvergaard [1991] have used this situation as a basis for an analysis of void or ductile crack growth.

7.1.7. Summary

Experimental techniques have been developed for conducting impact tests for the determination of ductile fracture processes. The damage appears as distributions of microvoids that appear to be nucleated at inclusions and to grow until they may coalesce and sever the sample. Quantitative post-test examinations of the partially fractured samples provide the nucleation and growth rates for the development of these voids. The data obtained are a basis for models describing the tensile fracture of ductile materials. The Nucleation-and-Growth model called DFRACT was developed to represent these rate processes.

7.2. Nucleation and Growth of Cracks and the Model for Brittle Fracture

This section focuses on the development of a computational model to represent brittle fracture under high-rate loading. Within the context of the forgoing discussion, "brittle" behavior is understood to represent a response mode characterized by cracking as the primary fracture mechanism as observed at a magnification of about 100X. Experimental evidence of brittle dynamic fracture indicates that, initially, the response is characterized by the formation of a large number of very small cracks, or microcracks. Under loading, the microcracks grow and coalesce to form a few larger cracks. Therefore, the model represents fracture as the development of microcracks. Here we begin with some experimental background based on the work of a few of the great many researchers who have studied the subject. Then we note some of the kinds of models that have been used to describe high–rate brittle fracture. To make the concepts more specific, we focus on the details of the processes in the SRI BFRACT model.

Many of the other models mentioned during the discussion have several characteristics in common with BFRACT. The discussion concludes with a summary of current work that will likely lead to a more complete, more complex, and probably more useful version of the brittle fracture model.

7.2.1. Experimental Aspects of Brittle Fracture

Various approaches have been used to provide the experimental basis for describing the response of materials to brittle fracture. The types of experiments performed depend on the type of information required and the level of detail pursued. The following are among the most commonly used approaches:

1. Study the pullback signals measured in an impact experiment to estimate the spall strength: such a study can occur without capturing and examining the target after the test.
2. Capture the damaged target from an impact experiment and examine cross sections of the target for evidence of the fracture morphology. This is the level of detail pursued for the BFRACT model.
3. Examine cross sections of the impacted target to relate the fracture processes to the grain structure and other metallurgical aspects of the material. This is the approach being pursued by Zurek et al. at Los Alamos (e.g., Zurek et al. [1998], Hixson et al. [1998], Tonks et al. [1995, 1998], and Thissell et al. [1998]).

Below, we discuss experimental studies from each of the three categories above. Later, the discussion focuses on the approach taken in the second category because this approach leads to the BFRACT fracture model described later in Section 7.2.3.

Romanchenko and Stepanov [1980] analyzed pullback signals to determine the critical stress for initiating spallation and found that the threshold increases strongly with strain rate. They sectioned specimens and examined them for damage to verify their conclusions based on the pullback signals. Speight and Taylor [1986] also studied spall behavior by interpreting free-surface velocity measurements. Dremin et al. [1992] used pullback signals in combination with metallographic examinations of impacted samples to examine damage in austenitic steel as a function of temperature. Kanel et al. [1994b] conducted spall experiments on aluminum, molybdenum, niobium, and alumina, varying the load duration down to nanoseconds and reaching 35% of the theoretical strength. They concluded that they could reach the theoretical strength with durations of 1 to 100 ps. Zaretsky and Kaluzhny [1995] measured free-surface velocity histories in stainless steel targets to infer a decrease in the spall strength with increasing temperature in the range from 170 K to 300 K. Gu and Jin [1998] found a similar decrease in spall strength in stainless steel using temperatures up to 900 K. Razorenov et al. [1995a] used the velocity pullback signal to study spall in high-purity titanium under loading by explosive and by a pulsed ion beam source. Razorenov et al. [1998] studied spall in aluminum, magnesium, and zinc over a wide range of temperatures, finding that the spall strength generally decreases significantly as the melting temperature is approached.[1] Using pullback signals, Arad et al. [1998] found in aluminum, copper, and a metallic glass that the spall strength increased rapidly with strain rates up to around 10^7 s^{-1}. Chang and Choi [1998] also used pullback signals to explore the increase in spall strength in tungsten heavy alloy as a function of strain rate and prior compressive stress level.

Grady [1988] collected spall data for many materials (ductile and brittle) and displayed these data as a function of strain rate. He noted that for materials like aluminum, in which both ductile spall and brittle spall are possible, brittle fracture occurs at lower rates and ductile fracture at higher rates. Gaeta and Dandekar [1988] performed spall experiments in single crystal and polycrystalline tungsten to examine the effect of shear strength on the spall behavior. Anderson et al. [1988] studied spallation in tungsten under explosive loading. Yaziv et al. [1986] performed spall experiments in BeO to examine the relationship between the apparent spall strength and the prior compressive stress above the HEL. Senf, Straßburger, and Rothenhäusler [1995] used high-speed photography to observe the fracture of glass under impact, and based on these observations, to delineate the nucleation processes. Yi Long Bai and his co-workers (Zhao et al. [1991]) have sectioned impacted samples, examined and quantified the crack (or void) damage, and developed governing equations for the nucleation and growth processes based on their observations.

Many studies have been performed to examine the complexity of the nucleation and growth processes, and to characterize the dependence of these processes on the inclusion content, heat treatment, temperature during the test, and the strain rate. Meyer and Aimone [1983] provide a large collection of data on the metallurgical aspects of spallation under both brittle and ductile conditions. Godse, Ravichandran, and Clifton [1989] examined crack propagation in steel with 15% upper bainite. They found that crack nucleation occurred primarily in the bainite phase, while crack propagation took place in the martensite phase. Elias, Rios, and Romero [1992] studied spall in high-strength, low-alloy steel. They found that the fracture mechanism and the spall strength depend significantly on the heat treatment. Zurek, Frantz, and Gray [1992] studied the effects of prestrain and prior compressive stress on the spall strength of 4340 steel. Buchar, Rolc, and Hrebicek [1992] found that spall strength varied with both fracture toughness and strain rate in several structural steels. Wang and Mikkola [1992] studied alpha-alumina under explosive loading, finding that fracture was related to the grain boundaries and orientations and to the plastic flow processes. Brar and Bless [1992] found that many aspects of the fracture phenomena under impact in alumina, glass, TiB_2, and SiC were material-dependent.

1. Results from this study were discussed in more detail in Chapter 5.

The experiments performed to support the development of the brittle fracture model described here were usually conducted by impacting a flyer plate against a thicker target plate of the same material such that fracture or spall occurred within the target plate (Barbee et al. [1972], Seaman et al. [1976], and Curran et al. [1987] give examples). Following impact, the targets were sectioned and polished, and the cross sections were examined for damage. The cracks were counted, and the resulting distributions were correlated with the stress durations to develop nucleation and growth rates for the damage. The materials tested include polymers, rocks, concrete, ceramics, propellant, steels, iron, and other metals.

Below, we summarize some of the experimental fracture observations; then we describe the method used for transforming areal crack density to density per unit volume; and finally, we provide a methodology for deducing nucleation and growth rates.

7.2.2. Determination of the Crack Size Distribution

The experimental methods used here are essentially the same as those described in Chapter 3 and in Section 7.1.1 above. They consist primarily of plate impact tests during which the target plate may have a sensor to record either the free-surface velocity at the rear surface of the plate, or the stress history at the interface between the rear surface of the plate and a softer buffer like epoxy or PMMA. These experiments are usually performed under controlled conditions and the targets are softly recovered for post-test inspection.

The recovered samples are sectioned and polished, and the cracks on each section are observed and quantified. Typical cross sections of spall-damaged Armco iron are shown in Figure 7.6. The figure shows that the cracks vary in number, size, and also in orientation with respect to the direction of wave propagation (vertical in the figure). As the cracks are counted, they are grouped according to length and orientation. For example, length groups of 0 to 5 μm, 5 to 10 μm, 10 to 15 μm, 15 to 25 μm, etc. might be used. Similarly, 10 angle groups of 9 degrees each may be used to represent the orientations. In this way a large matrix of counts is obtained for each region of the specimen. These cracks observed on a cross section must then be transformed to a volume basis (number per unit volume) to be considered material properties. The required transformation is outlined in the next section.

In this section, we have chosen to illustrate brittle fracture processes by a series of observations in Armco iron. In Chapter 8 we show observed fractures in a number of other brittle materials. Some representative sections from Armco iron targets originally examined by Seaman et al. [1976] and Barbee et al. [1970] are shown in Figures 7.6 and 7.7. In Figure 7.6 the direction of propagation is vertical; in Figure 7.7 it is horizontal. All the cross sections shown are characterized by a great number of small cracks at many different angles. In Figure 7.6, both the number and size of the cracks appear to increase with the tensile stress duration. Based on this observation, we postulate that both nucleation and growth processes are needed to adequately describe the evolution of the cracks during fracture.

The small cracks on some of the cross sections shown in Figures 7.6 and 7.7 appear to have intersected to form longer cracks. These longer cracks correspond with the macrocracks usually referred to in classical fracture mechanics. The photomicrographs shown in these figures indicate that the rough surfaces often seen on macrocracks really arise because the large crack is formed by the coalescence of many small cracks, and not by the irregular growth of a single large crack.

The grain structure of the Armco iron can be compared with the crack patterns in Figure 7.7. It appears that the microcracks are essentially straight and of two general types: intragranual microcracks propagating within the grain, and interfacial microcracks propagating along grain boundaries. Thus, the characteristic length of the microcracks is comparable to the characteristic grain size.

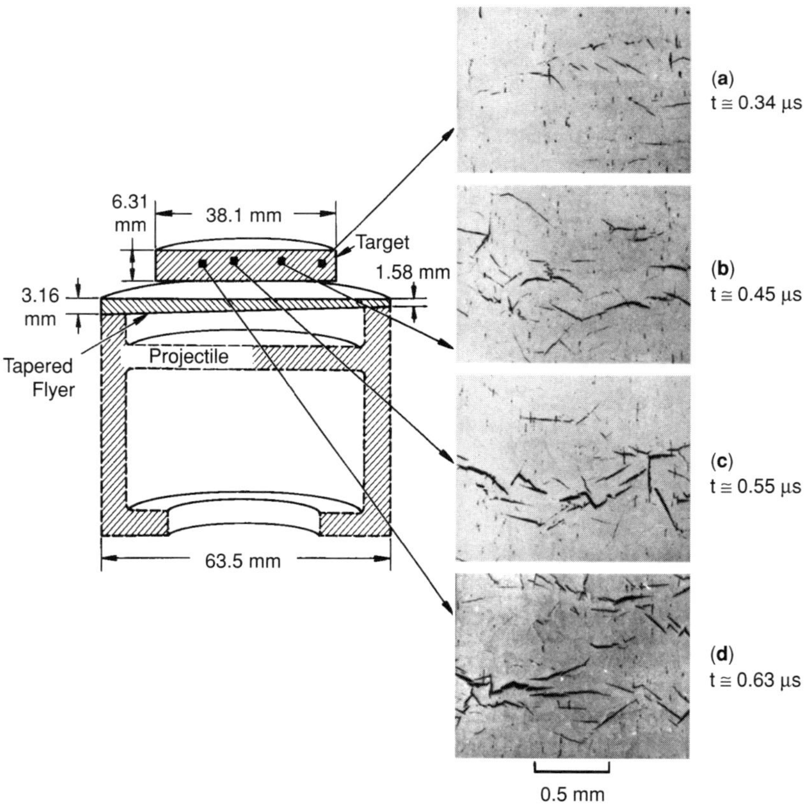

Figure 7.6. Configuration of a tapered-flyer impact experiment in Armco iron and observed damage on a cross section of the target.

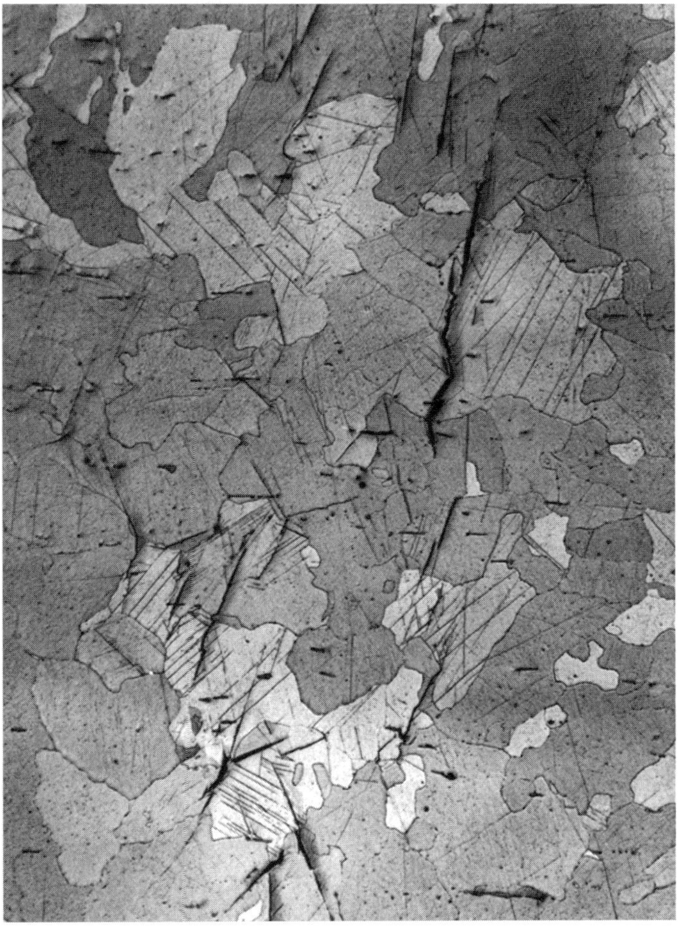

100 μm

Figure 7.7. Twins in the area of high crack density in Armco iron.

Figure 7.8 shows statistical areal distributions of cracks on the cross section of an Armco iron target impacted at a peak stress of about 4 GPa (Seaman et al. [1971, 1978b]). This figure does not distinguish between cracks of different orientations, so that the cumulative number of cracks shown represent the total number of cracks in all orientations. Each curve in Figure 7.8 refers to a different position through the target, and hence to a different tensile stress duration. The load duration that corresponds to the data in Figure 7.8 reaches a maximum near the center of the specimen and decreases with distance away from the center in either direction. The cumulative crack distributions shown in the figure reveal a trend of increasing number and size of cracks as a function of load duration (or distance away from the center of the specimen). The regularity of this trend suggests that the data may be uniform enough to be used as a basis for a model of microcracking.

Following the procedure developed for ductile fracture, the number of cracks is plotted as a function of stress duration to determine the nucleation rate. Similarly, the crack size is plotted as a function of stress duration to determine growth rates. The counting process is illustrated schematically in Figure 7.9. The cross section is divided into several strips such that all the material in each strip has about the same stress duration. Then, in each strip, the cracks are counted by size and assembled into size distributions for each orientation (not illustrated). These surface distributions (number per unit area) are then transformed, as outlined later, into number of cracks per unit volume. The size distributions on the planes of maximum damage from several impacts were fitted to the exponential form

$$N_g = N_0 e^{-(R/R_1)}. \tag{7.20}$$

Then the N_0 and R_1 values were used in plots with the parameters of the com-

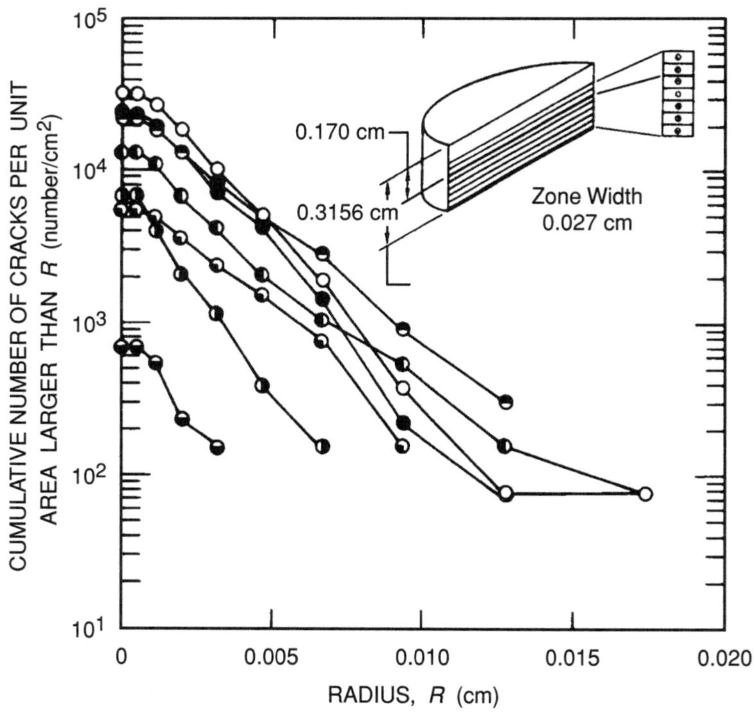

Figure 7.8. Observed cumulative crack concentrations on a cross section of a cylindrical target of Armco iron impacted at a nominal stress of 4 GPa.

puted stress wave (from a no-damage computation): the peak tensile stress σ_{max} and the stress duration Δt. An estimate of the nucleation rate is obtained by plotting $N_0 / \Delta t$ as a function of the maximum tensile stress on the planes of maximum damage from several impacts. A trend line through the data gives the nucleation function. Similarly the growth function is estimated from a plot of the distribution size parameter R_1 versus $(\sigma - \sigma_{g0})/\Delta t$.

The size distributions resulting from brittle fracture experiments have not been so easy to understand as they were in the ductile fracture case. There are several possible explanations for the complexity. First, the data exhibit more

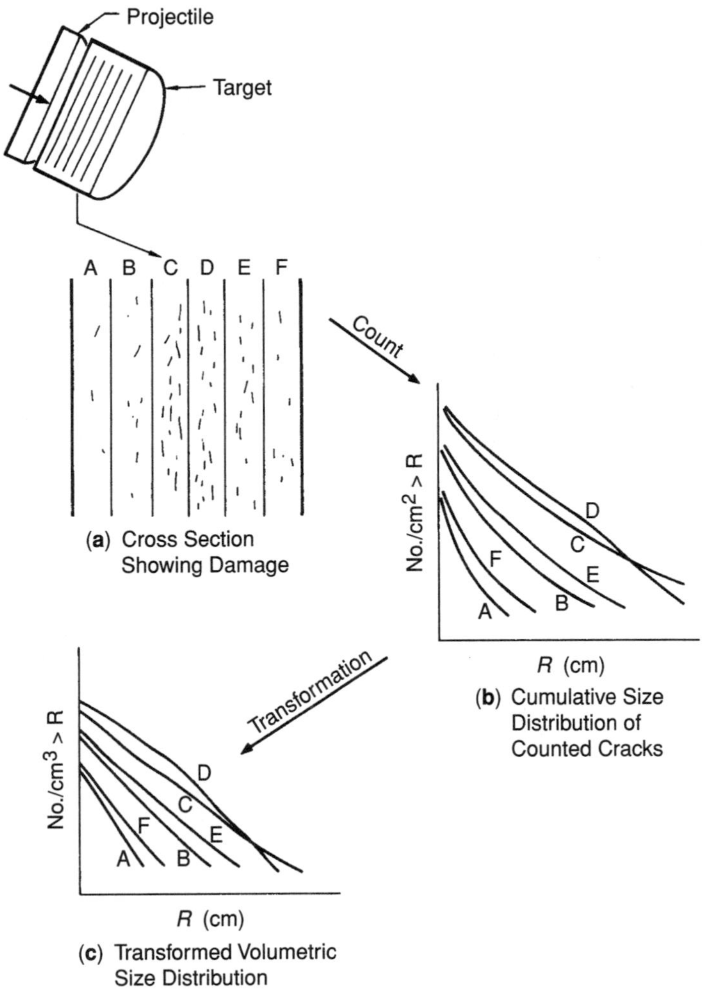

Figure 7.9. Acquisition and transformation of crack count data from cross sections of impacted cylindrical targets.

scatter so that major trends in the data are less readily observed. Second, the damage strongly affects the stress; hence, the nominal stress and stress duration obtained from a no-damage calculation are not closely related to the actual stress and duration that occurred in the specimen.

7.2.2.1. Transformations from Surface Counts to Volumetric Distributions

The study of brittle fracture is greatly aided if the number, size, and orientations of microcracks can be measured in test specimens. When such quantitative information is related to stress level and stress duration, nucleation and growth rates for the cracks may be obtained. For transparent materials it is possible to obtain this quantitative information directly by counting the cracks. However, for nontransparent materials the counting can only be conducted on a cross section. This method of counting captures only a slice through each of the cracks that intersect the cross section. As such, cross sectional counts alone cannot be expected to yield the true lengths, true orientations, or true numbers of cracks. But, if cracks are observed in large numbers, statistical analysis techniques can be used to calculate the true lengths, orientations, and number of cracks.

This section outlines a statistical analysis developed to transform surface (observed) crack size and orientation distributions to volumetric (true) distributions. The crack orientation distribution is assumed to be symmetric about one direction in the sample, and each crack is assumed to have a circular cross section. Thus, the analysis applies to spherical voids as well as to penny-shaped cracks. The analysis also applies to inclusions, grains, twins, new-phase platelets, and other metallographic features that obey the stated assumptions. No assumption is made about the form of the volumetric distributions, but the transformation works best when the number of cracks decreases exponentially with increasing crack size.

The transformation technique presented here is an extension of the work of Scheil [1931, 1935] and Saltykov [1958] who calculated true grain size distributions from apparent grain sizes on a cross section. Our approach is the same as Scheil's; however, our transformation is more complex because both size and orientation are considered, whereas Scheil considered only size.

For the crack counting operation, the shock-loaded specimen is sectioned on a plane that contains the direction of propagation of the initial compressive wave (Z direction in Figure 7.10). Each crack is measured to determine its apparent length $2c$, orientation angle α, width w, and position z through the thickness of the sample (see Figure 7.10). The ranges of length, orientation, width, and position are discretized into small intervals Δc, $\Delta\alpha$, Δw, and Δz for the calculations. We have chosen to consider either length or width separately to simplify the analysis. Also, the transformations for different Δz intervals are independent. Thus, for the transformation analysis all cracks are assigned to elements of a two-dimensional matrix of the form

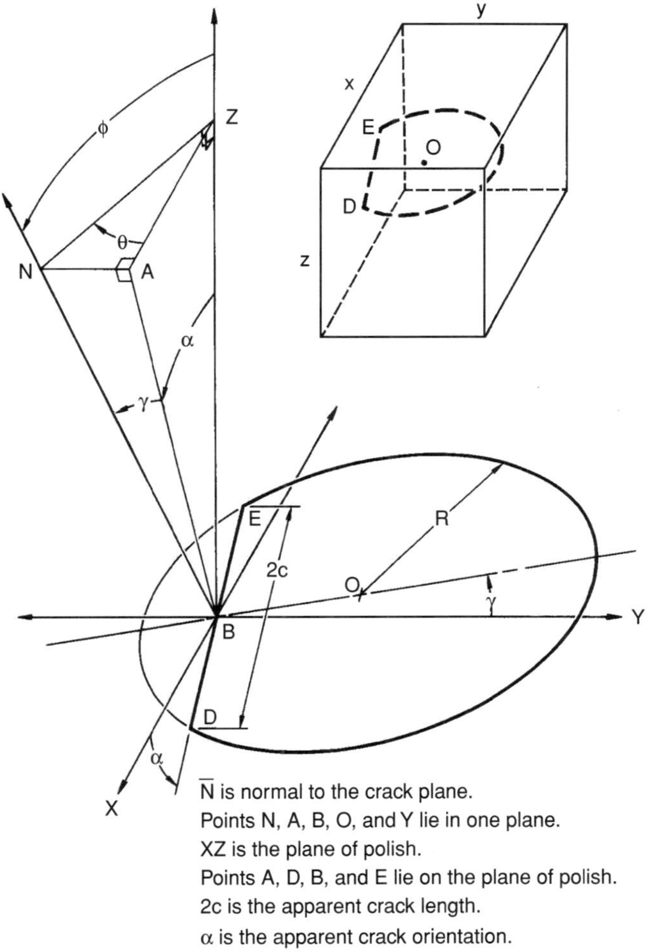

Figure 7.10. Circular crack intersecting the plane of polish.

$$n(\alpha, c) \qquad \text{or} \qquad n(\alpha, w), \tag{7.21}$$

where n is the number of cracks per unit area in the plane of polish in the orientation angle interval between α and $\alpha + \Delta\alpha$ and the size interval c and $c + \Delta c$ or the width interval w and $w + \Delta w$.

A sample crack size distribution observed in Armco iron is shown in Figure 7.11 for a 3.8 GPa impact. Here, the orientation is suppressed, and each curve represents $n(c)$ for a particular Z.

The crack distribution of interest is the volume distribution of cracks, specifically, $N(\phi, \theta, R, W)$, where ϕ is the angle that the normal to the actual crack surface makes with the Z axis, θ is the angle by which the normal to the crack plane is rotated around the Z axis, R is the true radius of the circular crack, and W is the true crack opening displacement (see Figure 7.10). Then N is the number of cracks per unit volume in the appropriate intervals of ϕ, θ, R, and W The following analysis shows the development of a procedure for performing the transformation from n, the number of cracks per unit area on the cross section, to N, the number of cracks per unit volume.

The analysis is based on the following simplifying assumptions:

1. The cracks are penny-shaped.
2. The crack distributions are independent of θ, the angle of rotation about the direction of wave propagation. Thus, it is assumed that there is an axis of symmetry for the orientation distribution.
3. The distributions do not vary in directions normal to the Z axis, that is, the wave propagation is one-dimensional and planar. Thus, a dominant direction of propagation and a reasonably homogeneous and isotropic material are assumed.
4. The crack length and width distributions may be handled separately.

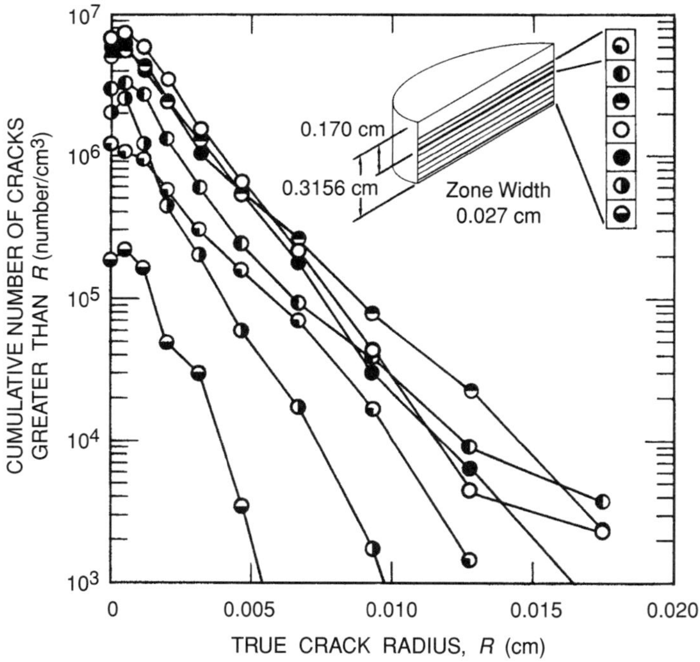

Figure 7.11. Crack size distribution in zones near the spall plane in an Armco iron target after a one-dimensional impact: Shot S25.

In the analysis, we consider only the length distributions. The same transformation may be applied separately for the width. We follow Scheil's method in proposing to divide the continuous functions n and N into discrete ranges and relate the values in those ranges by a matrix $A_{ij}^{\kappa m}$:

$$n(c_j, \alpha_i) = \sum A_{ij}^{km} N(R_m, \phi_k), \tag{7.22}$$

where $n(c_j, \alpha_i)$ is the number of cracks observed on the cross section with apparent radii between c_j and c_{j+1}, and with apparent inclinations between α_i and α_{i+1}; $N(R_m, \phi_k)$ is the number of cracks per unit volume with radii between R_m and R_{m+1}, and with true inclinations between ϕ_k and ϕ_{k+1}; and $A_{ij}^{\kappa m}$ is a fourth rank tensor.

The expression for $A_{ij}^{\kappa m}$ can be separated into the product of two matrices F_j^m and G_i^k (Seaman et al. [1978b]):

$$A_{ij}^{km} = F_j^m G_i^k, \tag{7.23}$$

where F_j^m contains only length quantities and G_i^k contains only angular quantities. F_j^m is identical to the transformation matrix used for spherical inclusions [See Scheil, 1931, 1935]. The evaluation of F_j^m and G_i^k as well as an algorithm for performing the inverse of the transformation in Eq. (7.23) has been written into the BABS2 computer program (Seaman et al. [1978b]).

The observed surface distributions shown in Figure 7.8 for Armco iron were smoothed and then transformed with the BABS2 computer code. The resulting distributions were summed over all orientations to reduce the numbers to functions only of crack radius and position in the specimen. These distributions, which are shown in Figure 7.11, indicate an exponential relation between cumulative number and crack radius.

The transformation described herein calculates true size and inclination of cracks or similarly shaped inclusions from counts of the cracks seen on a cross section. The transformation requires the calculation of two matrices: one that contains size terms only and is identical to Scheil's matrix for transforming spherical void counts; and a second one that contains only inclination terms. The accuracy of the method was demonstrated by comparisons to an exact analytical case.

7.2.2.2. Discussion

In brittle fracture under high-rate loading conditions, there is a range of fracture types. In Armco iron the microfractures occur across grains and growth is actually a coalescence of many microcracks to form a longer crack. Quite different behavior has been observed in the other materials described in Chapter 8. For most ductile materials, the cumulative crack size distribution has an exponential form similar to that which is seen with ductile void growth. This shape for the crack size distribution suggests that the crack growth rate is not strongly size dependent (the void growth threshold is independent of the void size). For more brittle materials such as the ceramic ZnS (exhibited in Figure 7.12), the size distribution has two exponential portions connected by a nearly horizontal section. These results were not obtained from a planar impact, so the conditions are more complex than those of a one-dimensional test. However, we believe that the

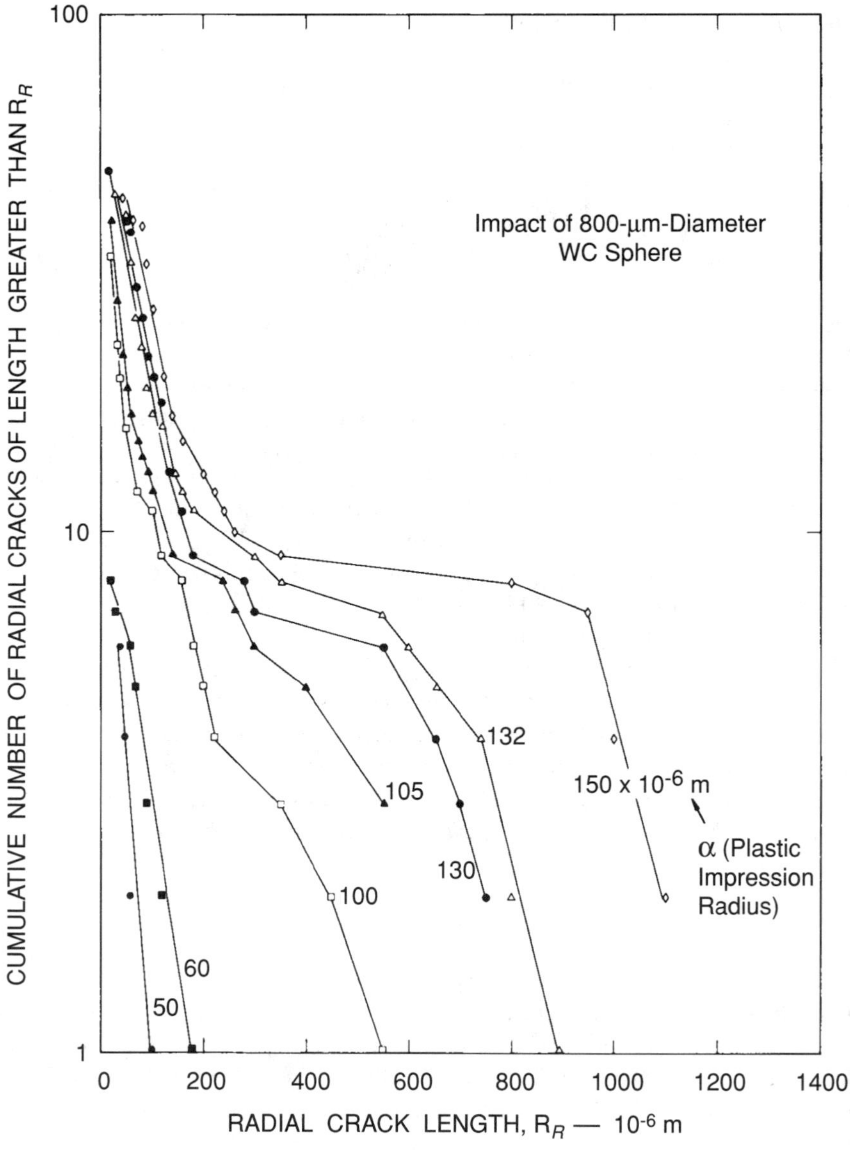

Figure 7.12. Measured size distributions of radial cracks in a zinc sulfide (ZnS) ceramic produced by 800-μm-diameter WC spheres at various impact velocities.

cracks with a radius less than about 200 μm were nucleated but did not grow because the stress reached was not sufficient to bring these cracks to critical conditions for growth (according to linear elastic fracture mechanics). But the larger cracks all grew at essentially the same speed, so the size distribution simply shifted to the right during the loading.

Our experience shows that it is difficult to obtain incipient damage from plate impact tests with brittle materials. This indicates that there is a narrow range in tensile stress between no damage and a fairly high level of damage. For this reason, the brittle fracture experiments are much harder to perform and to control than similar ductile fracture experiments.

7.2.3. Damage Processes in the Model

Models to represent brittle fracture have been constructed with many levels of detail. The simplest models account well for mechanics (energy and momentum) because these are common to all materials. But models are built based on various amounts of information about the specific material under study. A simpler model requires little, if any, experimental data. A complex model can require information that demands specific experiments just for the model development, but once developed and calibrated, the model can provide much more predictive detail. For example:

1. The model may account in detail for mechanics aspects of the fracture processes, such as the energy and momentum balance, and some macroscopic aspects of the material behavior such as the yield strength, toughness, and density. Grady et al. [1981, 1982, 1985, 1988, 1990] have developed a model of this type. The model can provide the average size of fragments after a fracture event.
2. A model, such as SRI BFRACT, accounts for mechanics features and also for the nucleation, growth, and coalescence processes characteristic of the fracture in a specific material. In this case, the results of the model calculations can provide the crack size distribution (detailed listing of the numbers and sizes of microcracks) and their orientations.
3. More metallurgically based models, such as that of Tonks et al. [1998] for tantalum, can be built to account for mechanics, nucleation and growth of microcracks, as well as for the specific grains and texture of the material under study. In this case, the computed results can provide even more detail about the fracture event than the previous two examples.

Many researchers have developed detailed microfracture models for a wide range of applications. Davison, Stevens, and Kipp [1977] derived a general approach to such models. Margolin [1983] has developed a model for an elastic material with distributions of pre-existing penny-shaped cracks in arbitrary orientations. The model accounts for both tensile opening (mode I) and shear sliding (mode II) of the cracks. The model, which provides an explicit solution for the stress state, does not account for crack nucleation, and it assumes that crack growth occurs at constant velocity. A similar elastic-brittle model has been derived by Krajcinovic and Fonseka [1981] for application to concrete fracture. Kipp, Grady, and Chen [1980] and Grady [1982] have also developed a brittle fracture model that has some micromechanical features. With this model, they were able to show that the fracture strength is proportional to the one-third power of strain rate (as observed experimentally) and to reproduce the measured dependence of fragment size on strain rate.

Evans, Ito, and Rosenblatt [1980] have given an example of a detailed study of crack formation in ceramic under impact using a combination of analytical and experimental efforts. A similar study was undertaken by Kumano and Goldsmith [1982] who also derived a model to represent their findings. Several Russian investigators (see, e.g., Dremin and Molodets [1980]) are actively pursuing the development of microfracture models, but they do not treat the cracks in detail.

Meyers and Aimone [1983] have reviewed fracture models for spalling in metals. Johnson [1981] and coworkers have been developing models for fracture in brittle materials (as well as ductile materials) and have a model with many features similar to the BFRACT model described here. Dienes [1981, 1985, 1996] continued the modeling effort initiated with Margolin [1980] and extended it from layered rock fracture to the fracture of other materials, including rocket propellant. This model is like BFRACT but has somewhat different nucleation and growth processes and has many crack orientations instead of the single orientation of BFRACT.

Wilkins, Streit, and Reaugh [1980] developed a cumulative-strain-damage model based on the analytical results of Rice and Tracey [1969] for the growth of spherical voids. But the model uses this growth relation to represent the gradual propagation of a macroscopic crack, yet without keeping track of the crack size or orientation. The model provides only a level of damage in each computational finite-element. They used the model to predict crack growth in 6061–T651 aluminum.

Perzyna [1986] and Wang and Lam [1995] have described a brittle fracture model in which the crack opening is proportional to the crack radius—a simplification which gives brittle fracture characteristics similar to those of the DFRACT ductile fracture model described in Section 7.1. In addition, Wang and Lam have made the growth process dependent only on the mean stress and have assumed that the orientations of the cracks are random, thus further minimizing the effects of crack and stress orientations. This approach is probably appropriate when the mean stresses are far above the threshold stress: in such cases the crack orientations may be uniformly distributed in solid angle as in Figure 7.6.

Espinosa [1995], Espinosa and Brar [1995], and Espinosa, Zavattieri, and Emore [1996] have presented a model for brittle fracture of ceramics in which they allow for nucleation and growth of cracks under tensile and shear loading and in many orientations simultaneously. This model appears to explain many of the characteristics of ceramic fracture under shock loading. The growth rate processes in the model are based on the elastic analysis of Freund [1990], rather than on observations of fracture data. This assumption of the crack growth rate is essential for work in ceramics because the rate is generally too high to be obtainable from post-test observations.

The SRI model, BFRACT, for dynamic tensile fracture and fragmentation, has been developed for the detailed simulation and study of brittle fracture processes in elasto-plastic materials (see Barbee et al. [1970], Seaman et al. [1976], Seaman et al. [1985], and Curran et al. [1987]). "Brittle" fracture is defined here as fracture occurring on the micro level with cleavage morphology rather than with a ductile void morphology. Included in the model are a threshold for the initiation of damage, a nucleation process for forming microcracks under tension, growth processes that are dependent on the tensile stress, coalescence and fragmentation processes with the resulting fragment size distribution, and stress relaxation associated with the developing damage. Cracks occur in a range of sizes at nucleation and throughout the calculation.

The processes in the present BFRACT model have represented, with reasonable fidelity, high-rate fracture in polycarbonate (Curran et al. [1973]; Seaman [1982], oil shale (Murri et al. [1977]), novaculite quartz (Shockey et al. [1974]), rocket propellant (Seaman et al. [1985], Seaman et al. [1998]), as well as beryllium and Armco iron (Barbee et al. [1972], Seaman et al. [1976], Shockey et al. [1973b]). Computations with the model agree satisfactorily with measured stress or particle velocity histories obtained during plate impact experiments and they also match the measured numbers and sizes of cracks observed in post-test examination of the impacted samples.

In this section, the analytical and experimental bases of the major equations of the model are presented. These equations are implemented in a subroutine BFRACT, which has been used with several wave propagation computer programs to simulate fracture processes. BFRACT acts as a stress–strain relation in such programs. BFRACT accounts for the following fracture processes, derived based on experimental observations:

1. *Initiation of fracture:* When the tensile stress normal to the crack plane exceeds a critical level, fracture begins.
2. *Nucleation of cracks:* Cracks are nucleated with a range of sizes and as a function of the tensile stress level.
3. *Growth of cracks:* The cracks grow or extend according to several different growth rate processes, depending on the ductility of the material.
4. *Coalescence and fragmentation:* As the number and sizes of cracks increase, fragments form until the entire material disintegrates into fragments.

Each of these processes is explored in the following paragraphs. The effects of the growing damage on the stress–strain relations are then examined.

7.2.3.1. Cracks and Crack Size Distributions

The microcracks in the model are penny-shaped: circular and flat. This shape was actually observed in polycarbonate (Curran et al. [1973]), a transparent material that lends itself to direct visual examination. Because a great number of cracks occur (usually 10^4 to 10^8 cm^{-3}), they are treated statistically, rather than individually. An indication of the form of a representative crack size distribution is shown in Figure 7.11 for a disk-shaped sample of Armco iron impacted by a flat plate to a shock stress of 3.8 GPa. The cracks observed in this sample varied in orientation as well as length, but the angular variation is suppressed in this plot. The several distributions shown represent different stages of growth (those further to the right have grown longer). To represent the observed crack sizes, the size distribution is usually given in the analytical form:

$$N_g = N_0 \exp(-R/R_1), \tag{7.24}$$

where N_0 and N_g are the total number of cracks per unit volume and the number per unit volume with a radius greater than R, and R_1 is the shape parameter for the distribution. For computer simulations, the distributions are represented by a series of line segments in the $N-R$ plane so they need not remain exponential throughout a given calculation.

The statistical nature of the crack distribution is indicated by the fact that N is a crack density (number per unit volume). Thus, the model assigns a crack density at each point in the material. The model is a continuum model, and the crack distribution acts as an internal state variable (i.e., it is an additional property of the continuum). This statistical approach is especially appropriate for finite-difference or finite-element calculations in which the material is represented by a number of small cells or elements. The crack density is independent of cell size, whereas the common treatment of individual cracks is highly dependent on cell size.

In the model, the cracks are presumed to open elastically under a stress σ_n normal to the crack plane (Sneddon and Lowengrub [1969]):

$$\delta = \frac{4\left(1-\nu^2\right)}{\pi E} R\sigma_n, \tag{7.25}$$

where δ is one half the maximum separation of the crack faces and E and ν are Young's modulus and Poisson's ratio. This elastic opening process only roughly approximates real material behavior in which cracks in fact open with a finite velocity and may remain open when the load is removed.

The crack faces form an ellipsoid with three semi-axes, δ, R, and R. Thus, the volume of a crack is

$$V_{1c} = \frac{4\pi R^2\delta}{3} = \frac{16\left(1-\nu^2\right)}{3E} R^3\sigma_n. \tag{7.26}$$

The volume of the entire crack size distribution can be obtained by combining Eqs. (7.24) and (7.26) into the following integral:

$$V_c = \int_{R_a}^{R_b} V_{1c} dN = \frac{16(1-\nu^2)}{3E} \sigma_n \int_{R_a}^{R_b} R^3 dN = \frac{16(1-\nu^2)}{3E} \sigma_n \tau_z = \varepsilon_n^c, \tag{7.27}$$

where R_a and R_b are minimum and maximum radii in the distribution. The quantity τ_z is introduced to represent the effect of the total crack distribution. The strain ε_n^c represents an average elongation strain normal to the crack plane.

In the current model, all cracks in the distribution in a given computational cell lie in a single orientation, although they may be considered to represent some range of orientation angles (in spherical coordinates) in the actual material. For each material element, the crack orientation is fixed normal to the maximum tensile stress at the time that any stress first exceeds the initiation stress criterion for nucleation.

7.2.3.2. Nucleation and Growth Processes

Two alternative types of nucleation and growth processes are provided in the model. Their characteristics are:

1. A constant threshold stress σ_{go} for growth. All sizes of cracks are nucleated and all grow when this tensile stress is exceeded. This process is appropriate for crack growth in ductile materials, in which considerable plastic flow accompanies cracking.
2. A fracture mechanics threshold for growth. Only cracks above a stress-determined size can nucleate and grow. This threshold seems to fit the more brittle materials, such as oil shale, ceramic, and novaculite (a quartzite rock).

In the model, somewhat different nucleation and growth relations are provided for these two types of growth threshold.

Nucleation in the model occurs as the new cracks are added to the existing set. These new cracks occur in a range of sizes with the size distribution given by Eq. (7.24). At nucleation, the parameter R_1 equals R_n, a material constant indicating the inherent flaw size. The rate at which cracks are nucleated is governed by the following function:

$$\frac{dN}{dt} = \dot{N}_0 \left\{ \exp\left(\frac{\sigma_n - \sigma_{n0}}{\sigma_1} \right) - 1 \right\}, \tag{7.28}$$

where dN/dt refers to the rate of increase of N_0, the total number of cracks per unit volume; N_0 and σ_1 are material parameters with the dimensions of number/m^3/s and Pa; and σ_n is the average stress normal to the crack plane during the time interval Δt. For the constant threshold stress, the size distribution nucleated at any time extends from $R = 0$ to $R_n^{\max}$, which is the maximum nucleated crack size (a material input parameter). For the fracture mechanics threshold, the size distribution nucleated in a time increment extends only from $R = R_c$ which is a critical size defined below, to $R = R_n^{\max}$. Although Eq. (7.28) has been successfully applied to the modeling of several materials, alternate forms should be considered. Experiments such as those of Senf, Straßburger, and Rothenhäusler [1995] with associated numerical simulations may lead to other nucleation functions.

The crack growth law derived from experimental data on cleavage crack growth in Armco iron and beryllium is

$$\frac{dR}{dt} = \left(\frac{\sigma_n - \sigma_{go}}{4\eta} \right) R, \tag{7.29}$$

where η is a growth factor with the dimensions of viscosity, and σ_{go} is a growth threshold stress. For σ_n less tensile than σ_{go}, no growth occurs. In our model, the crack velocity is not allowed to exceed the Rayleigh wave velocity in intact material. The growth law expressed by Eq. (7.29) was derived theoretically by Poritsky [1952] for the expansion of spherical voids in a viscous fluid and hence may be termed a viscous growth law. The fact that this growth law describes the growth of cleavage cracks in Armco iron and beryllium suggests that, in these materials, microcrack growth is accompanied by significant plastic flow in the high stress regions at the tips of the microcrack.

An alternate growth rate in which the growth rate is proportional to the shear wave speed C_s, can be expressed as

$$\frac{dR}{dt} = \begin{cases} fC_s & \text{for} \quad \sigma_n > \sigma_{go}\,, \\ 0 & \text{for} \quad \sigma_n \le \sigma_{go}\,, \end{cases} \tag{7.30}$$

Here f is subject to the condition that $0 < f \le 1$, thus ensuring that the crack propagation velocity is limited by the shear wave velocity in the material. Several investigators, including the present authors, have used this growth rate law, especially for very brittle materials. The growth law expressed by Eq. (7.30) is characterized by a sharp discontinuity in the damage variable just as the stress reaches the threshold value, σ_{go}. This discontinuity can cause difficulties during the course of numerical simulations. To alleviate the difficulties and facilitate the numerical solution, it is useful to allow the growth rate to increase from zero to fC_s over some narrow range of stress, rather than suddenly at the threshold.

Another alternate growth rate for elastic or quasi-elastic behavior is that derived by Freund [1973 and 1990] for the case of a semi-infinite crack in an elastic material subjected to an incident tensile stress wave. Freund's results indicate that the instantaneous value of the mode I stress intensity factor for a moving crack is equal to the product of the stress intensity factor of an equivalent stationary crack times a universal function of the crack tip velocity. Tsai [1973] derived the universal function for a penny-shaped crack moving at a constant velocity. Tsai's solution exhibits a trend of decreasing stress intensity factor with increasing crack velocity. The growth rate relations that stem from the results of Freund and Tsai were used by Gran and Seaman [1986] and by Antoun [1991] to represent dynamic crack propagation in concrete.

When an exponential distribution of flaws, represented by Eq. (7.24), grows according to the growth law expressed by Eq. (7.29), the distribution remains exponential. Thus, in any material where the crack size distribution is observed to remain exponential for a range in levels of damage, a viscous growth law is indicated. For fairly ductile materials such as Armco iron, in which considerable plastic flow accompanies crack growth, σ_{go} was observed to be a material constant; that is, the nucleation and growth process indicated by alternative (1) above was taken.

Growth with the constant threshold stress law is obtained by integrating Eq. (7.29) and taking σ_n as the time average of the solid stress normal to the crack plane:

$$R_{m+1} = \begin{cases} R_m \exp\left(\dfrac{\sigma_n - \sigma_{go}}{4\eta}\Delta t\right) & \text{for} \quad \sigma_n \geq \sigma_{go}\,, \\ R_m & \text{for} \quad \sigma_n < \sigma_{go}\,, \end{cases} \tag{7.31}$$

where Δt is the time increment, and R_m and R_{m+1} are radii at the beginning and end of the time increment, respectively. This expression shows that every radius in the distribution increases by the same ratio. Therefore, the new distribution obtained after growth is still an exponential, with the new value of the size parameter R_1 obtained from the old one by Eq. (7.31). Armco iron is an example of a material where crack size distributions have been observed to remain essentially exponential throughout their growth history (see Figure 7.11).

For more brittle materials, like rock, σ_{go} is the critical stress for crack growth according to linear elastic fracture mechanics, as in alternative (2) above:

$$\sigma_{go} = \sqrt{\frac{\pi}{4R}} K_{1c}, \tag{7.32}$$

where K_{1c} is the fracture toughness, a material constant. With Eq. (7.32) in the growth law, no growth occurs for cracks with a radius smaller than R_c,

$$R_c = \frac{\pi K_{1c}^2}{4\sigma_n^2}. \tag{7.33}$$

When the fracture mechanics form of the growth law (7.29) is integrated, accounting for the variation in σ_{go} according to Eq. (7.32), we obtain

$$\sqrt{R_{m+1}} = \left(\sqrt{R_m} - \sqrt{R_c}\right)\exp\left(\frac{\sigma_n \Delta t}{8\eta}\right) + \sqrt{R_c}\,, \tag{7.34}$$

where R_{m+1} and R_m are values of R at the end and beginning of the time increment Δt. Here the final size R_{m+1} is related in a complex way to R_m, so that the final distribution is not an exponential. To provide fidelity to the growth process in the model, the size distribution is represented by a series of exponential segments. Then the growth of the R values at the ends of the segments is computed by Eq. (7.34).

After nucleation and growth during a time increment, the nucleated crack size distribution is added to the growing distribution to form a single distribution representing the current state of damage in the material. This combined distribution may grow during the next time increment.

7.2.3.3. Coalescence and Fragmentation

"Coalescence" refers to the joining of two or more cracks to form a larger crack. "Fragmentation" represents the completion of the coalescence process and the formation of a spall plane. Coalescence and fragmentation occur during the late stages of damage development at which time fragments may form and separate from the remainder of the material. In the BFRACT model, coalescence acts like an enhanced growth rate, with the additional feature that the number of cracks decreases. When fragmentation occurs (a criterion for complete damage is reached), the crack size distribution is transformed to a fragment size distribution and the material cannot thereafter sustain any tensile stresses. Here we mention a few of the many researchers who have contributed to the theories of coalescence and fragmentation and then outline in detail the processes implemented in BFRACT.

For many years, Grady and his co-workers have made significant contributions to the development of fragmentation models (Grady [1981, 1982, 1985, 1988, 1990, 1995]; Kipp and Grady [1985, 1996]). The approach is mainly based on momentum and energy balances, but also includes some readily available material information. This analytical model describes dynamic fragmentation in both brittle and ductile materials. The model does not represent the details of the evolution of fracture processes and proceeds directly to full separation. The model results provide an average fragment size based on the energy induced in the sample under dynamic loading. The fragmentation model developed by Grady et al. has been validated through comparisons with several experiments and observations. Specifically, Grady and Kipp [1995] impacted metallic spherical samples onto nonmetallic plates and studied the breakup of the spheres with multiple flash radiography. Fragmentation of the samples correlated well with model predictions.

Glenn and Chudnovsky [1986] have also contributed to the theory of fragmentation. They developed a model similar to that of Grady et al., but added elastic strain energy to the energy balance equations. An alternate approach with more detail of the interaction of the cracks to produce fragments was given by Englman, Jaeger, and Levi [1984]. Zhang and Jin [1998] have provided an analytical approach for predicting the number and size of fragments obtained during fragmentation. Garrett, Rajendran, and Last [1996] have developed a coalescence and fragmentation model for steels in which ductile void growth is the dominant damage mechanism.

Tonks et al. [1998] have developed a model for growth and coalescence of brittle cracks based on the nucleation-and-growth concepts in the Johnson [1981] model. They fitted their model to spall experiments in uranium and determined an apparent spall strength for each sample in the range of 2 to 3 GPa. The results agree fairly well with the experimental results of Cochran and Banner [1977] and Grady [1986].

The BFRACT fragmentation processes were developed to represent observations in novaculite quartz (Shockey et al. [1974]). We begin by describing these observations qualitatively. Later, we derive equations to describe coalescence based on the experimental observations. In the novaculite experiments, the fragments occurred in a range of sizes from 35 μm to 5 mm, and all sizes had approximately the same aspect ratio, as shown in Figure 8.10 in Section 8.3. The physical process of fragmentation occurs when the cracks become so large that they begin to intersect other cracks. They may intersect in the same plane, thus forming larger cracks, and they may intersect at right angles, forming corners of fragments. Also, cracks in the same orientation, but on different planes, may coalesce by developing crack extensions out of the plane to join nearby cracks. Thus, a family of cracks in one orientation can coalesce and form a rough, multifaceted spall plane. These three coalescence possibilities are illustrated in parts (a) through (c) of Figure 7.13. Because cracks in only one orientation are permitted in the current version of the model, only coalescence of the first and third types are possible. At present, only the third type of coalescence is considered. In this coalescence pattern, fragments develop in each cell that fully fractures. Fragmentation in neighboring cells leads to the formation of a fragmented, or rubblized region, as shown in Figure 7.13(d). Fragmentation along a horizontal line or plane of cells leads to the formation of a spall plane, as shown in Figure 7.13(e), possibly with some loose fragments on the plane.

The physical processes by which cracks intersect and form fragments are not considered explicitly in the model; rather, the model provides only for a gradual transition from undamaged to fully fragmented material and an accounting of the fragment size distribution at the end. Let us consider the fully fragmented state first.

The fragments occur in a size distribution that we presume is related to the crack size distribution. Furthermore, we assume that for each size range, there is a constant ratio β between the number of cracks and the number of fragments:

$$\beta = \frac{N_0^f}{N_0}, \tag{7.35}$$

where N_0^f is the total number of fragments per unit volume. Thus, we expect small cracks to coalesce with small cracks to form small fragments, and large cracks to coalesce with large cracks to form large fragments. Chunky fragments are usually formed by a crack face above and below and by crack extensions on the vertical sides. Because each crack forms a side of two fragments and each fragment requires two crack faces, one crack is associated with each fragment;

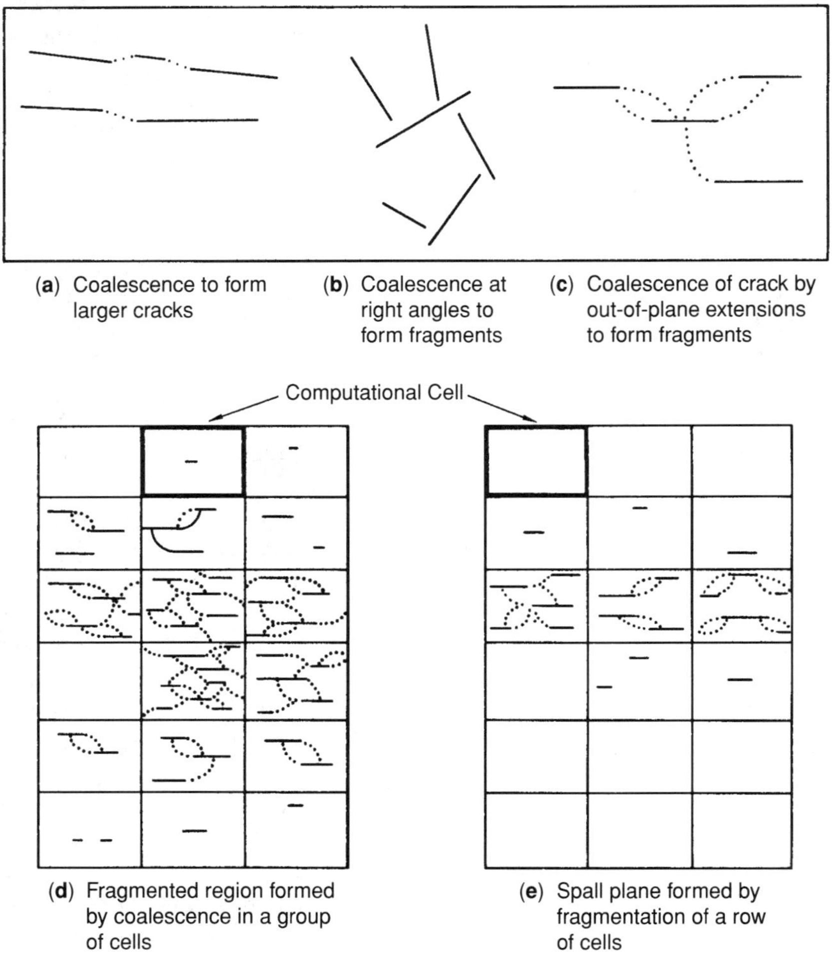

Figure 7.13. Coalescence and fragmentation processes envisioned for the model.

therefore, β is approximately one. Similarly, the fragment sizes are related to the crack sizes by a factor γ.

$$\gamma = \frac{R^f}{R}, \tag{7.36}$$

where R^f is the radius of the top or bottom of the fragments. The cracks forming the fragment top and bottom have about the same area as these fragment surfaces, so γ is about one.

Determining the fragment volume requires knowledge of the fragment shape. Assuming the top and bottom of the fragment to be approximately circular provides a means of calculating the fragment area as $\pi(R^f)^2$. The thickness of the fragment is assumed to be related to the radius R^f by the proportionality factor α such that the thickness is equal to αR^f. The fragment volume, V_1^f, can then be expressed as

$$V_1^f = \alpha\pi\left(R^f\right)^3 = T_F\left(R^f\right)^3. \tag{7.37}$$

Here the coefficients in the volume expression have been combined into a single constant T_F, which characterizes the aspect ratio of the fragments. Our experimental measurements indicate that T_F should be between 1 and 3. The relative volume of fragments, τ_f, is obtained by integration over N^f.

$$\tau_f = T_F \int_{R_a}^{R_b} \left(R^f\right)^3 dN^f = T_F \beta\gamma^3 \int_{R_a}^{R_b} R^3 dN = T_F \beta\gamma^3 \tau_z. \tag{7.38}$$

Thus, the fragment volume τ_z is closely related to the crack volume factors V_c and τ_z, as expressed in Eq. (7.27) and derived from the crack size distribution. When $\tau_f = 1$, the entire volume is fully fragmented. The process of gradual coalescence and fragmentation is treated in the model by requiring that τ_f always represent the portion of the material element that has fragmented.

After the fragmentation process was developed, a special coalescence process was constructed (Guéry and Seaman [1996]). This new coalescence process acts to produce enlarged cracks and also to reduce the number of cracks. It acts to alter the crack size distribution, but not to provide for the interaction of specific cracks. The coalescence rate process is assumed to have the same stress dependence as the growth equation; therefore, the form of the equation is

$$\left.\frac{dR}{dt}\right|_{coalescence} = B_c \left.\frac{dR}{dt}\right|_{growth}, \tag{7.39}$$

where B_c is a dimensionless coalescence factor with a magnitude on the order of 1. The number of coalesced cracks ΔN_c at any radius R is obtained by requiring that the surface area of the microcracks remain constant during coalescence:

$$\Delta N_c = \Delta N_0 \left(\frac{R_0}{R_c}\right)^2, \tag{7.40}$$

where R_0 is the radius after growth, R_c is the radius after coalescence, and ΔN_0 is the number at R before coalescence. The exponent of the radii in Eq. (7.40) reflects the assumption that a constant surface area is maintained. Trials with powers of 1, 2, and 3 showed only small differences among the resulting crack size distributions. The current BFRACT model uses this coalescence process in addition to the fragmentation process.

7.2.4. Stress Calculations Coupled with Damage

The stress calculations for intact material are conducted until the stresses exceed a threshold level and damage is initiated. At that time the plane of fracture is determined, as we will discuss later. For damaged material, the following sequence of calculations is performed:

1. Transform the average stresses in the global X, Y coordinate system to stresses in the solid material in directions normal and tangential to the fracture plane.
2. Compute the stresses in the solid intact material, accounting for the stress relaxation (material softening) due to the nucleation, growth, and opening of cracks.
3. Transform the stresses back to the global coordinate system, accounting for the reduction in area of intact material.

Each of these steps is outlined in the following sections, and a special calculation is provided for the case of full fragmentation resulting in a spall plane.

7.2.4.1. Stress–Strain Relations

The stress–strain (constitutive) relations include the usual elastic–plastic relations for intact material, plus a method to account for the effect of the developing damage on the stresses. We first present the stress–strain relations for undamaged material, and then proceed to those for damaged material.

The pressure P_s or average stress in the solid material is given by a Mie–Grüneisen relation that provides for linear thermoelastic behavior,

$$P_s = K_1\mu + K_2\mu^2 + K_3\mu^3 + \rho_s \Gamma E, \tag{7.41}$$

where

$$\mu = \frac{\rho_s}{\rho_{s0}} - 1$$

is the compressive volumetric strain; K_1, K_2, and K_3 are the bulk modulus series, Γ is the Grüneisen's ratio, ρ_s and ρ_{s0} are current and initial solid density, and E is the internal energy.

The deviatoric stress–strain relation is based on elasticity and von Mises plasticity with nonlinear work hardening, as discussed in Section 6.1.1.

For calculating stress–strain relations modified by damage, the damaged material is viewed as a composite of intact material with some separated fragments and material with reduced strength in the vicinity of active cracks. Two sets of stresses are considered: solid and continuum. The solid stresses are the average of those stresses appearing in the intact material; continuum stresses are the average stresses on a cross section of a combination of intact and fractured material. Continuum stresses are transmitted out of the subroutine to the calling program because they represent the stresses acting between computational cells.

The stresses are computed in two steps, and each step is affected by the amount of damage. In the first step, the imposed strain is separated into a portion taken by the intact solid material, ε_s, and a portion representing the crack opening displacement, ε_c. The separation of the total strains into solid and crack opening components is very important for small amounts of damage. The solid stresses are computed from the strain component ε_s. The average continuum stresses are then computed from the solid stresses, using the fraction of intact material as a reduction factor. This transformation from solid to continuum stresses dominates near the point of fragmentation or complete separation.

For the first step of calculating the stress in the solid material, the gross strain increments normal to the crack plane are separated into three portions following a suggestion of Herrmann [1971]:

$$\Delta\varepsilon_n = \Delta\varepsilon_n^s + \Delta\varepsilon_n^c + \Delta\varepsilon_n^{op}, \tag{7.42}$$

where the superscripts indicate strains taken by the solid material (superscript s), the crack opening (c), and the strain associated with the spall opening (op) that occurs in full separation. $\Delta\varepsilon_n^s$ is used in the elastic-plastic constitutive equations for the solid. The crack opening strain increment is computed as the change in crack volume from Eq. (7.27). The third term is simply the separation of the broken material, calculated on a true strain basis. Note that these are continuum strain increments—averages over the material in the cell—not strains defined at a point.

The strain increment $\Delta\varepsilon_n^s$ initially increases (in tensile magnitude) with increasing tensile loading. After cracks begin to nucleate and grow, $\Delta\varepsilon_n^c$ increases. With continued tensile straining, larger portions of the strain $\Delta\varepsilon_n$ are taken in crack opening $\Delta\varepsilon_n^c$. As the damage increases, the tensile solid strain $\Delta\varepsilon_n^s$ reaches a maximum and then begins to decrease. When separation occurs, the cracks close, so that $\Delta\varepsilon_n^c$ becomes zero. Simultaneously, $\Delta\varepsilon_n^{op}$ increases to fulfill Eq. (7.42).

The second step, the transformation from solid to continuum stresses, is accomplished using the damage quantity τ_f from Eq. (7.38):

$$\sigma_n^g = \sigma_n\left(1 - \tau_f\right), \tag{7.43}$$

where σ_n^g and σ_n are continuum and solid normal stresses on the plane of fracture. (No other stresses are modified to account for damage.) This relationship between the solid and continuum stress is borrowed from the analysis of Carroll and Holt [1972b] for a porous material:

$$P_g = \frac{\rho}{\rho_s} P_s, \tag{7.44}$$

where P_g and P_s are continuum and solid pressures, and ρ and ρ_s are continuum and solid densities.

7.2.4.2. Spallation

When large amounts of damage occur, the fractures are assumed to coalesce and form a local spall plane within the computational cell. Specifically, the criterion is that spall occurs when $\tau_f = 1$, where τ_f is the fragmented fraction from Eq. (7.38). Then the state of stress at the spall plane is taken to be one of plane stress, with zero stress normal to the spall plane. The spalled state route in the model contains the calculations of the stresses, plus an incrementing of the spall opening strain $\Delta\varepsilon_n^{op}$. The stress state is obtained from the usual elastic relations for the case of a free boundary. If yielding occurs, the stresses are computed using Mises plasticity and a normality rule.

7.2.5. Solution Procedure

To provide stress–strain relations that are sensitive to the growing damage, a method is required for determining the stress tensor for an imposed strain increment while accounting for the nucleation, growth, and coalescence processes. The governing equations that must be solved to determine the stresses are

1. growth law, Eq. (7.29);
2. nucleation rate law, Eq. (7.28);
3. crack opening relation, Eq. (7.27);
4. pressure–volume-energy relation for intact material, Eq. (7.41), and a corresponding deviator stress relation from Section 6.1.1; and
5. strain decomposition, Eq. (7.42).

The solution method is essentially an integration of the nucleation and growth rate laws, combined with the auxiliary algebraic equations giving the stress–strain relations. Because of the complexity of the stress–strain and damage relations, a direct solution for the stresses is not possible in general, even for small time steps. The equations constitute a "stiff" system, which means that very small time steps may be required to obtain accurate and stable results. Three different methods with varying degrees of sophistication have been used to solve the BFRACT equations; each can be adjusted to give satisfactory results where the method is appropriate:

1. An implicit solution combining *iteration and subcycling*, using the solid strain as the basic unknown. This method is appropriate when there is basically one independent unknown; hence, only cracks in one plane and cracking from either tensile or shearing strain, but not both. The process begins with an estimate of the solid strain increment normal to the crack plane—the one independent variable. With this esti-

mate of the strain increment, the stresses are computed, followed by all the damage factors. After computation of the crack opening strain, we can test the validity of our estimate with the strain decomposition equation, Eq. (7.42). If Eq. (7.42) is not satisfied, another estimate is made and the process is repeated, iterating until Eq. (7.42) is satisfied within an acceptable pre-specified tolerance. In performing the iterations, the second and subsequent estimates are made using the Newton-Raphson method until the estimates bracket the correct value. Thereafter, a *regula falsi* scheme is used until a converged solution is obtained. If the iterations are unsuccessful after a prescribed number of attempts, a subcycling process is initiated whereby the time step is reduced and the process is repeated until convergence.

2. A *direct integration* of the differential equations for nucleation, growth, coalescence, and energy change using a standard method such as Runge-Kutta. The method used must include checks on the accuracy because the stable time step cannot be readily computed in advance. If insufficient accuracy is obtained at any step, the time step is reduced, and the integration is repeated. In this approach more powerful solution techniques such as the semi-implicit method, the Rosenbrock-Kaps-Rentrop method (e.g., Press et al. [1994]), or a similar method appropriate for stiff differential equations should be employed. Many crack orientations can be used: the basic unknowns are the numbers and sizes of the cracks on each plane.
3. A direct solution in which a *compliance* tensor is derived for the cracked material, and the stress is obtained by inverting this compliance tensor to form a stiffness tensor, and then multiplying it times the imposed strain increment.

To illustrate this latter approach, let us consider the solution for an elastic-brittle material for which the crack opening strain ε^c for a penny-shaped crack is proportional to the applied normal stress σ_n as follows:

$$\varepsilon^c = \frac{16\left(1-\nu^2\right)}{3E} NR^3\sigma_n = C^c\sigma_n, \tag{7.45}$$

where ν and E are Poisson's ratio and Young's modulus, N is the number of cracks per unit volume, R is the crack radius, and C^c is a compliance modulus. Two additional relations are now introduced. First, we introduce the consistency condition requiring that the total strain be equal to the sum of the elastic and cracking strains (ε^e and ε^c) in any orientation:

$$\varepsilon^t = \varepsilon^e + \varepsilon^c. \tag{7.46}$$

Next, we introduce the relation for the elastic compliance modulus C^e:

$$\varepsilon^e = C^e\sigma_n. \tag{7.47}$$

Substituting from Eqs. (7.45) and (7.47) into Eq. (7.46), and taking the time derivative of (7.46), we obtain

$$\frac{d\varepsilon^t}{dt} = C^e \frac{d\sigma_n}{dt} + \frac{dC^c}{dt}\sigma_n + C^c \frac{d\sigma_n}{dt} = \left(C^e + C^c\right)\frac{d\sigma_n}{dt} + \frac{dC^c}{dt}\sigma_n. \quad (7.48)$$

Solving for the stress rate, we find

$$\frac{d\sigma_n}{dt} = \left(C^e + C^c\right)^{-1}\left[\frac{d\varepsilon^t}{dt} - \frac{3C^c}{R}\frac{dR}{dt}\sigma_n\right]. \quad (7.49)$$

With this relation, we can solve for the change in stress that corresponds to an imposed strain increment and the current state of cracking. We note here that the term containing $C^e + C^c$ gives the apparent stiffness assuming no change in crack size, that is, the elastic response of the cracked material. The negative term containing the crack growth rate provides for the decrease in stress caused by crack growth during the time step. For stable and accurate computations, the time step must be limited such that (a) the negative term in the brackets does not cause the stress to become compressive, and (b) R and C^c do not change significantly during the time step.

Because this method accounts for the change of damage during the time step, it can be stable and fairly accurate even in highly nonlinear problems. The apparent stiffness from this equation is always positive for no change in damage (accurately reflecting the usual elastic modulus formulations for a cracked material), but because of the term reflecting the changing damage, it also provides for the negative stress–strain slopes expected when damage is being accumulated during softening. This implicit method described above can be used as a standalone solution procedure, in which the damage-dependent modulus reflects merely the initial state of damage for the time step, or it can be used as the first part of a predictor–corrector approach.

Experience indicates that the set of equations comprising the fracture model is very nonlinear, especially near the threshold, thus requiring a very powerful method to obtain a converged and stable solution. Usually, the natural time step provided by the wave propagation code for an explicit solution procedure is somewhat longer than the stable time step for the damage solution. Therefore, a direct integration method (such as 2 above) requires some subcycling during critical periods of fast damage growth. During these same periods, methods 1 and 3 above may also require shorter time steps to provide adequate accuracy.

7.2.6. Current Studies for Amplifying the Brittle Fracture Treatment

In the sections above, we have outlined a fairly simple model to represent the nucleation, growth, and coalescence processes associated with the dynamic fracture of a brittle material. Naturally, there are ongoing efforts to develop models that fit better special materials and provide a more detailed representation of materials in general. Here we mention several approaches that are now being taken to improve brittle fracture models.

The *Mechanics* approach emphasizes the stresses, strains, and observable features of the fractures. Espinosa's [1995, 1996] multiple-plane method is notable here: his model allows for several orientations of cracks to be active simultaneously and treats both shearing and tensile damage. A similar model was constructed by Seaman et al. (Seaman et al. [1981, 1983, 1985, 1992], Seaman and Dein [1983], Shockey et al. [1984], Curran and Seaman [1985, 1986], Curran et al. [1987]) for shear banding in material undergoing large shear strains under high-rate conditions such as in projectile impacts and in the steel cylinders of exploding rounds. This model, termed SHEAR4 or SHEAR3D, has 9 active orientations for bands or cracks. The "bands" are shaped like penny-shaped cracks, but are formed when a region of the material undergoes excessive shearing strain. These bands also nucleate and grow, but the nucleation and growth are functions of shearing strain and strain rate. Currently there is a need to develop a model that combines shear banding and tensile crack opening because both these processes occur in projectile penetrations and in fragmenting cylinders. In many situations, shear banding occurs first, producing a shattered region. Then when tension occurs, the bands are opened and grown as cracks; eventually, the material is fragmented.

An important result in this mechanics approach is the simulation by Needleman and Tvergaard [1991] of the propagation of a crack by growth and coalescence of microvoids. This is an indication of our current state of knowledge about the growth of cracks in fairly ductile material. More work is required along this line to be able to use this knowledge to solve practical problems.

Metallurgical and experimental approaches continue to illustrate the complexity of the fracture processes and also show that fracture differs from one material to another and can also depend on details such as grain size, heat treatment, and other metallurgical features. The FRASTA (FRActure Surface Topographic Analysis) method of Kobayashi (Kobayashi et al. [1991, 1994, 1997], Kobayashi and Shockey [1991a, 1991b], Schmidt et al. [1997]) examines the cracked surfaces of a broken specimen to determine the pattern of cracking, the fracture toughness, the proportion of cleavage or more ductile types of fracture, and other features of the surface. Tonks [1993, 1995, 1998] is showing how to describe ductile fracture in tantalum based on a very detailed knowledge of the processes of plastic flow on grains in this complex material. Razorenov et al. [1998] noted the differences in spall behavior between single crystals and polycrystalline materials, and for differences in grain orientations.

Many studies have shown the complexity of the nucleation and growth processes, indicating the dependence of these processes on the inclusion content, heat treatment, temperature during the test, and the strain rate. Meyer and Aimone [1983] show a large collection of data on the metallurgical aspects of spallation in both brittle and ductile conditions. Senf et al. [1995] studied the nucleation of fractures in glass during high-speed impact, leading to a better understanding of the nucleation process in this very brittle material. Gaeta and Dandekar [1988] examined spall processes in single crystal and polycrystalline tungsten to examine how the spall strength may be related to the shear strength. Godse, Ravichandran, and Clifton [1989] examined crack propagation in steel with 15% upper bainite. They found evidence of crack nucleation in the bainite, but propagation through the martensite; hence, an important difference in fracture behavior based on metallurgical aspects of the material.

These studies are leading in two markedly different directions:

1. A general model applicable to a wide range of brittle materials and requiring a minimum of specific information about the material. This is the desired result of the mechanics approach.
2. Recognition that an accurate treatment of fracture in any material requires a large amount of knowledge about that material in the temperature range and under the specific prior treatment and loading. The appropriate model must then account for the multiplicity of phases and their interaction, the change of behavior with temperature, and the alterations caused by its loading and temperature history.

It is reasonable to expect continued progress in the modeling of dynamic fracture to advance in both directions, providing both more accurate general models and also more models for very specific material applications.

7.2.7. Summary

A basic model for representing the brittle microfracturing and fragmentation processes under high-rate tensile loading has been developed. It contains nucleation, growth, fragmentation, damage-caused stress-relaxation processes, and a stress solution procedure suited to highly nonlinear behavior. The model handles multiple loading and unloading paths, and reconsolidation. The model has been used to represent brittle fracture in a wide range of materials. However, it is not expected to match all materials. The current model is an extension of earlier models into the high-damage range where full separation and fragmentation occur. Because the model has provisions for all the basic brittle damage processes, it is suitable for use initially to represent fracture in any new material. The initial attempts to match experimental data would indicate any needed modifications in the processes. These processes, which are isolated in the model so that each can be altered independently of the others, can be changed in the computational model and the experiments re-simulated until a satisfactory match is obtained. Within this framework, the model can be viewed as one element in a flexible experimental–computational approach to fracture problems.

8 Applications of the Nucleation-and-Growth Fracture Method

In this chapter, we present a few illustrative samples of the application of the NAG (Nucleation-and-Growth) fracture approach. The chapter spans the range of materials that have been considered, noting the two basic types of fracture: ductile and brittle. In some cases, the data are well defined and appear to fit the general assumptions of the model and hence are well represented. In others the data are limited, of poorer quality or only estimated, and do not fit the predictions of the model. In most cases, the tests considered are plate impacts.

As we have seen in earlier chapters, the NAG model is complicated. Its application requires an extensive set of carefully controlled experiments followed by a demanding process of data analysis. Under what circumstances is the NAG approach to be used, as opposed to simpler methods that only indicate the level of damage?

1. When there is a need to represent fracture over a wide range of strain-rates or sizes. Then, knowledge of the underlying physics makes the interpolation or extrapolation more reliable. For example, in current rocket propellant research, fracture is being examined over strain-rates from 10^{-2} s^{-1} to 10^5 s^{-1} and scales from grams to megagrams, so a detailed treatment of the physics aids in the formulation of the model.
2. When a detailed end result such as the fragment size distribution is desired. In many situations it is necessary to know the size and velocity distribution of fragments because of subsequent damage that they may cause in other materials.
3. When multiple stress waves are to be represented so the behavior of partially damaged material is needed, or when the subsequent strength of the material after fracture is desired.

We begin our set of examples with 1145 aluminum to illustrate fracture in ductile materials. The experimental results here indicate the variation of damage with distance through the sample and from sample to sample, and also the reliability we may expect from our model. Then we describe fracture in a polycarbonate, a transparent material, so that we can examine the damage in three dimensions as well as on a cross-section. With the study of novaculite quartzite, we continue the fracture study to full fragmentation. The study of fracture in solid rocket propellant provides a sample of the NAG approach in a very complex composite material with significant rate effects in the undamaged material. Some results in a structural grade of beryllium are presented to indicate that the same kind of fracture occurs under electron-beam, X-ray, and other radiation loading as under impacts. Finally, we show some results in two steels and in Armco iron.

8.1. Ductile Fracture of Commercially Pure Aluminum

Commercially pure aluminum is a good example of a very ductile material. We use this material as a means for introducing the nucleation-and-growth approach in detail. The damage in this material appears initially as nearly spherical voids that enlarge and increase in number with evolving damage. In this section we wish to illustrate how fracture processes actually occur, especially the likely scatter in data, and also to show which aspects of fracture are represented by the model assumptions and which are not. The nucleation-and-growth (NAG) model used here is called DFRACT (for Ductile FRACTure). When the model does not fit the data well, we expect that the model assumptions are inappropriate or that phenomena not taken into account are occurring. For example, there may have been impurities randomly distributed in the material or even bands of impurities, giving preferential locations for the voids and also weakening the bonding so that lower thresholds were more appropriate in one sample than in the other samples.

To study fracture in commercially pure aluminum, we undertook a series of plate impact tests in 1145 aluminum. A summary of the configurations and important parameters of these tests is provided in Table 8.1. After the tests the samples were sectioned following the procedures outlined in Chapter 3 (Section 3.4). The numbers of voids were counted by sizes and then transformed according to the statistical procedure of Scheil as described in Chapter 7 (Section 7.1.2). These data on void volume were the basic data on which the DFRACT model is based. Figures 8.1 and 8.2 contain plots of the variation of the relative void volume as a function of distance through the specimens. Generally, the void volume plots show a fairly sharp peak at the region of maximum damage; whereas, except for test 939, there is only a broad peak for the number density. We may conclude from the variation of the numbers of voids that nucleation occurs early during the tensile pulse while the stress amplitudes are little affected by the adjacent damage; but growth continues throughout the period of tension and is therefore strongly affected by the adjacent damage.

Table 8.1. Configurations for 1145 Al impacts.

Test no.	Configuration V (m/s)	a (cm)	b (cm)	Peak stress C† (GPa)	T‡ (GPa)	Void volume fraction	Crack density (cm^{-3})	Damage state
847	128.9	0.236	0.635	9.8	8.8	6.0×10^{-3}	4.7×10^{6}	Incipient
873	132.0	0.236	0.635	10.0	9.0	1.5×10^{-2}	2.2×10^{7}	Incipient
849	142.6	0.236	0.635	10.8	9.8	6.2×10^{-2}	2.3×10^{7}	Intermediate
872	154.2	0.236	0.635	11.7	10.7	8.4×10^{-2}	2.7×10^{7}	Intermediate
939	185.6	0.114	0.317	14.1	13.1	1.4×10^{-1}	8.1×10^{7}	Incipient
857	207.9	0.236	0.635	15.7	14.7	No count	—	Intermediate
798	250.5	0.173	0.632	19.0	18.0	No count	—	Full separation

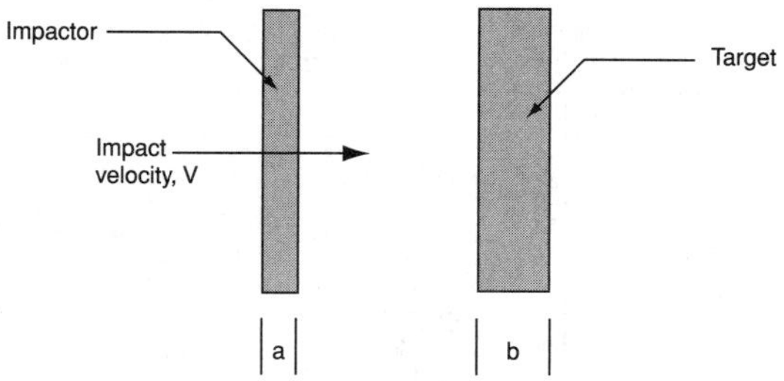

†C indicates compressive stress.
‡T indicates tensile stress.

From these plots, we also get an indication of the scatter in the damage quantities. For example, in test 847 the relative void volume in the region of maximum damage (0.38 cm from the impact interface) varies from 0.003 to 0.013, or by a factor of 4. Similarly, the number density for test 847 varies by a factor of 2. An indication of the scatter of the data from test to test was obtained by plotting the peak void volumes and number densities as a function of impact velocity as shown in Figures 8.3 and 8.4. Test 939 was conducted with the target and impactor each about half as thick as those for tests 847, 873, 849, and 872; so the damage for 939 is expected to be about half what it would be for full scale. The average void volume data for the full-scale tests shown in Figure 8.3 fit together fairly well, with test 849 being probably too high. Within tests the void volume values vary within a factor of 2 to 3. The average number density shows more scatter: tests 873 and 849 are probably high by a factor of 2 or 3 and 847 is 30% low. These plots give some indication of the accuracy with which we can expect to represent a suite of data with the model. The computed results for both relative void volume and number density should be within a factor of two of the experimental observations.

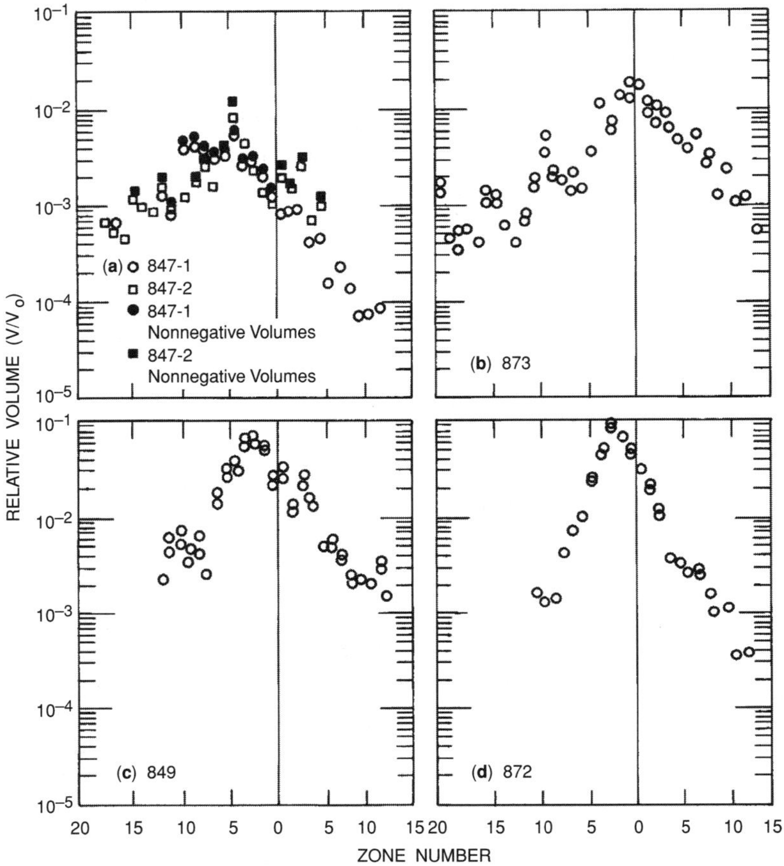

Figure 8.1. Void volume distributions as a function of position in Al 1145 targets 847, 873, 849, and 872.

Figures 8.3 and 8.4 also show the damage computed on the planes of maximum damage for the set of fracture parameters listed in Table 8.2. An additional simulation was made for a pseudo-test 872 with the dimensions of test 939. With the results of this pseudo-test we were able to plot the trend lines between the velocities for tests 872 and 939. It is evident that the model can represent the main trends of the data, but also that there is considerable variation from test to test. Matching the model to the data for the highest damage levels is especially difficult. The cross-sections showing the highest damage also showed signs of significant amounts of coalescence, a phenomenon that is not yet represented in the model.

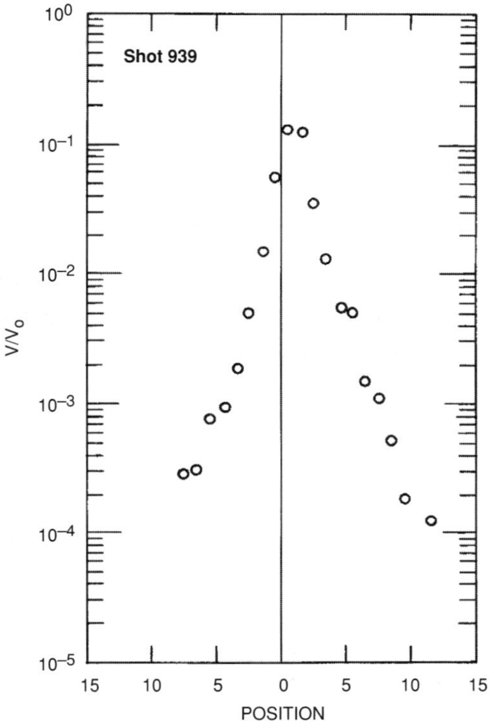

Figure 8.2. Void volume distribution as a function of position in Al 1145 target 939.

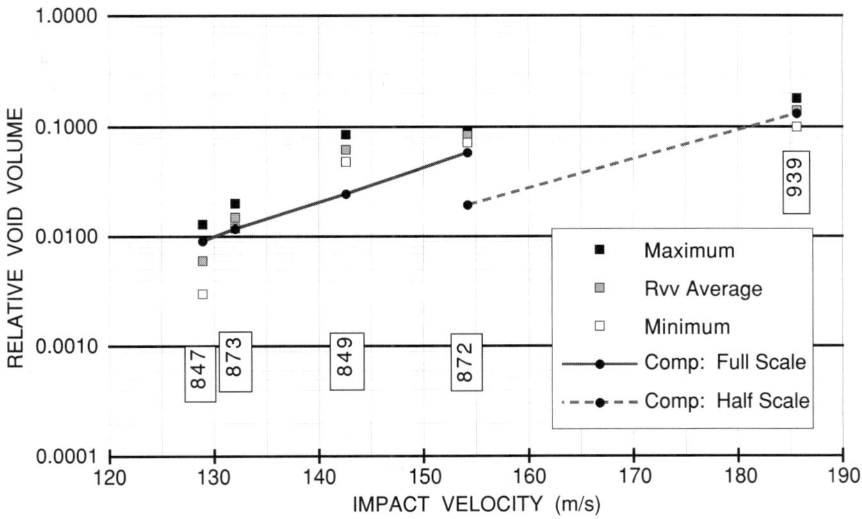

Figure 8.3. Variation of void volume with impact velocity in Al 1145.

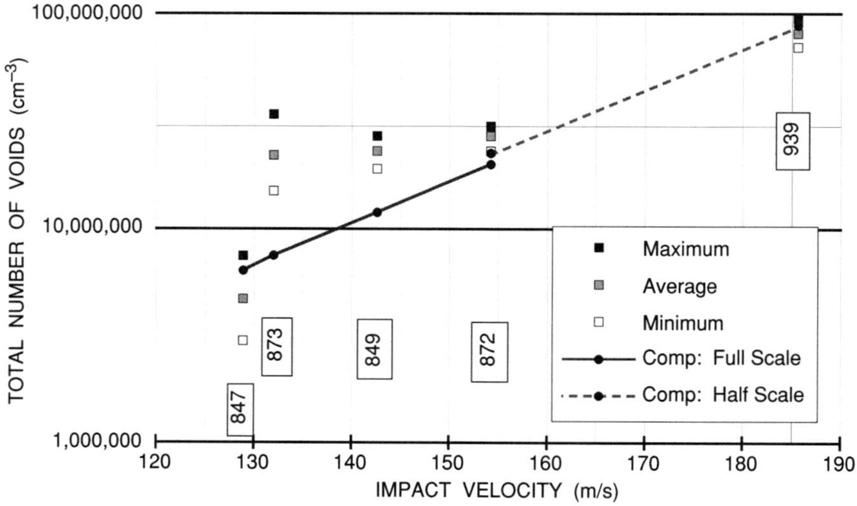

Figure 8.4. Variation of the number of voids with impact velocity in Al 1145.

Table 8.2. Fracture parameters for Al 1145.

Symbol	Units	Value	Description
T_1	$(\text{Mpa} \cdot \text{s})^{-1}$	1.0×10^5	Growth rate coefficient
σ_{go}	MPa	450	Threshold for growth
R_n	cm	1.0×10^{-4}	Mean radius at nucleation
$\dot{N}_o$	Number/m^3/s	5.0×10^{16}	Nucleation rate coefficient
σ_{no}	MPa	300	Threshold for nucleation
σ_1	MPa	65	Stress sensitivity for nucleation

8.2. Brittle Fracture of Polycarbonate

The first illustration of the use of BFRACT, the NAG model for brittle, crack-like behavior, is with polycarbonate; a tough, transparent plastic used as bullet-proof glass. Under impacts it showed crack-like fractures which could be readily seen because the material is transparent. Several tests were made spanning a good range of damage and the results were consistent with each other (suggesting that the material behavior is repeatable from sample to sample) and with the assumptions of our model. This study of polycarbonate showed that with a few carefully conducted experiments and post-test examinations one could determine appropriate parameters for the NAG brittle fracture model. Because of the transparency of the material, this study allowed us to verify the assumptions behind our crack counting and transformation methods.

A series of plate impact experiments was performed on polycarbonate (Curran, Shockey, and Seaman [1973]) to determine the shock compression parameters, yield strength under shock, and spall behavior. The spall tests considered here, listed in Table 8.3, show a range of impact velocities and computed stress levels that provided damage levels from just above a threshold level to almost sufficient to produce a complete separation in the target.

The heavy damage observed in Test 5 is illustrated in the cross-section shown in Figure 8.5 and in the crack size distributions of Figure 8.6. As we see in Figure 8.5, there is a main region of damage which could be described as a single macro crack, but we note that it consists of many short cracks at many distances through the specimen. These microcracks are essentially all horizontal, although they sometimes bend out of this plane to intersect with other cracks. Figure 8.6 contains the quantitative information about the damage: the location of each zone of damage that was counted and the cumulative size distribution of cracks in each zone. These size distributions have already been transformed from a surface count on the cross-section to a volumetric size distribution using the method of Scheil as noted in Chapter 7 (Section 7.2.2). The variation of damage from zone to zone suggests that the sample contained a narrow band of heavy damage with only narrow zones of collateral damage, and then was undamaged outside this band.

Table 8.3. Spall tests in polycarbonate.

	Thickness (mm)		**Impact velocity (m/sec)**	**Peak tensile stress† (MPa)**	**Observed damage**
Test no.	Flyer‡	Target‡			
5	3.05	6.58	152.2	183	Near separation
6	3.17	6.45	142.8	171	Near separation
7	3.17	6.53	137.2	164	Minor, threshold

† Peak tensile stress was computed on the assumption of no damage.

‡ Both the flyer and target were of polycarbonate.

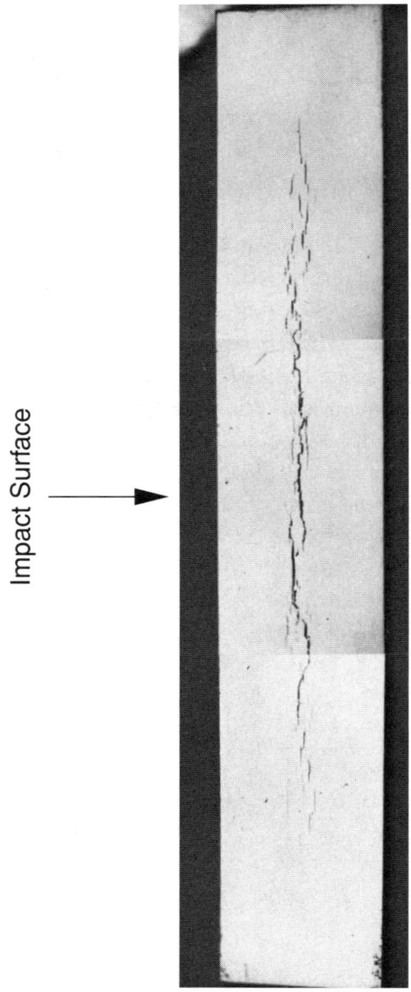

Figure 8.5. Polished section through polycarbonate specimen 5, showing the distribution of crack traces intersecting the surface.

From these fracture data and from the other data collected during the study, a complete set of (elastic–plastic) constitutive relations and fracture parameters was assembled. With the BFRACT model in a one-dimensional wave propagation computer program, simulations were undertaken of the experiments and the fracture parameters were adjusted to give a good fit to the experimental data. Figure 8.7 contains comparisons of the computed and measured crack-size distributions on the planes of maximum damage. The comparison is reasonably good, if not exceptionally so. The model comparisons for Test 7 suggested that this test had a stress level just above the threshold for nucleation of cracks. Very little nucleation occurred (very few cracks), but the nucleated cracks grew significantly, producing the unusually shallow slope to the observed crack size distribution. The fracture parameters determined for this material are listed in Table 8.6 at the end of this chapter. The range in experimental data is not sufficient to ensure that we have a unique set of parameters, so special caution should be exercised in extrapolating outside the domain within which these parameters have been verified.

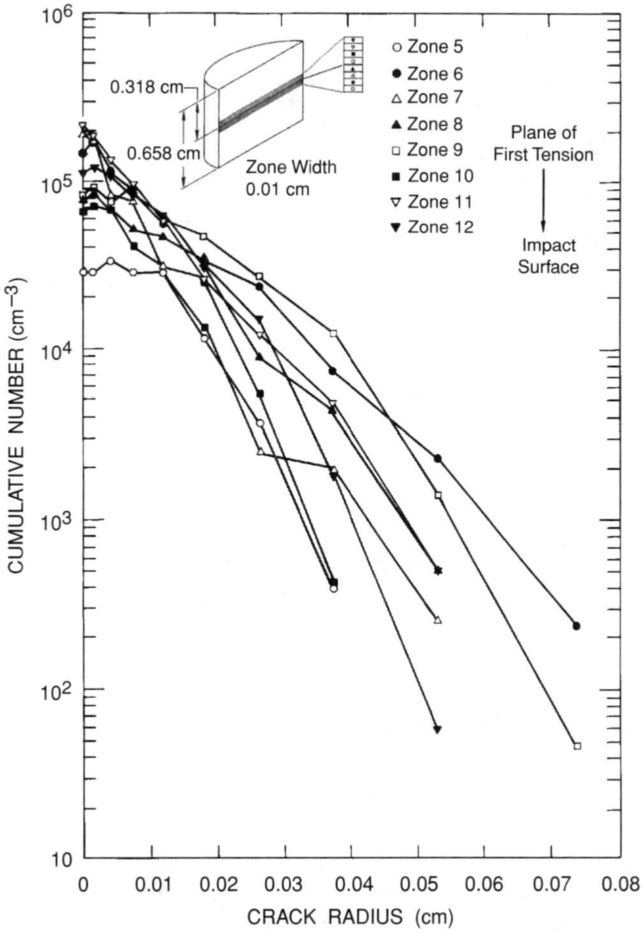

Figure 8.6. Volumetric crack size distributions in polycarbonate target 5 at several distances through the target.

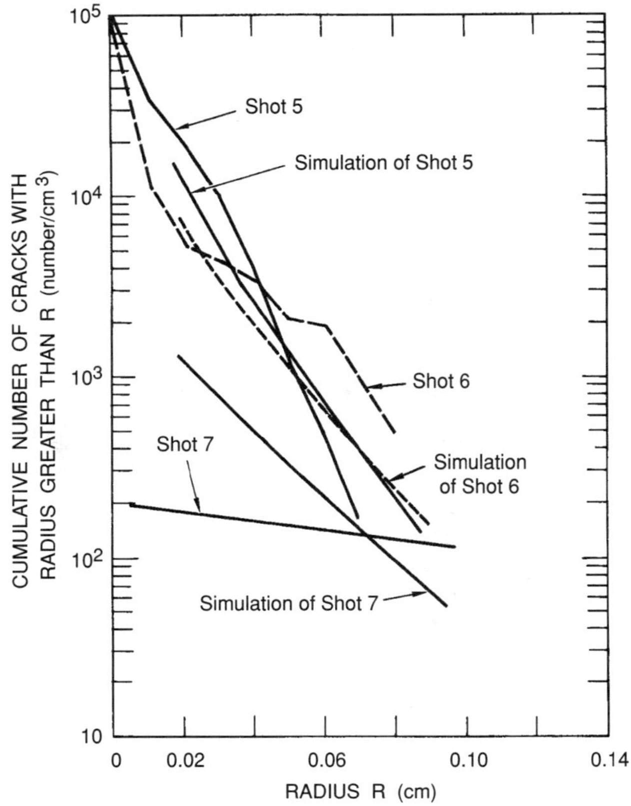

Figure 8.7. Comparison of computed and observed crack sized distributions on planes of maximum damage in three polycarbonate target disks.

8.3. Fracture and Fragmentation of Rock (Quartzite)

A series of impact experiments were conducted by Shockey et al. [1974] to determine the fracture and fragmentation behavior of fine-grained quartzite when subjected to high-rate loading. The quartzite, called Arkansas Novaculite, is a naturally occurring polycrystalline mineral. It is pure, dense, homogeneous, and consists of equisized, equiaxed, and randomly oriented quartz grains having an average diameter of about 10 μm. A population of flat flaws preexists on planes roughly parallel to each other. For one sample these flaws were counted on cross-sections and transformed to the three-dimensional crack size distribution displayed in Figure 8.8.

Controlled impact experiments on novaculite were carried out with a light gas gun using a flat projectile impact technique, so that fracture and fragmenta-

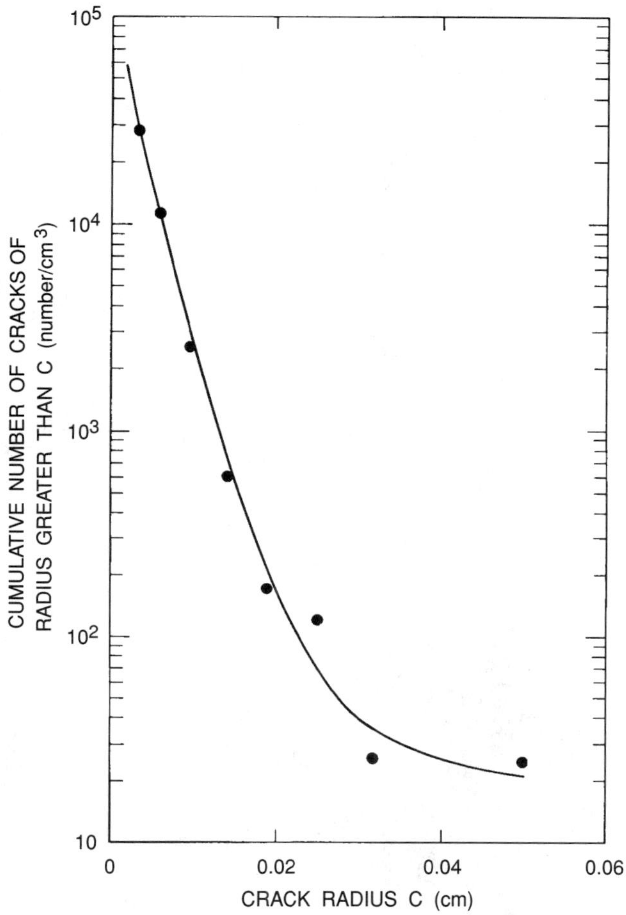

Figure 8.8. Size distribution of inherent flaws in Arkansas novaculite.

tion occurred under one-dimensional strain conditions. The cylindrical specimens were 6.3-mm thick and 13 to 38 mm in diameter. They were fitted into constraining rings of aluminum (which has approximately the same impedance as quartzite) to permit recovery of the sample after the impact with minimal subsequent damage.

The 53 impact tests resulted in some samples with no damage, some with a few cracks and others with fragments. Some samples were sectioned, as shown in Figure 8.9, to reveal the cracking. The impacts were from the top in the figure and we see that the cracks are horizontal, that is, normal to the direction of the maximum tensile stress. As we see in the figure, the number of cracks, the length of the cracks, and the general level of damage increase with increasing impact stress.

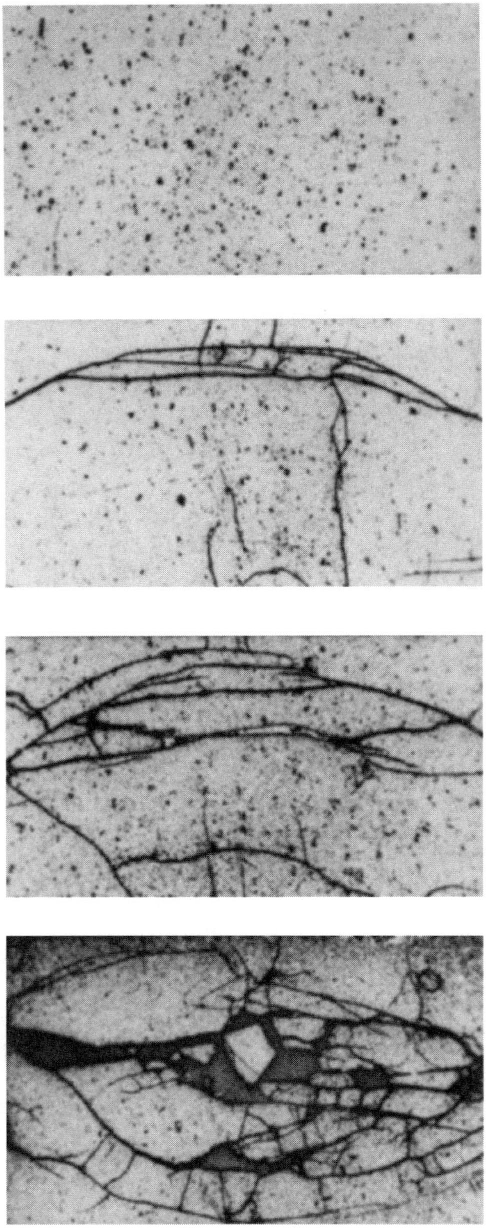

(**a**) Experiment 48
41.4×10^6 Pa

(**b**) Experiment 47
45.8×10^6 Pa

(**c**) Experiment 46
53.1×10^6 Pa

(**d**) Experiment 52
138×10^6 Pa

Figure 8.9. Polished cross-sections of Arkansas novaculite specimens, showing the extent of fracture damage produced at increasing levels of dynamic tensile stress.

Microscopic observations of the markings on crack surfaces in some samples indicated that the cracks were approximately circular and that their growth occurred by radial extension. The initiation sites could not be determined, but we presumed that cracks grew from the small crack-like flaws that existed in the rock before the application of any external load.

Sample 53 was impacted at 48.9 m/s (essentially the same impact velocity and test conditions as sample 52 shown in Figure 8.9(d), but this sample was separated from the constraining ring and allowed to disintegrate. The fragments were sieved into separate size groups and samples of these fragments are shown in Figure 8.10. The individual fragments are roughly equiaxed with six to eight major facets regardless of the fragment size. From these observations, we concluded that each facet had been formed from a single crack; hence, six to eight cracks had participated in forming each fragment. Because each facet seemed to be formed by a single crack, we concluded that large cracks had formed large fragments and (for mathematical convenience we assumed) that small cracks had formed small fragments. Large fragments tended to contain cracks that had not yet coalesced; therefore, not all the cracks participated in forming the fragments.

Based on the foregoing observations, we formulated a quantitative description of the fragmentation process; this description is the basis for the fragmentation process in the BFRACT fracture model. After a certain degree of crack growth and coalescence, fragments of various sizes form. The fragment sizes reflect the crack size distribution that led to the fragmentation. To compute the fragment size distribution, we have assumed relationships between the numbers of cracks and the numbers of fragments, the crack sizes and the fragment sizes, and also assumed the shape of the fragments. The fragment radius (radius of an equivalent sphere) is given by

$$R_{if} = \gamma R_{ic}, \tag{8.1}$$

where R_{if} and R_{ic} are the ith radii in the size distribution table for the fragments and cracks respectively, and γ is a constant with a value of about 1. Similarly, the numbers of cracks and fragments are related by

$$N_{if} = \beta N_{ic}, \tag{8.2}$$

where N_{if} and N_{ic} are numbers of fragments and cracks corresponding to the ith radii in the size distribution table. β is a constant with a value of 1/3 to 1/4 to agree with the fact that 6 to 8 cracks form each fragment (each crack forms facets of two fragments). Here, we are neglecting the fact that not all cracks may coalesce an\d form fragments.

Figure 8.10. Photomicrographs of various-sized fragments from experiment 53.

The volume of the fragments at the *i*th radius is given by

$$\Delta V_f = T_f\left(-\frac{\partial N_f}{\partial R_f}\Delta R_f\right)R_f^3. \tag{8.3}$$

Here,

$$-\frac{\partial N_f}{\partial R_f}\Delta R_f \tag{8.4}$$

is the number of fragments in the radius interval and T_f is a material constant to account for the bulkiness of the fragments ($T_f = 4\pi/3$ for spheres).

A new variable τ was introduced to represent the progress toward fragmentation. At full fragmentation τ is the total volume of fragments per unit volume, that is, one. Using Eq. (8.3) we can define τ as

$$\begin{aligned}\tau &= T_f\int_0^{R_{\max}}\left(-\frac{\partial N_f}{\partial R_f}\right)R_f^3\,dR_f\\ &= T_f\beta\gamma^3\int_0^{R_{\max}}\left(-\frac{\partial N_c}{\partial R_c}\right)R_c^3\,dR_c\,.\end{aligned} \tag{8.5}$$

Thus, the progress variable τ can be computed from the crack size distribution.

With the foregoing data and observations, we formulated a fracture model to represent both the cracking and the fragmentation seen in the quartzite. The nucleation aspect was treated by allowing a gradual activation of the initial flaw size distribution shown in Figure 8.8. Growth was treated with the usual relation for growth as a function of stress level above a size-related threshold stress. The material constants are shown in Table 8.6 listed at the end of this chapter. The constants that did not come directly from the observations were determined by multiple simulations of the impact experiments illustrated in Figure 8.9. Then experiment 53 was simulated and the computed and observed fragment size distributions were compared, as shown in Figure 8.11. Evidently, the model can be used to represent many aspects of the fracture and fragmentation processes.

8.4. Fracture and Fragmentation of a Solid Rocket Propellant

Solid rocket propellant is a soft, rubbery composite made by adding grains of a high-energy explosive to a polymer with adhesive and viscous properties. The present discussion concerns a prototype propellant designated SRI-A. The equation-of-state data for the propellant listed in Table 8.4 were obtained from impacts, ultrasonic tests, and other measurements. Fracture may occur by debonding of the polymer from the grains because of excessive shear or tensile stress.

The material appears to be very ductile, because its elongation under tensile strain can be 50% to 200% before fracture becomes apparent. But the appearance of the damage is crack-like. During tensile fracture, voids probably open around the explosive particles and other inclusions and enlarge both viscoelastically and viscoplastically until some critical rupture strain is reached around the stretching periphery—then tearing occurs through the rubbery matrix. When the stress is removed, the void closes and looks like a crack.

Four planar gas-gun impacts were conducted on the prototype propellant with projectile velocities from 44 to 191 m/s (computed impact stresses from 80 to 370 MPa). No fracture occurred in the lowest velocity impact, but various levels of spall and fragmentation took place in the other three. Unfortunately, no low-damage results were obtained, although such data are needed

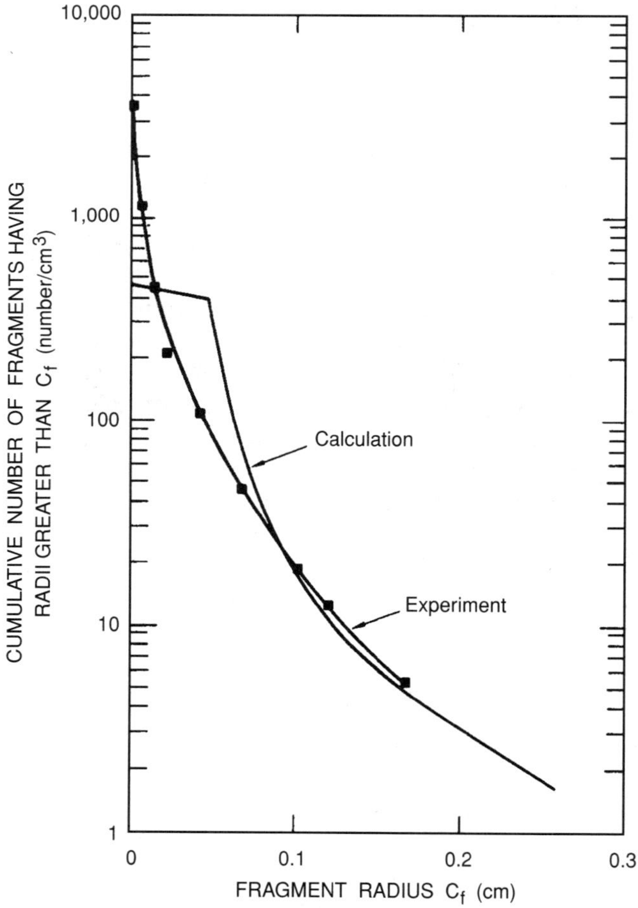

Figure 8.11. Comparison of experimental and computed fragment size distributions for experiment 53.

Table 8.4. Equation-of-state parameters for propellant SRI-A.

Description	Symbol	Value	Units
Density	ρ_{so}	1.85	Mg/m^3
Yield stress	Y	50.0	MPa
Shear modulus	G	1.24	GPa
Bulk modulus	K_1	5.67	GPa
Second bulk modulus coefficient	K_2	39.1	GPa
Third bulk modulus coefficient	K_3	182.0	GPa

for an accurate determination of the nucleation and growth parameters. Only the larger cracks appear to grow under dynamic loading; therefore, we selected the fracture mechanics-based growth threshold approach, using a fracture toughness to govern the stress threshold. The fracture parameters were obtained from iterative simulations of the four experiments; these parameters are given in Table 8.6 at the end of this chapter. The damage computed with these parameters generally agrees well with the observed damage, as detailed in the following paragraphs.

A cross-section of a propellant specimen impacted by a tapered flyer having a velocity of 73 m/s is shown in Figure 8.12. The spall plane appears well defined in this test, although there is actually considerable damage in the form of

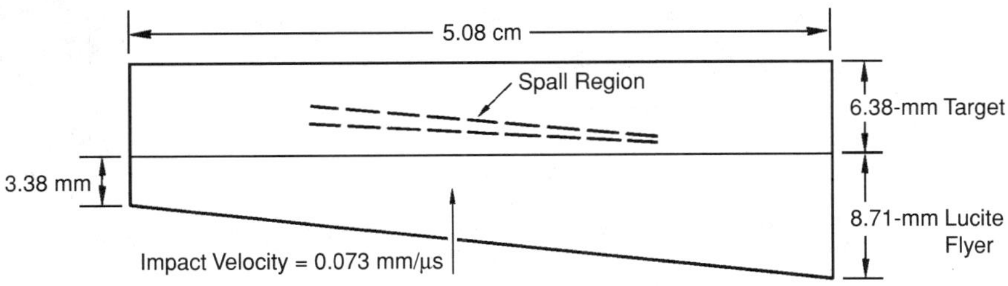

(**a**) Initial configuration and calculated width and location of spall plane

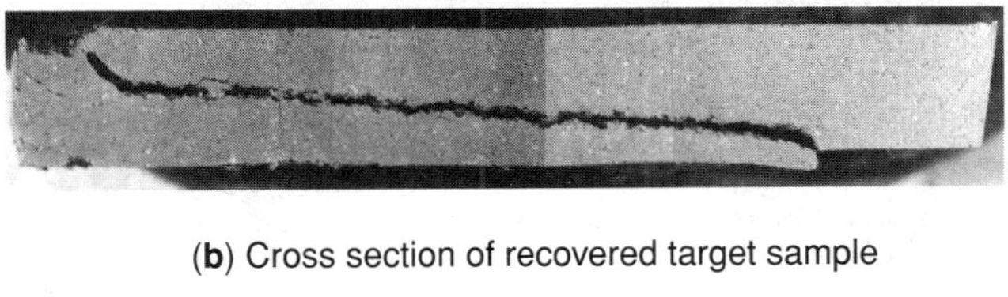

(**b**) Cross section of recovered target sample

Figure 8.12. Comparison of computed and observed spall plane locations in experiment 1458 in propellant SRI-A (Murri et al. [1982]).

small cracks in the material adjacent to this plane (seen in higher magnification views). The figure shows good agreement between computed and observed spall locations and widths of the spalled region. The location of the spall plane is readily predicted by most dynamic fracture models because it corresponds to the plane of first tension, which can be identified even without a fracture model. The thickness of the fractured region producing the separation, however, depends on the damage-induced stress relaxation processes in the model and is therefore a more severe discriminator between models.

A specimen impacted at 160 m/s (Figure 8.13) showed evidence of a double spall. According to the simulation of this test, two rarefaction waves collided to form a tensile stress and then to cause a broad region of damage, including the uppermost spall, during the first period of tensile stresses following impact. This spall caused the formation of recompression waves that propagated away from the spall region and toward the free-surfaces of the specimen. The recompaction waves caused more rarefaction waves as they reflected from the free-surfaces. These secondary rarefaction waves were strong enough to cause a second period of tensile damage and even a second spall closer to the impact plane.

An impact of the propellant at 191 m/s resulted in complete fragmentation of the sample. The fragment size distribution for the sample is shown in Figure 8.14. The "average radius" given on the abscissa is the radius of an equivalent sphere of the propellant with the same mass as the actual fragment. Both the

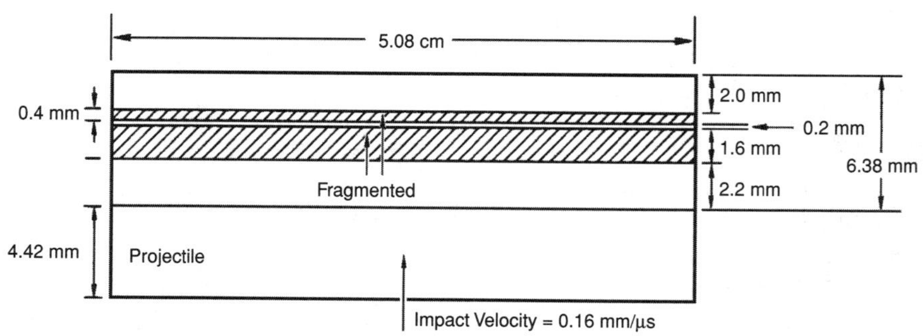

(a) Configuration and computed results showing two fragmented regions

(b) Cross section of specimen recovered from experiment 1460 showing "double" spall

Figure 8.13. Comparison of computed and observed fragmented regions in experiment 1460 on propellant SRI-A.

computed and observed distributions show a change of slope around 0.1 to 0.2 cm; this effect is probably associated with the fracture mechanics concept that allows only cracks larger than a critical size to grow.

The results of the four experiments ranged from no fracture to spallation to fragmentation. All three levels of damage were well represented by the model simulations, which used a single set of fracture parameters. Unfortunately, no low-damage impacts were available; hence, the nucleation and growth parameters for the model cannot be considered unique.

Examination of these experimental data and comparison of the data with the results obtained with the model indicated two areas where further developments are needed:

1. Results for low and intermediate damage are needed to determine the nucleation and growth parameters more accurately.
2. A more physically descriptive model of the fragmentation process should be developed. The current treatment, although it represents the fragmentation data quite well, includes no detailed steps from the cracked state to the fragmented state.

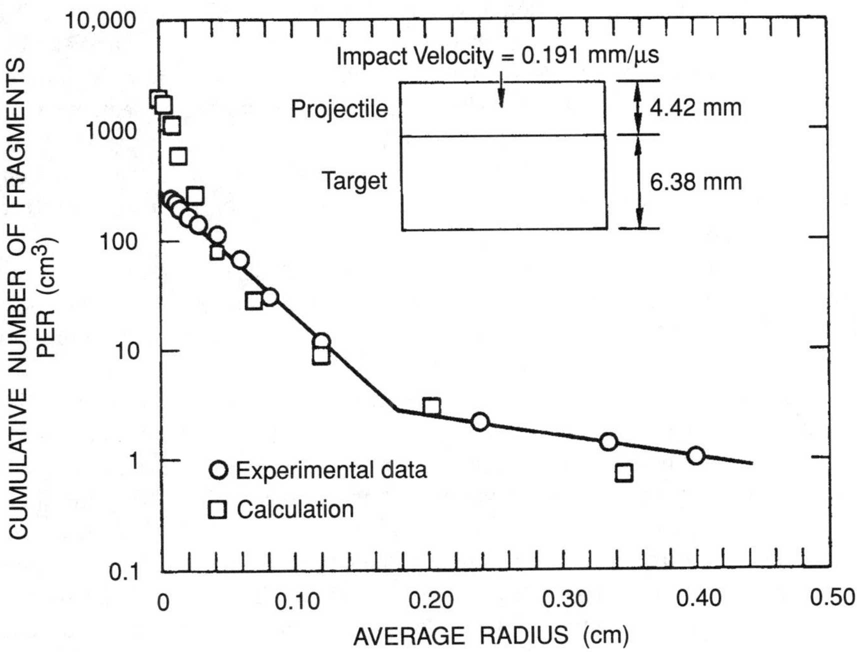

Figure 8.14. Comparison of computed and experimental fragment-size distributions for experiment 1462.

8.5. Fracture of Beryllium Under Impact and Thermal Radiation

A study was made of several grades of beryllium subjected to impact loading and also to thermal stress produced by radiation deposition. Here, we emphasize the results for S-200 beryllium, a structural grade. The impact and the initial modeling work were reported by Shockey et al. [1973] and the radiation results were described by Shockey et al. [1979]. We used plate impacts to produce a range of fracture levels and to derive from these the relevant fracture parameters. The S-200 beryllium is strongly strain-rate dependent; therefore, we used a deviator stress model from Read and Cecil [1972] based on the concepts of dislocation dynamics. For beryllium, this model replaced the standard elastic–plastic model for the deviator stresses in the brittle fracture model. Here, we examine the use of the model and parameters to represent some thermal radiation deposition experiments.

The extent of fracture damage produced in an S-200 beryllium specimen subjected to a burst of electron beam radiation was calculated from the measured nominal fluence and the dynamic fracture parameters determined earlier from the results of plate impact experiments. A plate specimen 3.86-mm thick was irradiated at a fluence level of 98 ± 10 cal/cm^2 (a peak dose of 85 cal/g at the front), then sectioned and polished to reveal the internal cracks shown in Figure 8.15(b). The brittle fracture model was used to simulate this experiment, using the fracture parameters from the plate impact experiments. The comparison between the computed and measured numbers and sizes of cracks as a function of position in the specimen (shown in Figure 8.16) shows satisfactory agreement in view of the uncertainties in electron beam fluence. Also, there is clearly a significant amount of scatter in the observed numbers of cracks in the specimen. The disagreement is most noticeable for the largest cracks: the model tends to predict too few large cracks.

The 74- and 124-cal/cm^2 dose experiments illustrated in Figure 8.15 were also simulated, but the agreement with the observed damage was not as good. The reasons for the disagreement indicate several of the current shortcomings of the model. These shortcomings are listed below.

The cracks we see in Figure 8.15 are very close together, have been influenced by their neighbors, and some of them remain open after the test. Yet our counting procedure and modeling technique were developed for isolated cracks only. In the present case the counting procedure led to essentially the same crack size distributions on the central planes for all three samples. However, we have the impression from the cross-sections that there is a marked increase in damage with increasing fluence levels. By contrast, the simulations led to distinctly different crack size distributions for the three doses: incipient, intermediate, and full fragmentation levels were predicted.

The brittle fracture model in use at the time of these simulations provided for elastic opening only, although the surrounding intact material can behave

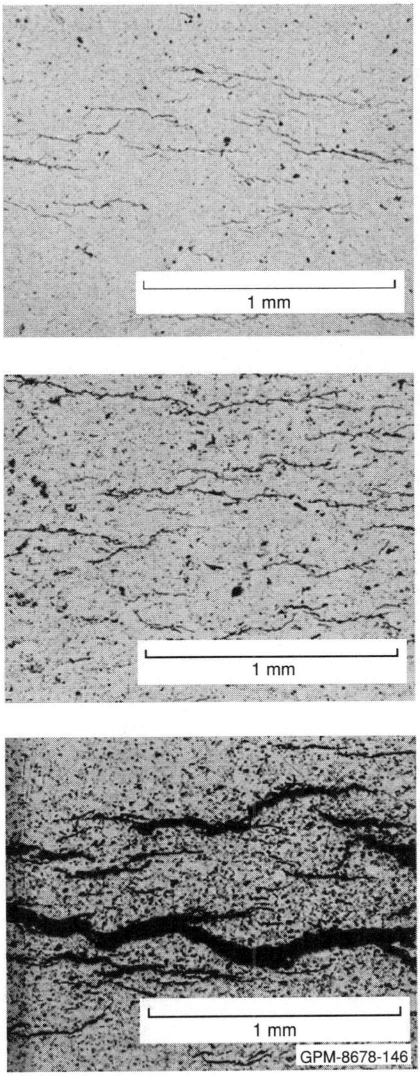

(a) 74 cal/cm^2

(b) 98 cal/cm^2

(c) 124 cal/cm^2

Figure 8.15. Polished and etched cross-sections through plate specimens of S-200 beryllium, showing internal brittle cracks induced by irradiation with an electron beam at three intensities.

plastically, and even exhibit rate-dependent behavior. A plastic crack opening procedure should have been provided. The model also could have been provided with a limit on the number of nucleation sites for cracks. Then we could have represented the fact that the same numbers of cracks were observed in all three tests.

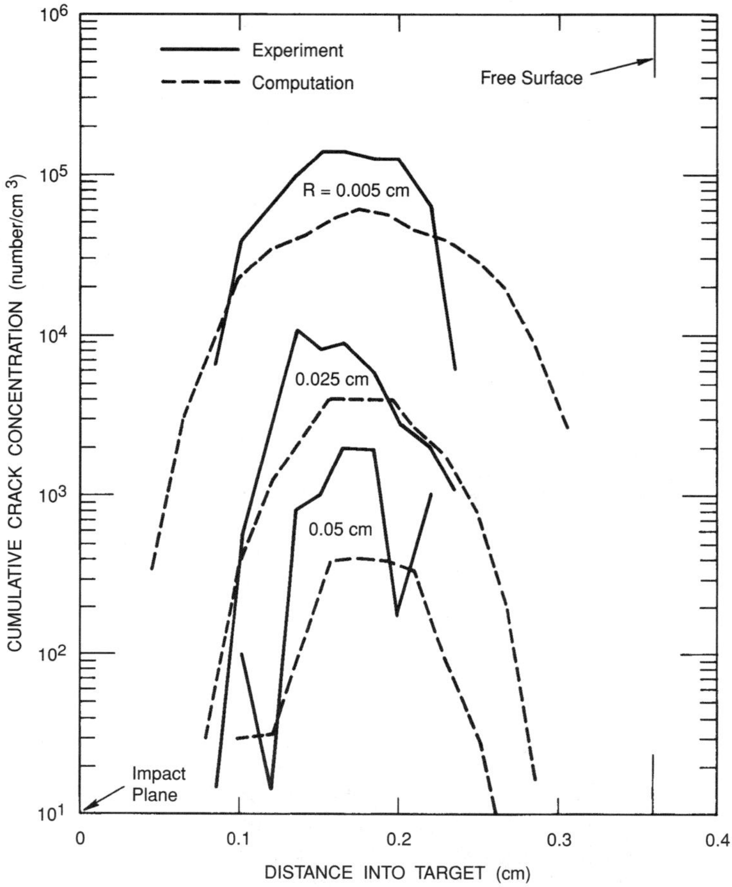

Figure 8.16. Comparison of computed and measured fracture damage in a beryllium specimen irradiated with an electron beam at a fluence of 98 cal/cm^2.

The main purpose of this study of beryllium was the examination of the inherent differences between fracture under thermal radiation (in this case, by electron beam) and fracture under mechanically induced stresses alone. The material undergoing fracture was not heated significantly by the beam. The observations were that the fractures which occur under thermal radiation appeared to be the same as in impact tests. Also, the simulations indicated that the cracks nucleated and grew under the same levels of tensile stress that had caused these fracture processes under impacts. Hence, it seems evident that the fracture modeling approach is appropriate under thermal radiation as well as under impact loading: the cracks respond only to the stress levels they experience and there is no special response to radiation.

No fracture parameters for beryllium are listed in the tables at the end of this chapter because this material required a special form of the model, including the rate-dependent deviator stress model from Read and Cecil and somewhat different nucleation and growth processes.

8.6. Fracture of Steel and Iron Under Impact

Here, we examine how fracture occurs in practical engineering materials: steel and iron with a range of ductilities. Two of the materials are strong, tough steels which are also termed "rolled homogeneous armor" or RHA because of their military use: Mil-S-12560B and XAR30. The third material is Armco iron. The appearance of cracks in these materials is shown and we suggest the processes by which they were formed. In each case, we attempt to represent the damage by the NAG brittle fracture model BFRACT.

XAR30 Armor steel. Dynamic fracture of XAR30 (a type of rolled homogeneous armor steel) was undertaken with a series of flyer plate impact tests, which caused partial spall of varying levels. The yield strength is 1.45 GPa, the ultimate tensile strength is 1.6 to 1.8 GPa, the Rockwell C hardness is 52, and the elongation is 14.5%. Figure 8.17 shows cross-sections of two disks that were impacted at 200 and 259 m/s respectively. In each case the flyer plate was about half the thickness of the sample (12.1 and 2.54 mm), but tapered as indicated in Figure 8.18. The taper causes a variation in the duration of loading and in the location of the fracture plane from the impact plane. Both specimens show many cracks in a region of fracture near their centers, not a single crack or plane of fracture. In Figure 8.17(a), the cracks do not follow the taper, but lie parallel to the plate surfaces (rolling plane) indicating that the material has a lower threshold for fracture in the through-the-thickness direction. The sample in Figure 8.17(b) has an almost completely formed fracture plane, which has probably occurred by coalescence of many small cracks.

The crack counts along the cross-section with the longest duration of loading (because the flyer is thickest at this end) in sample 5 are shown in Figure 8.18. This impact was simulated using a stress-wave propagation code with the BFRACT model for brittle fracture. The resulting crack size distributions for both cross-sections A-A and B-B are shown in Figure 8.19. In this case the BFRACT version had only an exponential crack size distribution (rather than the table of number vs. radius now in use), but the correspondence with the experimental data is quite good.

MIL-S-12560B Armor steel. A similar series of plate impact spall tests was performed on MIL-S-12560B; another rolled homogeneous steel with a much lower hardness. The yield strength is 1.03 GPa, the ultimate tensile strength is 1.12 GPa, the Rockwell C hardness is 38, and the elongation is between 15% and 25%. Figure 8.20 shows a cross-section of a disk impacted at 201 m/s. The larger cracks are composed of many cracks about 1 mm long that have coalesced to form the long cracks. As shown in Figure 8.21, the section in Figure 8.20 is just one of three sections taken through a target impacted by a tapered flyer. Figure 8.21 also shows the crack concentrations on section A-A.

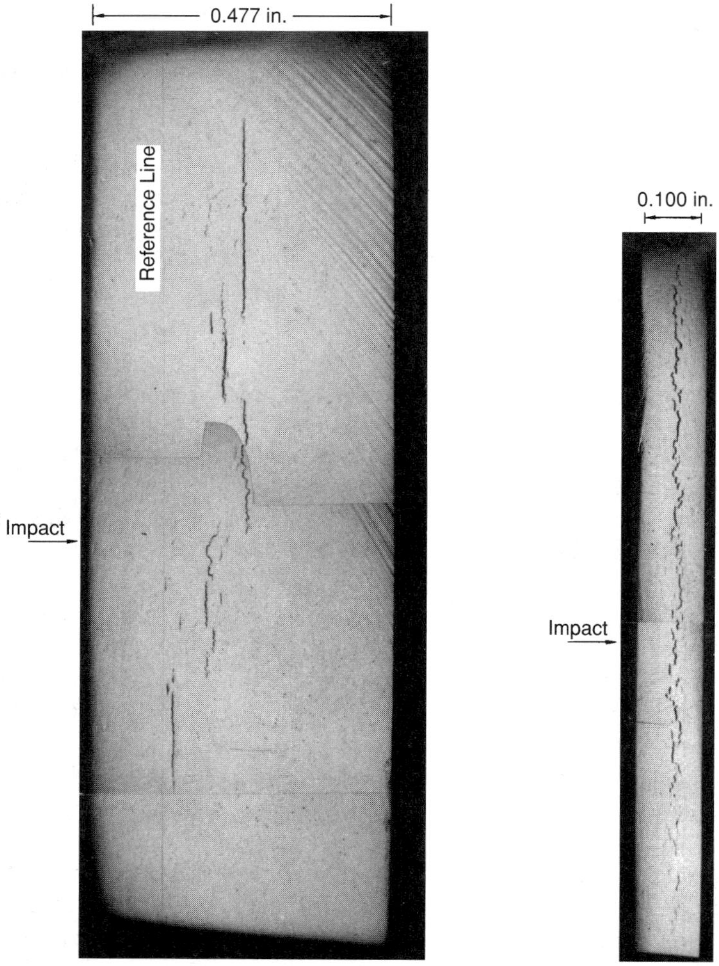

(**a**) Specimen 2024-4, Section A-B (surface is parallel to taper of flyer)

(**b**) Specimen 2024-5, Section A-A (surface is perpendicular to taper of flyer)

Figure 8.17. Polished cross-sections of XAR30 samples, showing that cracks are parallel to the rolling direction.

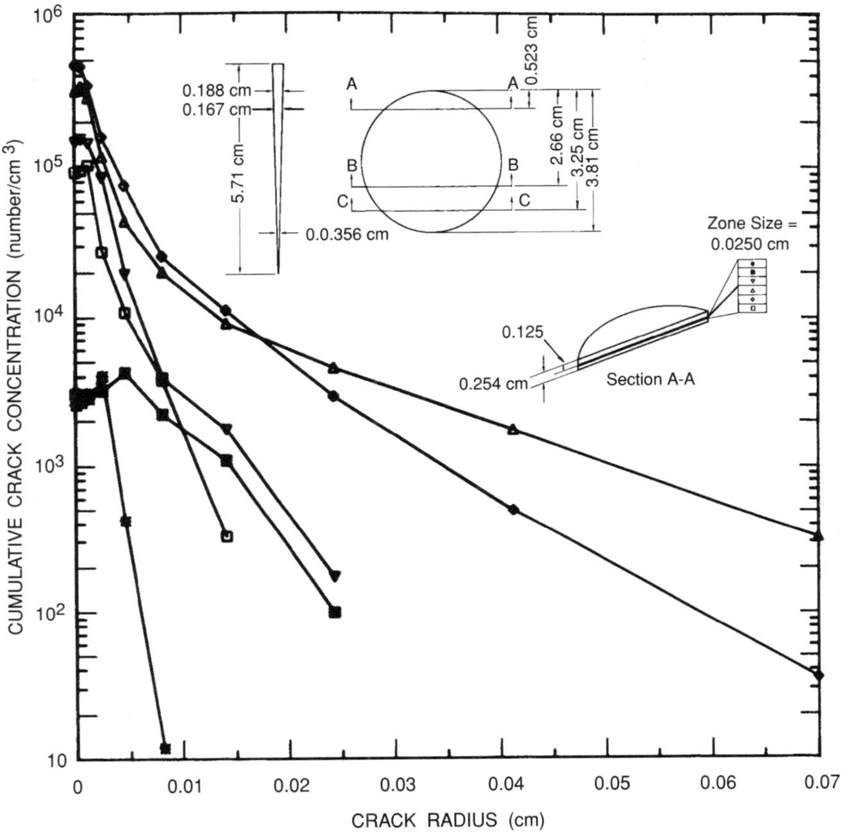

Figure 8.18. Measured crack-size distribution in XAR30 steel on section A-A of Test 5.

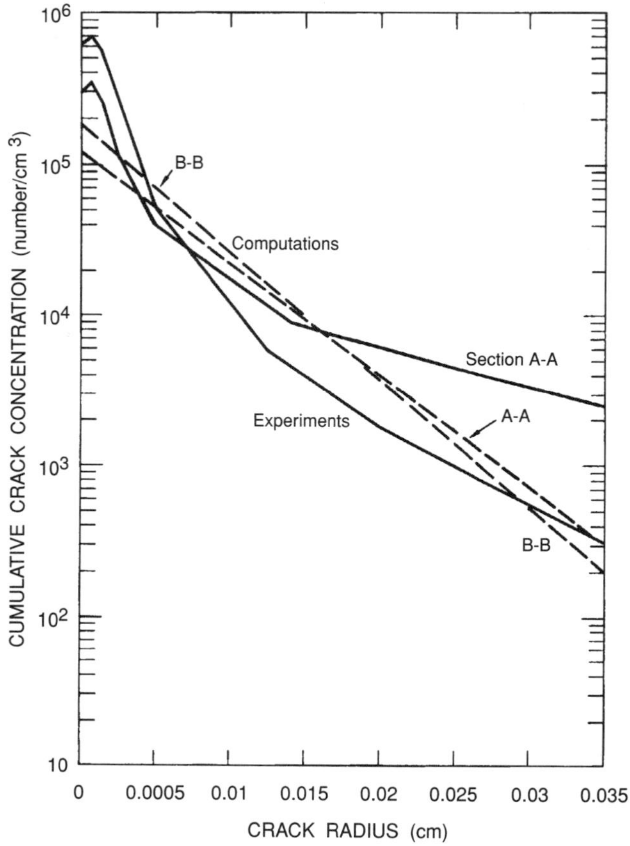

Figure 8.19. Comparison of measured and computed crack-size distributions on the planes of maximum damage for tapered flyer impact 5 in XAR30.

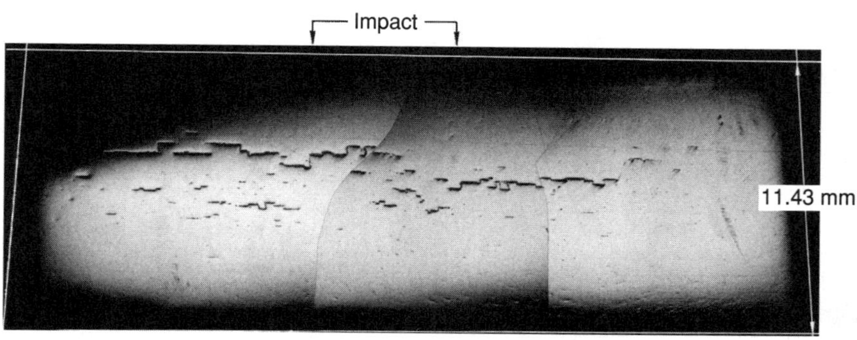

Figure 8.20. Polished cross-section of section A-A in experiment 2 on MIL-S-12560B steel, showing fractures on the rolling plane.

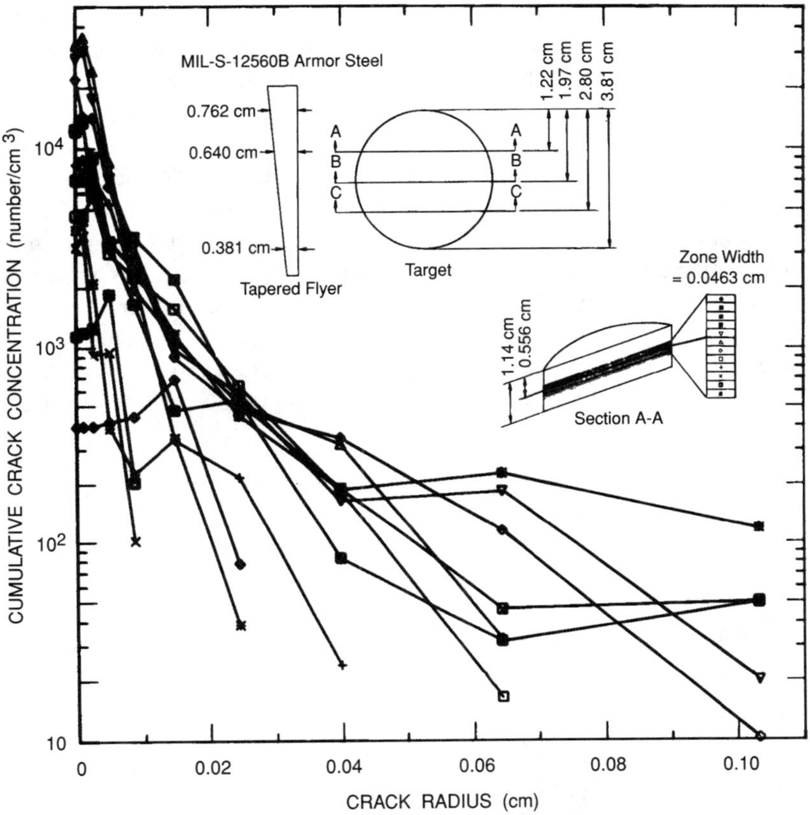

Figure 8.21. Measured crack-size distribution on section A-A of experiment 2 on MIL-S-12560B steel.

This impact was simulated using BFRACT and the resulting crack-size distributions for cross-sections A-A, B-B, and C-C are shown in Figure 8.22. This BFRACT version had only an exponential crack size distribution, but the correspondence with the experimental data is quite good. An alternate comparison is made in Figure 8.23 in which we have plotted the variation of the number of cracks per unit volume as a function of position through the target. This result demonstrates that the simulations are providing approximately the correct breadth of damage.

Armco iron. Armco iron is a much softer material than the steels discussed above, so the fracture behavior is somewhat different. The yield strength is 193.5 MPa, the ultimate tensile strength is 262 MPa, and the elongation is 35%. Again, a series of tapered flyer experiments were conducted to survey the damage to be obtained. For one of these (test S1), conducted at an impact velocity of 103 m/s (impact stress of about 2.0 GPa), a section was made in the target parallel to the direction of taper in the flyer. Portions of this section are shown in Figure 8.24: in correspondence to the thickness of the flyer at that point are indications of the duration of the tensile stress. In this material, brittle cleavage cracks are nucleated and traverse individual metal grains, being arrested at grain boundaries. These cracks then coalesce by extension of ligaments between noncoplanar cracks, that is, by a ductile tearing along the grain boundaries. Therefore, the fracture process involves both brittle and ductile mechanisms.

To examine how the appearance of fracture changes between quasi-static loading and plate impact loading, we examined the fracture surfaces as indicated in Figure 8.25. The surfaces are formed by the coalescence of a large number of dimples, and often the dimples contain the inclusion that we presume is the site of the crack initiation. For both loading rates, the largest dimples are 10 to 15 μm in diameter. However, for the dynamic loading, there is a great range of

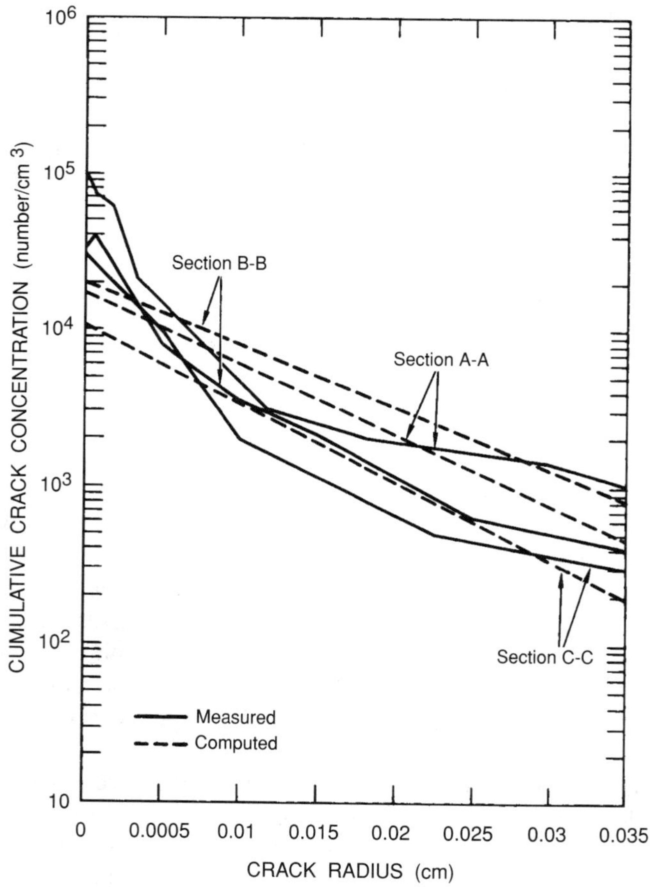

Figure 8.22. Comparison of measured and computed crack size distributions on the planes of maximum damage at three locations in specimen 2 of MIL-S-12560B steel.

dimple sizes, but for the slow loading, the dimples are nearly the same size. We may conclude from this observation that in the high-rate test there was sufficient stress (13 GPa) to debond most of the inclusions, but in the slow test only the largest inclusions were debonded.

In our study of Armco iron, we focused especially on test S25, which showed an intermediate level of damage spread over the central third of the specimen. The crack-size distributions (obtained by counting and then transforming from areal to volume distributions with the BABS2 computer program) are shown in Figure 8.26 for the 3.156-mm thick specimen. The impactor, also of Armco iron, was 1.138-mm thick and had an impact velocity of 196 m/s.

In this experiment (S25), there was a manganin gauge between the Armco iron target and a PMMA buffer behind the target. The record from the Manganin gauge is compared with the computed stress history in Figure 8.27. To aid in understanding the modification of the stress history caused by the damage, a history from a no-damage simulation is also shown.

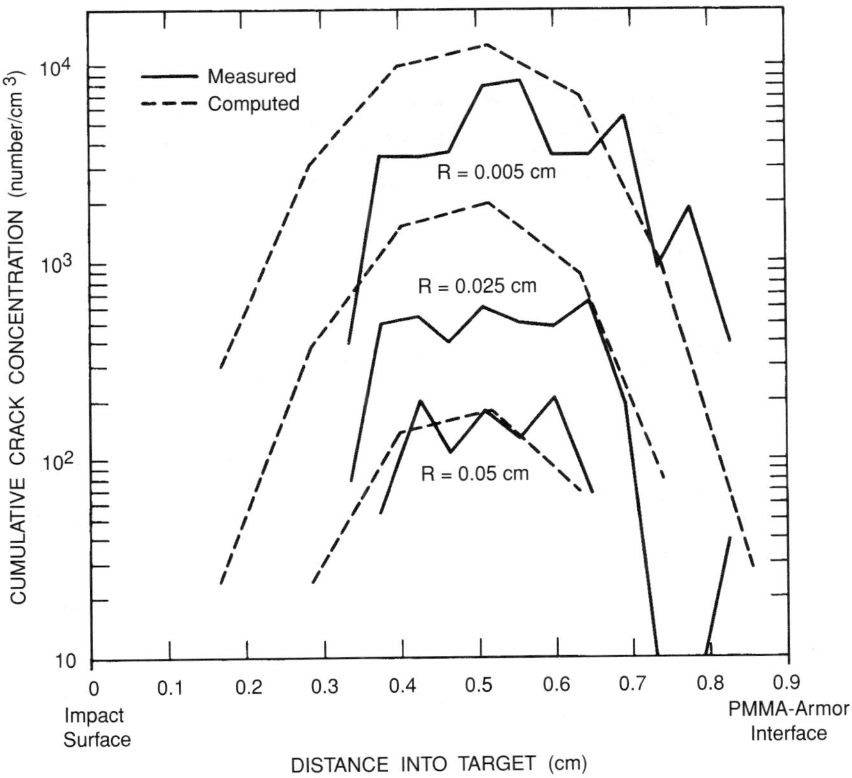

Figure 8.23. Comparison of measured and computed crack size distributions for three crack sizes as a function of depth in specimen 2 of MIL-S-12560B steel.

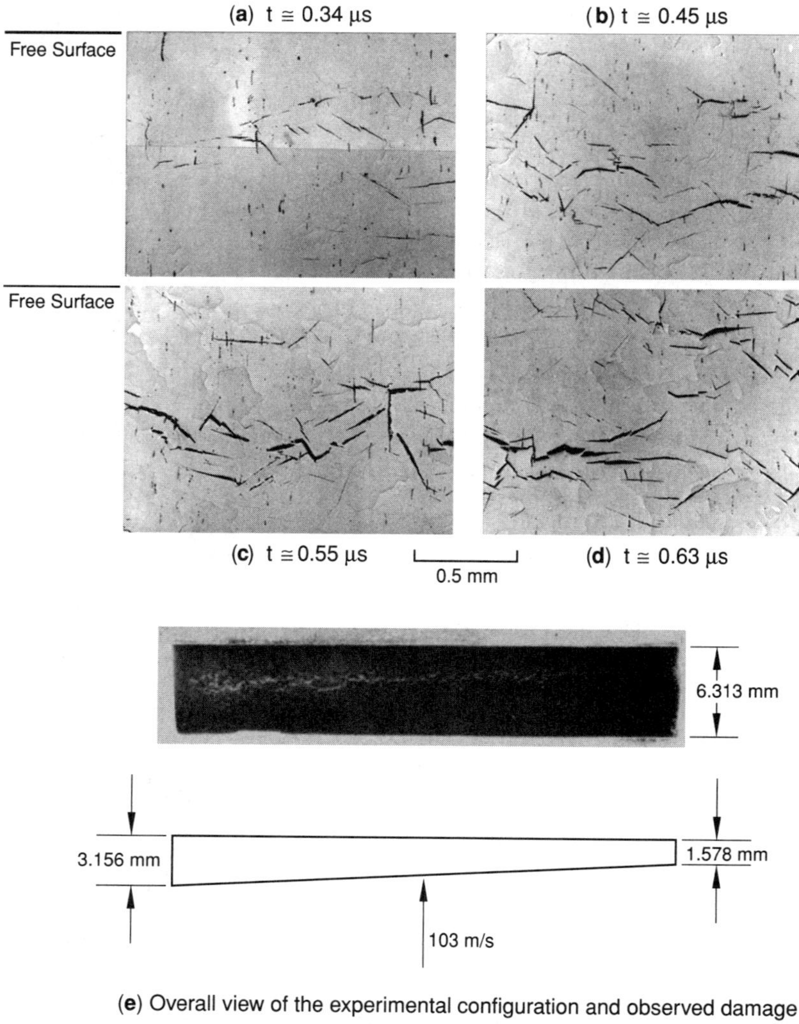

Figure 8.24. Damage distributions observed in cross-sections of the target in a tapered flyer test S1 on Armco iron.

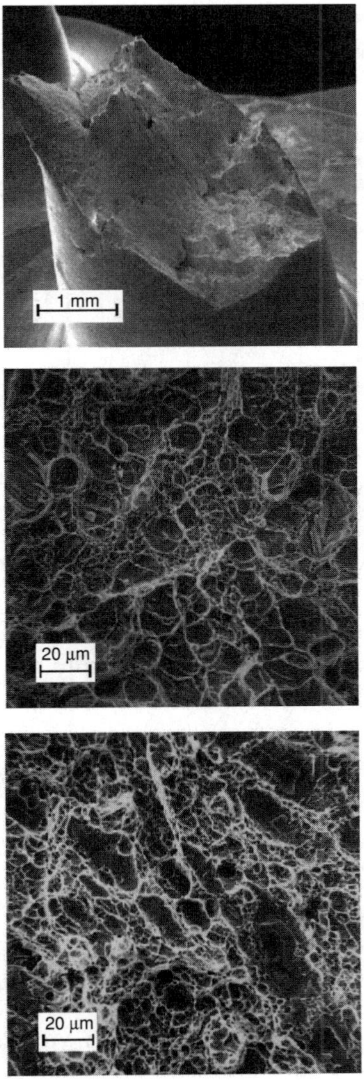

(**a**) Fracture surface produced at a strain rate of about 10^{-3} s^{-1} in a short-transverse tensile specimen

(**b**) Higher magnification view of (a), showing angular particles in the dimples

(**c**) Fracture surface of impact-loaded specimen 2151-4, showing similar dimple pattern as the quasi-static specimen in (b)

Figure 8.25. Scanning electron micrographs of fracture surfaces in MIL-S-12560B steel produced by quasi-static and dynamic loading.

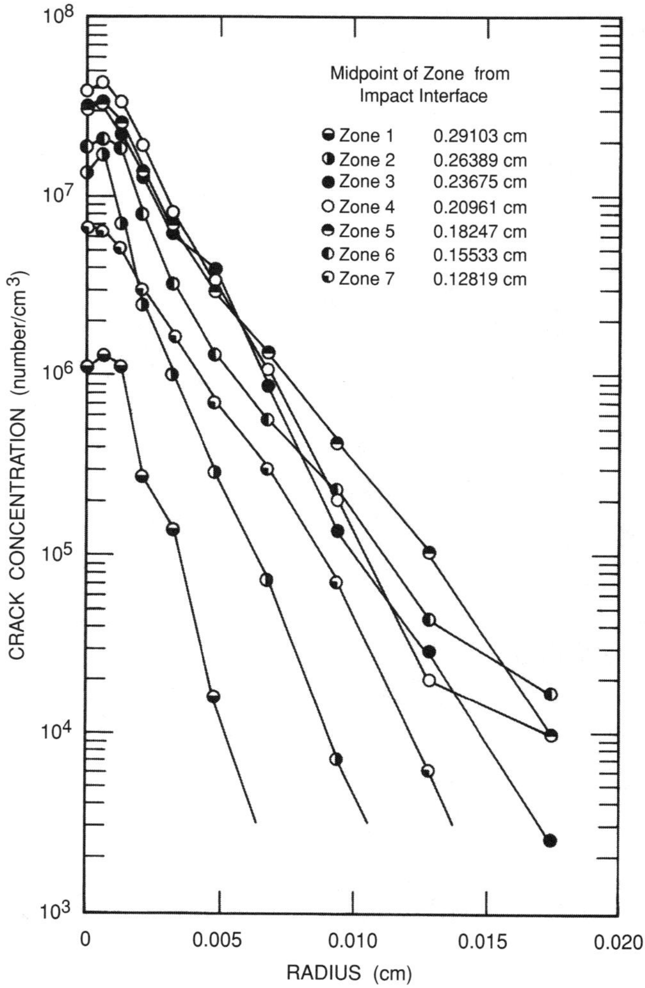

Figure 8.26. Crack-size distributions for test S25 in Armco iron from measurements (counted and transformed by BABS2).

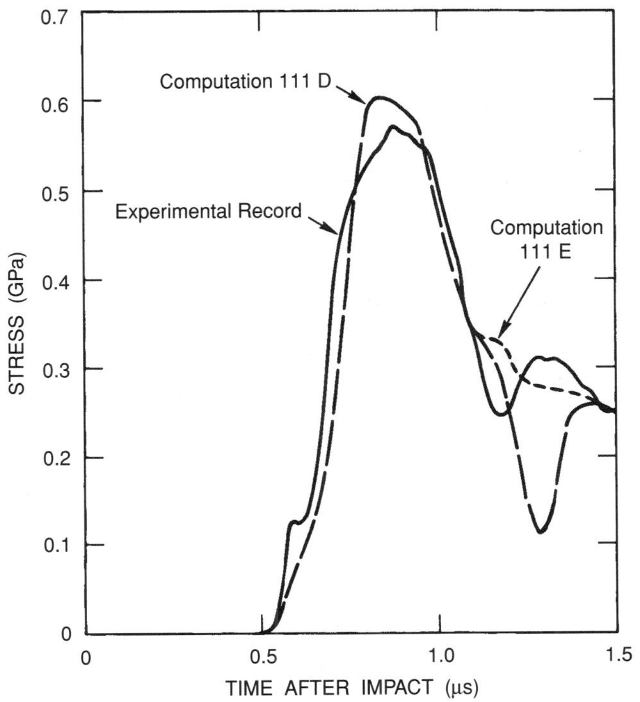

Figure 8.27. Comparison of the measured stress record behind the Armco iron target in test S25 with a simulated stress history computed using BFRACT.

8.7. Discussion

In these samples of fracture obtained under high-rate loading conditions, we have seen a range of fracture types. In Armco iron, the microfractures occurred across grains and growth was actually a coalescence of many microcracks to form a longer crack. The polycarbonate behaved in a more glass-like manner, forming long cracks that probably extended by growth at the tips.

For the most ductile materials, there is an exponential cumulative crack-size distribution. This suggests that the crack growth rate is not strongly size dependent. For more brittle materials such as the zinc sulfide ceramic ZnS (Shockey et al. [1981]), the size distribution has two exponential portions connected by a nearly horizontal section: this shape is expected from linear elastic fracture mechanics.

When sufficient data were available, the number and size of cracks were measured and related to the imposed stresses and stress durations. In this way, we obtained approximate nucleation and growth rates for the cracks.

In several materials, we had difficulty in causing incipient damage by plate impacts. This difficulty indicates that there may be a narrow range in tensile stress between no damage and a fairly high level of damage. For this reason, the brittle fracture experiments are much harder to perform and to control than similar experiments in more ductile materials.

The fracture parameters for most of the materials described in this chapter are listed in the following tables. Table 8.5 summarizes the fracture parameters for the two steels XAR30 and MIL-S-12560B, and for armco iron; and Table 8.6 summarizes the fracture parameters for polycarbonate, novaculite, and propellant.

Table 8.5. Fracture parameters for MIL-S-1256B steel, XAR30 steel, and Armco iron.

Symbol	Units	MIL-S-1256B	XAR30	Armco iron	Description
T_1	$(\text{Mpa} \bullet \text{s})^{-1}$	900	550	6000	Growth rate coefficient
K_{Ic}	MPa $\sqrt{\text{cm}}$	—	—	3.4	Fracture toughness
or, σ_{go}	MPa	200	10	—	Threshold for growth
R_n	cm	3.0×10^{-3}	4.0×10^{-3}	1×10^{-4}	Mean radius at nucleation
$\dot{N}_o$	Number/m^3/s	2.5×10^{14}	4.0×10^{14}	1×10^{19}	Nucleation rate coefficient
σ_{no}	MPa	1120	250	300	Threshold for nucleation
σ_1	MPa	100	179	950	Stress sensitivity for nucleation
$R_{\max}$	cm	—	—	—	Maximum radius at nucleation
$\beta\gamma T_f$	—	1.0	—	—	Crack-to-fragment factor

Table 8.6. Fracture parameters for polycarbonate, novaculite, and propellant.

Symbol	Units	Polycarbonate	Novaculite†	propellant	Description
T_1	$(\text{MPa} \bullet \text{s})^{-1}$	5.0×10^3	3×10^6	1.5×10^5	Growth rate coefficient
K_{Ic}	MPa $\sqrt{\text{cm}}$	3.6	11.0	1.28	Fracture toughness
R_n	cm	2.0×10^{-3}	5.7×10^{-3}	5.2×10^{-3}	Mean radius at nucleation
$\dot{N}_o$	Number/m^3/s	7×10^{16}	3.86×10^3	6×10^{13}	Nucleation rate coefficient
σ_{no}	MPa	160	0	50	Threshold for nucleation
σ_1	MPa	14.1	4×10^{-3}	10	Stress sensitivity for nucleation
$R_{\max}$	cm	1.0×10^{-2}	.05	2.0×10^{-2}	Maximum radius at nucleation
$\beta\gamma T_f$	—	4.0	1.0	1.33	Crack-to-fragment factor

† The nucleation rate is given by the gradual activation of the known size distribution of preexisting flaws.

9
Concluding Remarks

9.1. Conclusions

Investigations of the spall phenomenon provide important information about the fracture properties of materials under short duration loading conditions unattainable with conventional quasi-static testing techniques. The one-dimensional strain state that prevails in spall experiments lends itself to straightforward interpretation, and the short-duration, high-amplitude stress pulses that can be produced with shock wave loading techniques make it possible to investigate material behavior at stresses much higher than those attainable using other methods.

These factors contributed to the popularity of spall experiments as a means of investigating dynamic fracture properties, and numerous spall investigations have been performed. Unfortunately, contradictory results of spall strength measurements have been reported in several instances, partly because all methods of measuring the dynamic tensile stress in materials during spalling are indirect. This lack of a direct technique for measurement of the spall strength has led to several indirect methods. Each of the methods uses a different approach to determine the dynamic tensile stress; and, in some cases, the measurements are based on primitive schemes that do not account for the wave dynamics during spall.

Meanwhile, detailed analyses have shown that the development of spall fracture influences the wave dynamics, and in turn, the wave dynamics influence the development of fracture. Thus, the choice of the measurement technique and the method of analyzing the experimental data are crucial in spall investigations. For this reason, we have devoted a great deal of attention to the theoretical background, experimental technique, and sources of errors associated with spall strength measurements.

Time-resolved measurements made during shock wave loading provide the most accurate data about fracturing stresses under spalling. The free-surface velocity profiles contain information not only about the spall strength of materials, but also about stress relaxation during fracture and the kinetics of damage evolution. More detailed information about the kinetics of the fracture process can be obtained by performing post-test microscopic examinations of impacted spall samples. Not only do such examinations provide information about the nature of damage, the damage nucleation sites, and the mode of fracture (i.e., brittle crack propagation versus ductile void growth), they also provide microstatistical data that can be used to develop quantitative estimates of the damage nucleation and growth processes.

The experimental and analytical tools for investigating the spall phenomenon are versatile and can be applied to study the response of practically all solids and liquids. The experiments described throughout the book illustrate this broad capability with applications to ductile and brittle materials, metals, ceramics, metal and inorganic single crystals, glasses, elastomers, and liquids in a wide range of temperatures, load intensities, and pulse durations.

One of the main goals of spall fracture investigations is to construct constitutive models for analyzing such phenomena as high velocity impacts, explosions, and laser and particle beam interactions with condensed targets. In this regard, two constitutive modeling approaches were discussed. In the first approach, the derivation of empirical constitutive relations is based on measurements of free-surface velocity profiles over a wide range of load durations and is guided by an analysis of the influence of damage evolution on the wave profile during spalling. In the second approach, constitutive relationships of the nucleation and growth (NAG) type are constructed based on extensive microstructural observations of recovered spall samples subjected to a variety of predetermined loading conditions. NAG models are more complex than empirical models, and their calibration is more demanding. The trade-off to this relative complexity is that NAG models provide added insight into the damage process, and because the model is based on elementary physical processes, the material parameters are identifiable from experimental data and the results can be extrapolated with higher confidence outside the domain for which the model is calibrated.

Both types of constitutive descriptions are evolutionary and are computationally oriented. Choosing a constitutive model for a particular application depends to a large extent on the desired output. For applications where it is enough to characterize the response of the material in terms of kinematic variables (e.g., stress and strain), empirical models provide an adequate description of the resistance to fracture. Constitutive models of the NAG type, however, should be used when more detailed information about the state of the material is needed (e.g., flaw or fragment size distribution, or response of partially damaged material).

Spall fracture is different from fracture under quasi-static loading conditions. Conventional quasi-static strength tests are always accompanied by complicating factors such as plastic flow preceding fracture, transition from a uniaxial stress state to a triaxial stress state with the development of necking, and surface and environment effects. In contrast, experiments with plane shock waves provide a unique base of information about the strength of solids under one-dimensional strain, and stress states close to three-dimensional tension. Neither the surface of the body nor isolated coarse defects contribute to the main development of the spall fracture. The short duration of the applied stress makes possible the creation of large overstresses in the material near the minimum of the potential curve, $p(V)$, thus making it possible to study the conditions of elementary fracture events on a structural level close to that of the ideal crystal structure. In this sense, spall testing under shock wave loading conditions provides an opportunity for measuring the fundamental strength properties of matter.

9.2. New Applications

Spall strength measurements can also be useful for many engineering applications. For example, new technologies for processing materials using pulsed lasers are now being developed. For the success of such technologies, it is important to characterize the material behavior under conditions similar to those experienced during the laser treatment. In this way, the process can be optimized and undesirable fractures can be prevented. Spall studies can be used to aid in understanding the behavior of the material. When a pulsed laser is used as generator of shock load, spall strength measurements can be performed on very small samples obtained from different locations along the work piece. In this way a strength map can be constructed using a procedure similar to that used in microhardness mapping.

Spall studies can also be used to study the adhesion properties of coatings and to optimize processes ranging from laser cutting and drilling to laser surgery. Clinical studies have demonstrated that, for some applications, surgical lasers are superior to conventional surgical procedures. However, a fundamental understanding of the failure process in biological tissue, particularly when the tissue is irradiated by a laser, is not yet available. Such an understanding can be gained through carefully conceived spall experiments. The information could be used to develop and calibrate constitutive models for tissue failure, which could then be implemented in finite-element and finite-difference computer programs and used to optimize the laser surgery procedure.

Spall studies also have potential applications in the mining industry. The economy of many mining operations depends strongly on the fragment size distribution achieved during explosive blasting. Since fragmentation is the culmination of a crack nucleation, growth, and coalescence process, it is reasonable to expect that a fundamental understanding of this process would lead to practical algorithms that could be used to optimize explosive fragmentation and attain better control of the fragment size distribution. Spall studies could play a significant role in developing such algorithms because in a spall experiment, it is possible to produce damage under controlled conditions and characterize the damage in terms of flaw and fragment size distributions and in terms of material properties that affect the fragmentation process. The experiments could then be simulated using a suitable constitutive model. Once the fragmentation process is understood and characterized, appropriate fragmentation algorithms can be developed for use in mining operations.

Spall studies can be applied in still more areas. Potential areas of research where spall studies can be applied in the near future include studying the resistance to fracture in nanosecond and subnanosecond ranges of load duration, direct observations of fracture dynamics in transparent materials, and investigations of the dynamic fracture of composite materials, reinforced concrete, biological tissues, and other nonhomogeneous and nonmonolithic materials. Other areas of future research include investigating the tensor properties of the dynamic strength of anisotropic materials. This includes inherently anisotropic materials like composites as well as materials that start out being isotropic, then become anisotropic as a result of cold work or directional deformation and damage.

Perhaps the most important area of future work is in establishing the link between classical fracture mechanics and microstatistical fracture mechanics. The existence of this link has been established by careful observations of the 'process zone' ahead of the macrocrack tip. In the process zone, microflaws nucleate and grow under the action of the amplified tensile stress field near the macrocrack tip, and the macrocrack grows by the coalescence of these microflaws. Thus, the processes that control macrocrack propagation are similar to those observed in spall experiments, and this fact provides the essential link between the classical fracture mechanics approach used to describe macrocrack propagation and the microstatistical fracture mechanics approach used to describe spall damage which is governed by the near-simultaneous nucleation and growth of many microcracks.

Bridging the gap between the classical fracture mechanics approach and the microstatistical approach offers the exciting possibility of relating material properties to underlying micromechanisms and hence material response to processing variables. If established, such a relationship could be used to design materials with superior fracture properties and response characteristics.

Appendix
Velocity Histories in Spalling Samples

This Appendix contains a compilation of spall experiments performed by Genady Kanel and his co-workers at the Russian Academy of Sciences. Each of the experiments reported is described in sufficient detail to allow the reader to attempt to reproduce the results or simulate the experiment. To maximize the utility of the data, we also provide references to the original publications where the reader can obtain additional information about the techniques and procedures used to produce the data. The information provided includes (1) a description of the material investigated, including its density and elastic properties, (2) a schematic of the experiment, (3) the dimensions and conditions of the material investigated, (4) the technique used to perform the measurement and the associated experimental error, and (5) the experimental results, which take the form of a particle velocity history recorded at the free surface of the sample or at the interface between the sample and a softer material.

Table A.1 serves as a roadmap for the data presented in the appendix. It provides the density and sound velocities, both bulk and longitudinal, for all the materials included in the compendium, and points to the location where data for these materials can be found within the appendix.[1] Table A.2 shows the experimental configurations of the spall experiments documented in the appendix. Associated with each configuration is a set of dimensions, but since dimensions varied between experiments, the actual dimensions used in various experiments are provided in the captions of Figures A.1 through A.136 where the experimental spall data are presented.

1. Material compositions and western equivalents of FSU Metal alloys were provided earlier in Table 5.1.

Table A.1. Summary of the materials included in this compendium of spall data, their densities, sound velocities, and the location within the appendix of the figures containing their spall data.

Material	Density (g/cm^3)	Sound speed (km/s)		Figures
		C_l	C_b	
Aluminum AD1	2.71	6.40	5.25	A.1–A.2
Aluminum D16	2.78	6.40	5.34	A.3–A.6
Aluminum AMg6M	2.61	6.40	5.25	A.7–A.13
Steel 3	7.85	5.97	4.65	A.14–A.15
Steel 45	7.78	5.98	4.65	A.16–A.17
Steel Kh18N10T (stainless)	7.90	5.74	4.65	A.18–A.22
Steel KhVG	7.95	5.85	4.65	A.23–A.28
Titanium	4.50	6.51	5.11	A.29–A.35
Titanium VT5-1	4.45	6.15	5.11	A.36–A.38
Titanium VT6	4.43	6.15	5.11	A.39–A.43
Titanium VT8	4.45	6.15	5.11	A.44–A.46
Copper M2	8.93	4.60	3.96	A.47–A.52
Copper single crystal	8.93	4.60	3.96	A.53–A.58
Nickel	8.86	5.63	4.57	A.59–A.60
Molybdenum (polycrystal)	10.21	6.44	5.14	A.61–A.63
Mo single crystal<100>	10.21	6.44	5.14	A.64–A.68
Mo deformed single crystal <100>	10.21	6.44	5.14	A.69–A.71
Mo single crystal <110>	10.21	6.44	5.14	A.72–A.78
Mo single crystal <111>	10.21	6.44	5.14	A.79–A.83
Niobium single crystal <110>	8.59	5.03	4.44	A.84–A.86
Nb deformed single crystal <110>	8.59	5.03	4.44	A.87–A.91
Magnesium Ma1	1.75	5.61	4.50	A.92–A.94
Armco iron	7.80	5.97	4.65	A.95–A.99
Lead	11.35	2.25	2.03	A.100–A.102
Tin	7.29	3.43	2.61	A.103–A.104
Epoxy	1.20	2.62	—	A.105
PMMA	1.186	2.71	2.65	A.106–A.112
Rubber, white (grade 7889)	1.34	—	1.50	A.113–A.117
Propellant Simulant	1.60	—	1.85	A.118–A.122
Alumina (ruby)	3.99	11.20	8.00	A.123–A.127
Alumina (z-cut spphire)	3.99	11.20	8.00	A.128–A.129
Quartz, x-cut	2.65	5.57	3.69	A.130–A.131
Titanium carbide (with Ni binder)	5.28	9.15	7.00	A.132–A.136

Table A.2. Configurations used to perform the spall experiments documented in this Appendix.

<table>
<tr><th rowspan="2">Designation</th><th colspan="2">Test description</th><th rowspan="2">Configuration schematic</th></tr>
<tr><th>Configuration</th><th>Sensor type</th></tr>
<tr><td>A</td><td>Flyer plate</td><td>Capacitor gauge</td><td>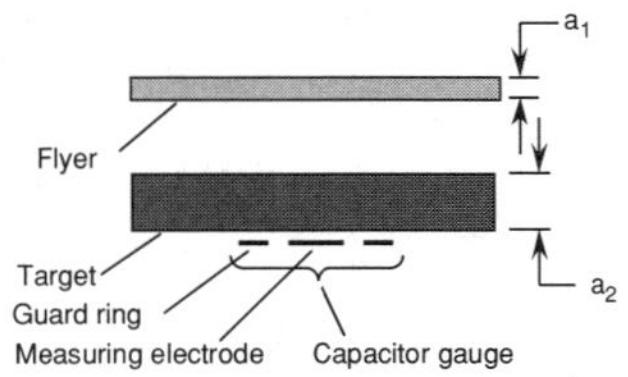
</td></tr>
<tr><td>B</td><td>Flyer plate</td><td>VISAR</td><td>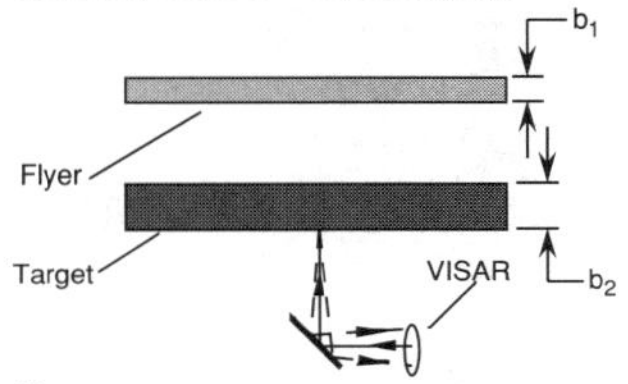
</td></tr>
<tr><td>C</td><td>Flyer plate (a base plate separates the target from the flyer plate)</td><td>VISAR</td><td>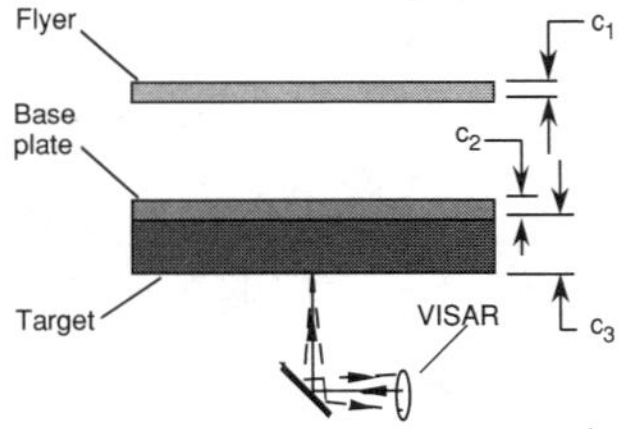
</td></tr>
<tr><td>D</td><td>Flyer plate</td><td>VISAR (reflected off a thin Al foil bonded to the back surface of the target)</td><td>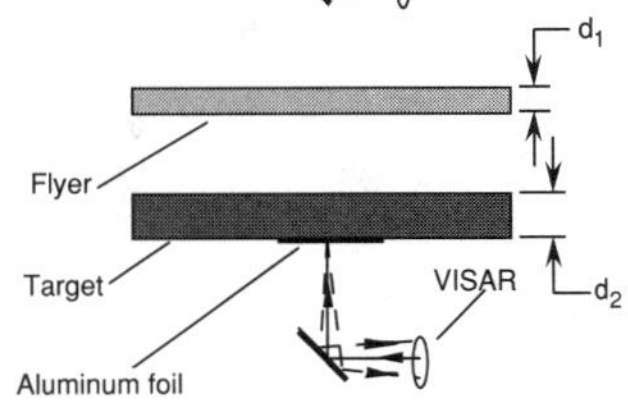
</td></tr>
<tr><td>E</td><td>Flyer plate</td><td>VISAR (reflected off a thin Al foil sandwiched between the target and a Hexan window)</td><td>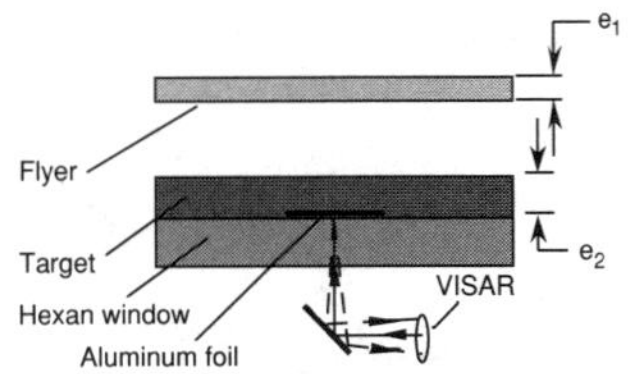
</td></tr>
<tr><td>F</td><td>Flyer plate (a base plate separates the target from the flyer plate)</td><td>VISAR (reflected off a thin Al foil bonded to the back surface of the target)</td><td>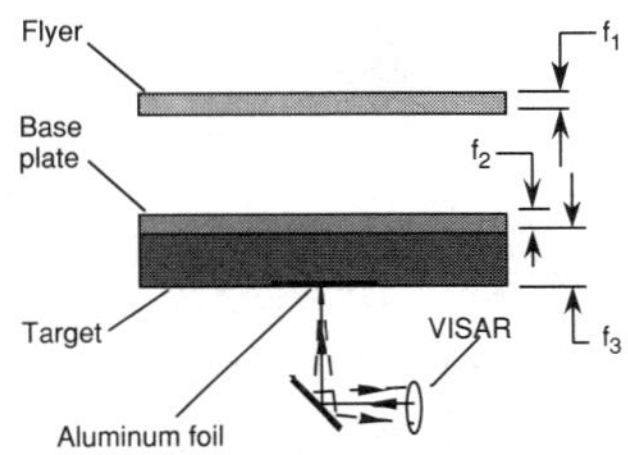
</td></tr>
</table>

Table A.2 (*continued*). Configurations used to perform the spall experiments documented in this Appendix.

Designation	Test description: Configuration	Test description: Sensor type	Configuration schematic
G	Flyer plate (a base plate separates the target from the flyer plate)	VISAR (reflected off a thin Al foil sandwiched between the target and a lexan window)	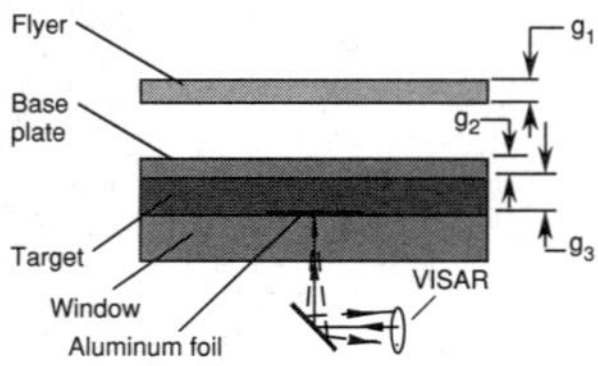
H	In-contact explosives	Capacitor gauge	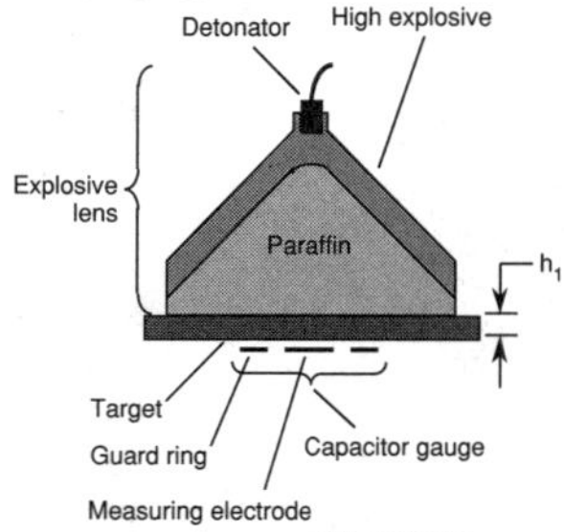
I	In-contact explosives (a buffer plate separates the target from the explosive package)	Capacitor gauge	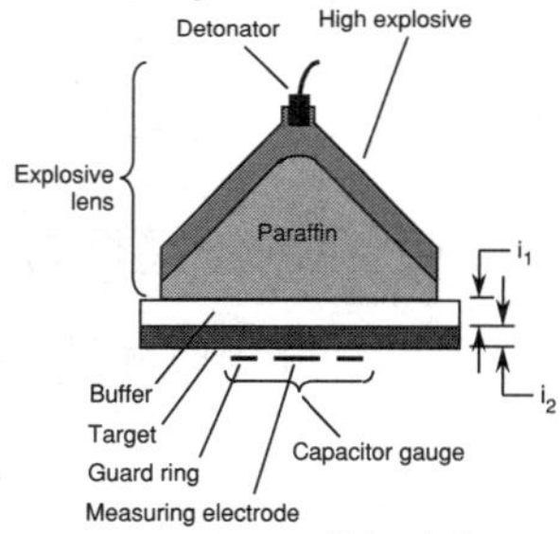
J	In-contact explosives	VISAR	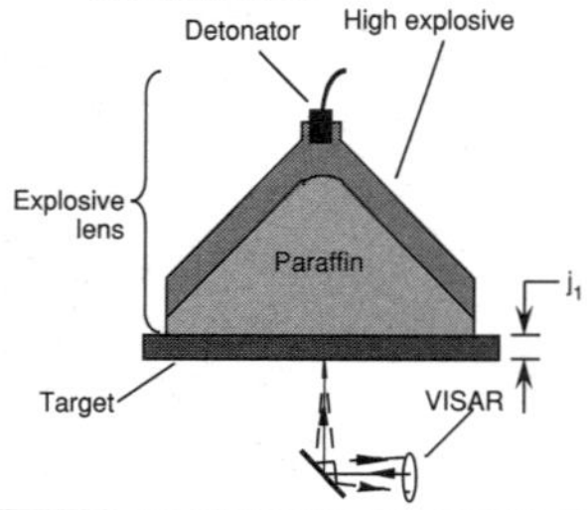

Table A.2 (*continued*). Configurations used to perform the spall experiments documented in this Appendix.

Designation	Test description: Configuration	Test description: Sensor type	Configuration schematic
K	Ion beam (energy deposited directly into the front surface of the target)	ORVIS	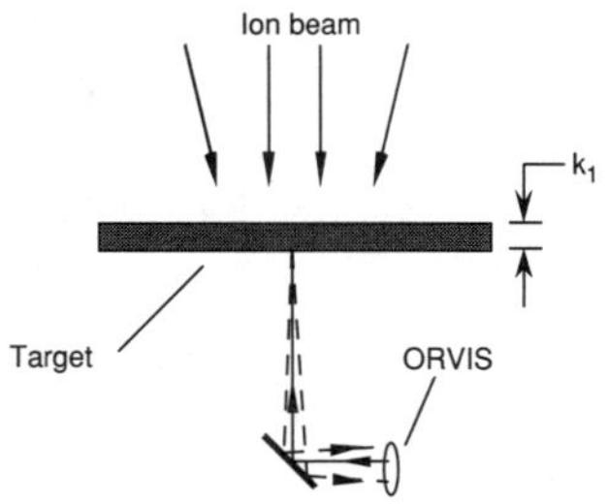
L	Ion beam (energy from the beam is used to accelerate a flyer plate)	ORVIS	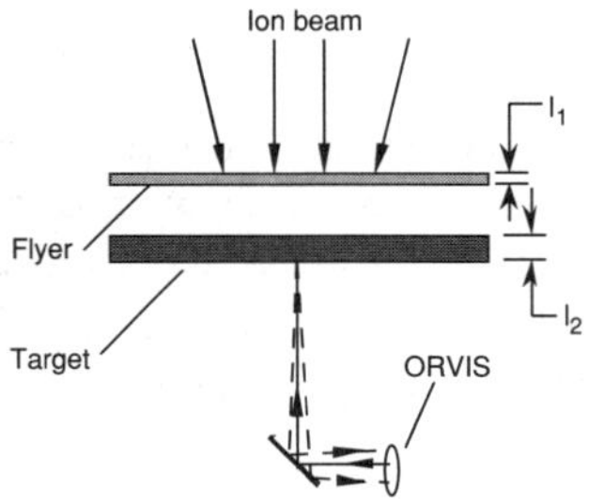
M	Ion beam (energy deposited directly into the front surface of the target)	ORVIS (measurement made at the interface between the back surface of the target and a window material)	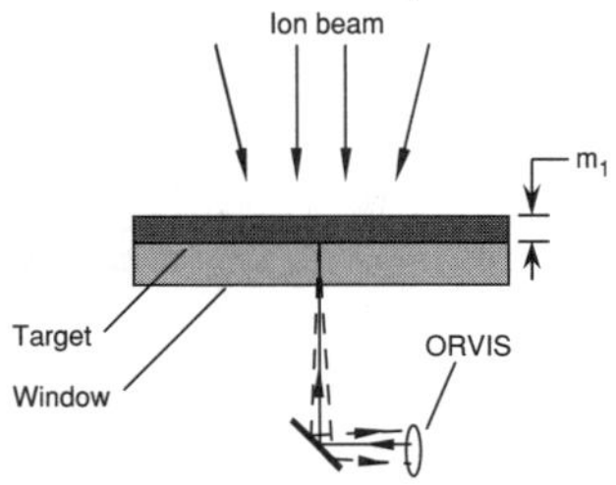

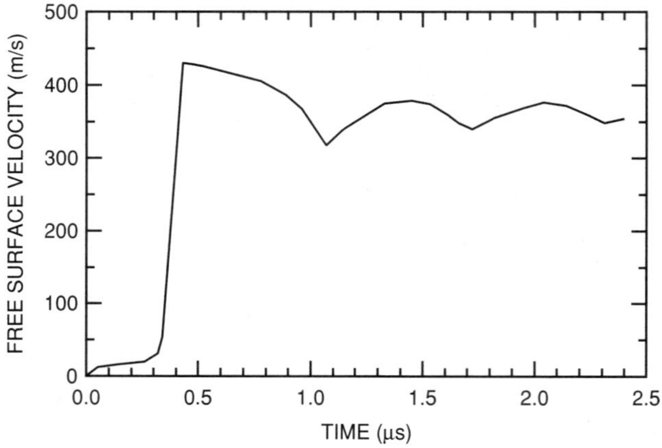

Figure A.1. Velocity history at the free surface of an aluminum AD1 target impacted by an aluminum flyer plate at an impact velocity of 450 ± 20 m/s. The experiment was performed using configuration A (see Table A.2) with a_1 = 2 mm and a_2 = 14.9 mm. The data were recorded using a capacitor gauge with an electrode diameter of 20 mm and a measurement accuracy of ±4%. The spall strength was 0.87 ± 0.05 GPa, and the spall thickness was 1.59 mm (Kanel [1982a]).

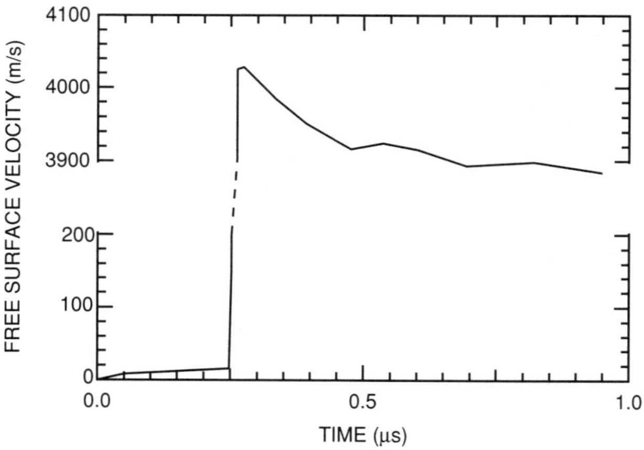

Figure A.2. Velocity history at the free surface of an aluminum AD1 target impacted by an aluminum flyer plate at an impact velocity of 4000 ± 150 m/s. The experiment was performed using configuration B (see Table A.2) with b_1 = 2 mm and b_2 = 10.0 mm. The data were recorded using a VISAR with a measurement accuracy of 3 to 5 m/s. The spall strength was 0.87 ± 0.1 GPa, and the spall thickness was 0.68 mm (Kanel [1987]).

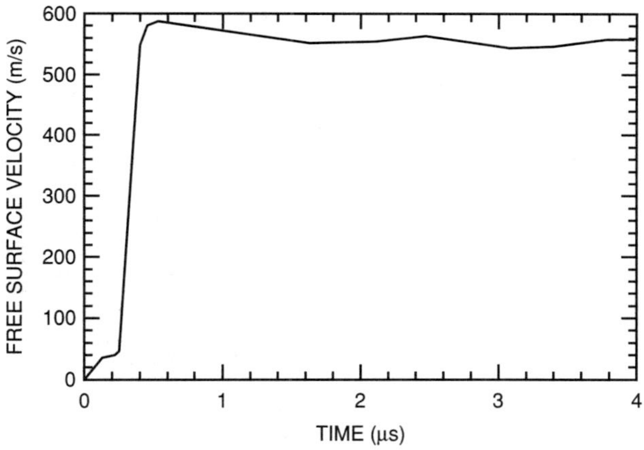

Figure A.3. Velocity history at the free surface of an explosively loaded aluminum D16 target. The experiment was performed using configuration I (see Table A.2) with $i_1 = 10$ mm and $i_2 = 10.0$ mm. The data were recorded using a capacitor gauge with an electrode diameter of 20 mm and a measurement accuracy of ±4%. The spall strength was 0.60 ± 0.05 GPa, and the spall thickness was 4.0 ± 0.5 mm (Kanel [1982a]).

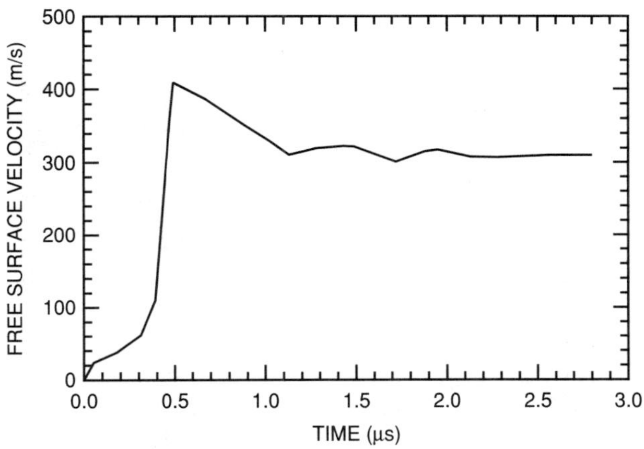

Figure A.4. Velocity history at the free surface of an aluminum D16 target impacted by an aluminum flyer plate at an impact velocity of 450 ± 20 m/s. The experiment was performed using configuration A (see Table A.2) with $a_1 = 2$ mm and $a_2 = 15.0$ mm. The data were recorded using a capacitor gauge with an electrode diameter of 20 mm and a measurement accuracy of ±4%. The spall strength was 0.78 ± 0.05 GPa, and the spall thickness was 1.85 (±10%) mm (Kanel [1982a]).

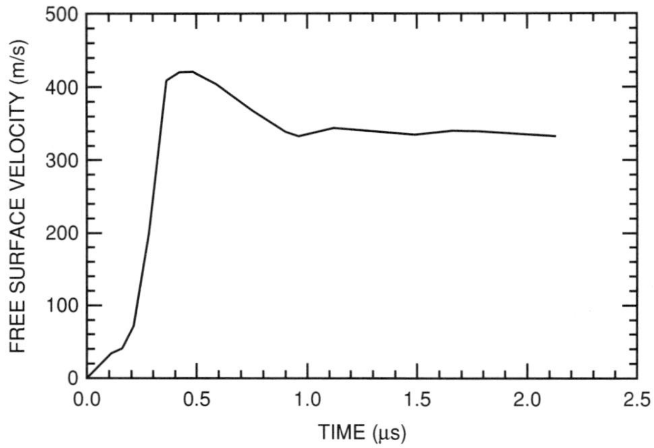

Figure A.5. Velocity history at the free surface of an aluminum D16 target impacted by an aluminum flyer plate at an impact velocity of 450 ± 20 m/s. The experiment was performed using configuration A (see Table A.2) with a_1 = 2 mm and a_2 = 10.0 mm. The data were recorded using a capacitor gauge with an electrode diameter of 20 mm and a measurement accuracy of ±4%. The spall strength was 0.72 ± 0.05 GPa, and the spall thickness was 1.73 (±10%) mm (Kanel [1982a]).

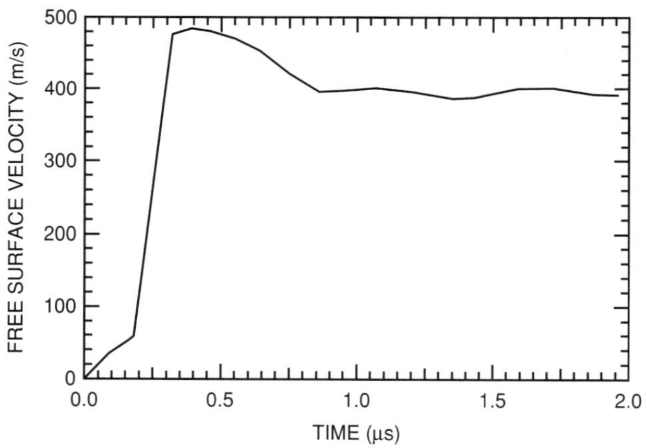

Figure A.6. Velocity history at the free surface of an aluminum D16 target impacted by an aluminum flyer plate at an impact velocity of 700 ± 20 m/s. The experiment was performed using configuration A (see Table A.2) with a_1 = 2 mm and a_2 = 10.0 mm. The data were recorded using a capacitor gauge with an electrode diameter of 20 mm and a measurement accuracy of ±4%. The spall strength was 0.66 ± 0.05 GPa, and the spall thickness was 1.54 (±10%) mm (Kanel [1982a]).

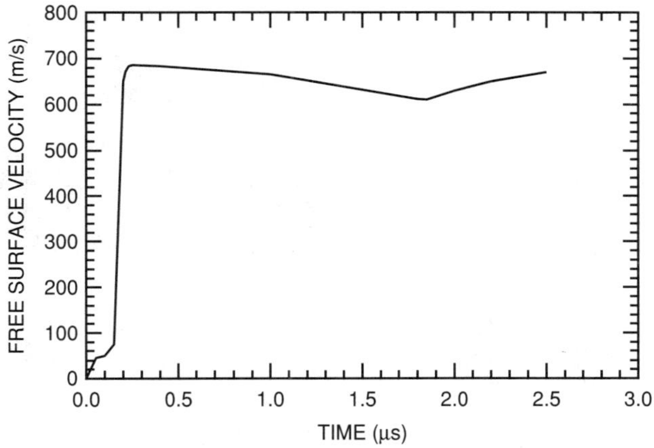

Figure A.7. Velocity history at the free surface of an explosively loaded aluminum AMg6M target. The experiment was performed using configuration H (see Table A.2) with h_1 = 10 mm. The data were recorded using a capacitor gauge with an electrode diameter of 5 mm and a measurement accuracy of ±4%. The spall strength was 0.57 ± 0.1 GPa, and the spall thickness was 4.7 (±10%) mm (Kanel et al. [1984a]).

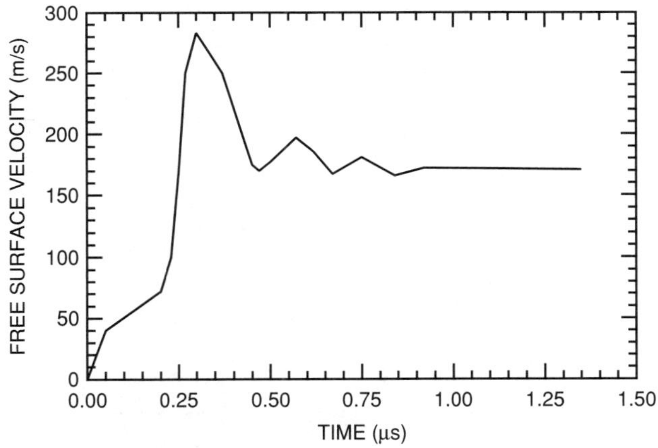

Figure A.8. Velocity history at the free surface of an aluminum AMg6M target impacted by an aluminum flyer plate at an impact velocity of 660 ± 20 m/s. The experiment was performed using configuration A (see Table A.2) with a_1 = 0.4 mm and a_2 = 9.6 mm. The data were recorded using a capacitor gauge with an electrode diameter of 20 mm and a measurement accuracy of ±4%. The spall strength was 0.83 ± 0.08 GPa, and the spall thickness was 0.61 (±10%) mm (Kanel et al. [1984a]).

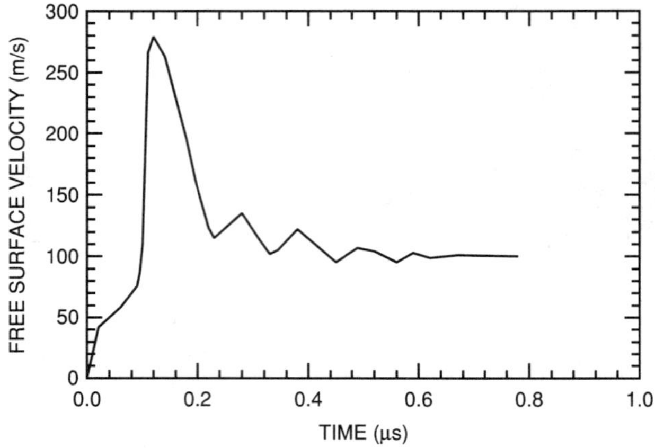

Figure A.9. Velocity history at the free surface of an aluminum AMg6M target impacted by an aluminum flyer plate at an impact velocity of 660 ± 20 m/s. The experiment was performed using configuration A (see Table A.2) with a_1 = 0.19 mm and a_2 = 4.4 mm. The data were recorded using a capacitor gauge with an electrode diameter of 20 mm and a measurement accuracy of ±4%. The spall strength was 1.15 ± 0.05 GPa, and the spall thickness was 0.34 (±10%) mm (Kanel et al. [1984a]).

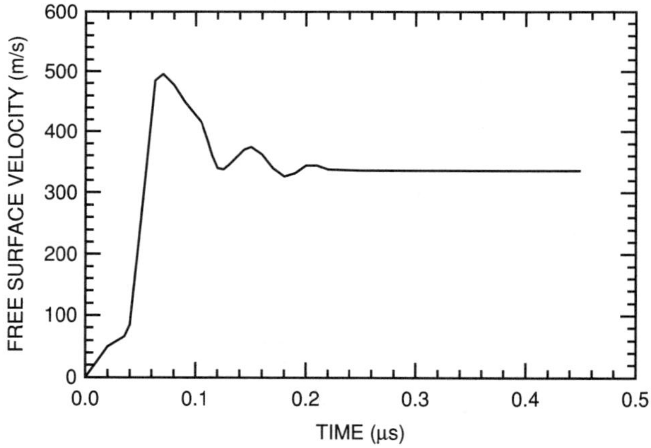

Figure A.10. Velocity history at the free surface of an aluminum AMg6M target impacted by an aluminum flyer plate at an impact velocity of 660 ± 20 m/s. The experiment was performed using configuration A (see Table A.2) with a_1 = 0.19 mm and a_2 = 1.8 mm. The data were recorded using a capacitor gauge with an electrode diameter of 20 mm and a measurement accuracy of ±4%. The spall strength was 1.2 ± 0.12 GPa, and the spall thickness was 0.18 (±10%) mm (Kanel et al. [1984a]).

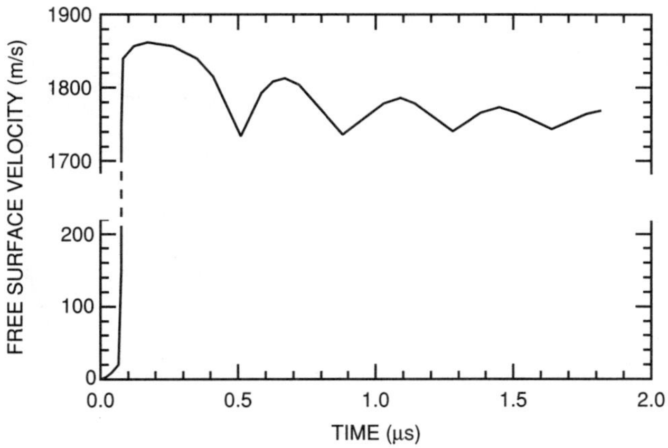

Figure A.11. Velocity history at the free surface of an aluminum AMg6M target impacted by an aluminum flyer plate at an impact velocity of 1900 ± 70 m/s. The experiment was performed using configuration B (see Table A.2) with b_1 = 2 mm and b_2 = 7.0 mm. The data were recorded using a VISAR with a measurement accuracy of 3 to 5 m/s. The spall strength was 0.93 ± 0.04 GPa, and the spall thickness was 1.23 (±10%) mm (Razorenov and Kanel [1986]).

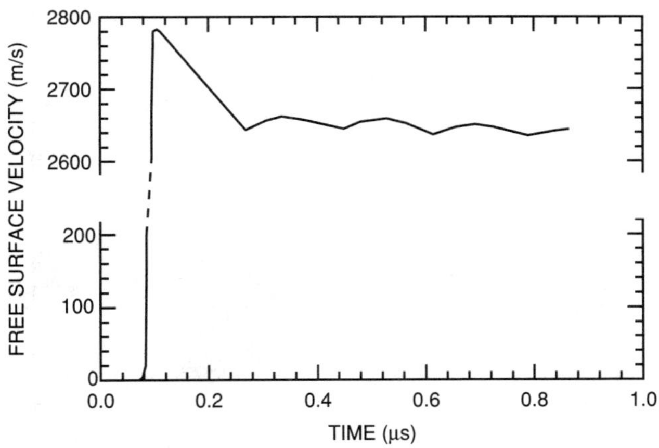

Figure A.12. Velocity history at the free surface of an aluminum AMg6M target impacted by an aluminum flyer plate at an impact velocity of 3000 ± 100 m/s. The experiment was performed using configuration C (see Table A.2) with c_1 = 2 mm, c_2 = 4.0 mm (copper), and c_3 = 10.0 mm. The data were recorded using a VISAR with a measurement accuracy of 3 to 5 m/s. The spall strength was 0.93 ± 0.04 GPa, and the spall thickness was 0.54 (±10%) mm (Razorenov and Kanel [1986]).

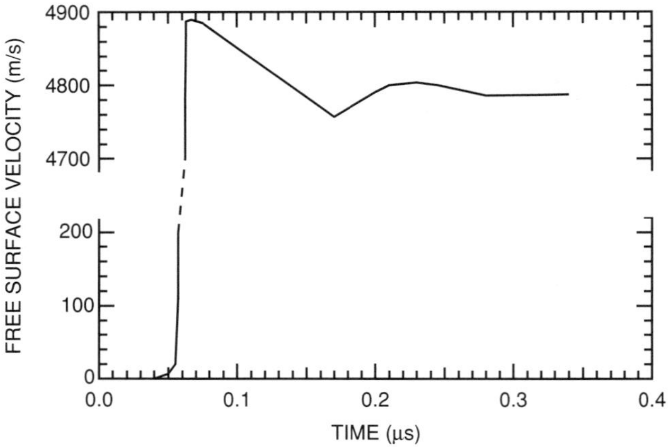

Figure A.13. Velocity history at the free surface of an aluminum AMg6M target impacted by an aluminum flyer plate at an impact velocity of 5300 ± 150 m/s. The experiment was performed using configuration C (see Table A.2) with c_1 = 2 mm, c_2 = 4.0 mm (copper), and c_3 = 4.5 mm. The data were recorded using a VISAR with a measurement accuracy of 3 to 5 m/s. The spall strength was 0.93 ± 0.04 GPa, and the spall thickness was 0.35 (±10%) mm (Razorenov and Kanel [1986]).

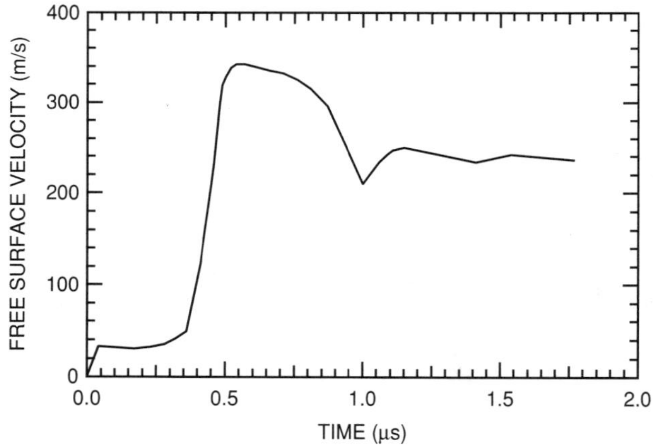

Figure A.14. Velocity history at the free surface of a steel 3 target impacted by an aluminum flyer plate at an impact velocity of 660 ± 20 m/s. The experiment was performed using configuration B (see Table A.2) with b_1 = 2 mm and b_2 = 10.0 mm. The data were recorded using a VISAR with a measurement accuracy of 3 to 5 m/s. The spall strength was 2.9 ± 0.1 GPa, and the spall thickness was 1.16 (±10%) mm (Kanel et al. [1987b]).

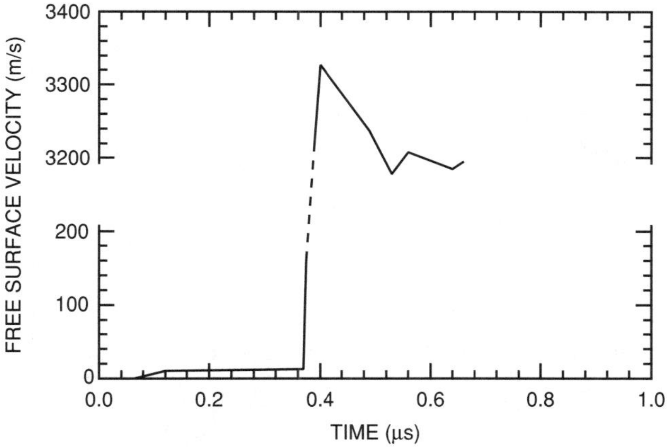

Figure A.15. Velocity history at the free surface of a steel 3 target impacted by an aluminum flyer plate at an impact velocity of 5300 ± 150 m/s. The experiment was performed using configuration B (see Table A.2) with b_1 = 2 mm and b_2 = 5.0 mm. The data were recorded using a VISAR with a measurement accuracy of 3 to 5 m/s. The spall strength was 2.9 ± 0.1 GPa, and the spall thickness was 0.31 (±10%) mm (Kanel et al. [1987b]).

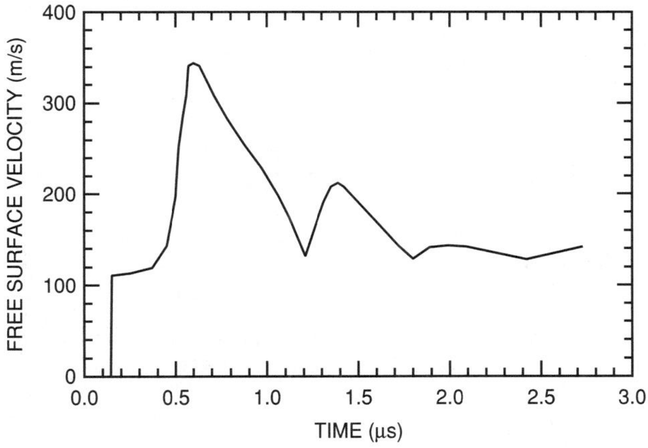

Figure A.16. Velocity history at the free surface of a steel 45 target impacted by an aluminum flyer plate at an impact velocity of 700 ± 20 m/s. The experiment was performed using configuration B (see Table A.2) with b_1 = 2 mm and b_2 = 11.1 mm. The data were recorded using a VISAR with a measurement accuracy of 3 to 5 m/s. The spall strength was 4.25 ± 0.1 GPa, and the spall thickness was 1.81 (±10%) mm (Razorenov et al. [1992]).

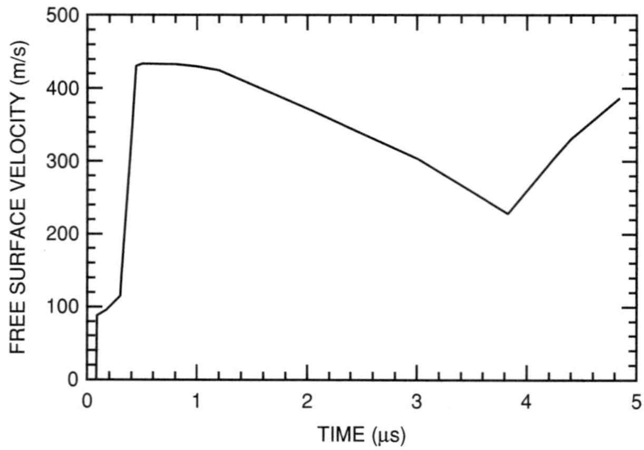

Figure A.17. Velocity history at the free surface of an explosively loaded steel 45 target. The experiment was performed using configuration J (see Table A.2) with j_1 = 11.1 mm. The data were recorded using a VISAR with a measurement accuracy of 3 to 5 m/s. No spall signal was observed (Razorenov et al. [1992]).

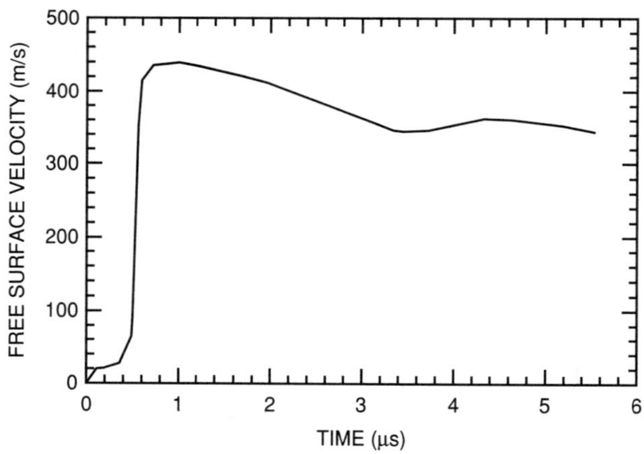

Figure A.18. Velocity history at the free surface of an explosively loaded stainless steel (Kh18N10T) target. The experiment was performed using configuration H (see Table A.2) with h_1 = 15 mm. The data were recorded using a capacitor gauge with an electrode diameter of 20 mm and a measurement accuracy of ±4%. The spall strength was 1.73 ± 0.11 GPa, and the spall thickness was 6.9 (±10%) mm (Kanel [1982a]).

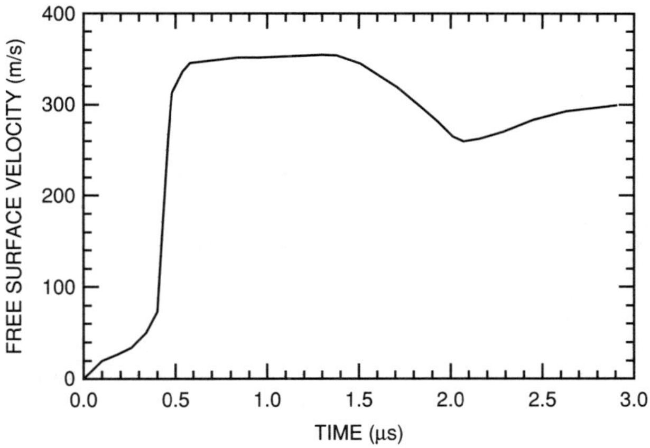

Figure A.19. Velocity history at the free surface of a stainless steel (Kh18N10T) target impacted by an aluminum flyer plate at an impact velocity of 600 ± 10 m/s. The experiment was performed using configuration A (see Table A.2) with a_1 = 5.0 mm and a_2 = 10.0 mm. The data were recorded using a capacitor gauge with an electrode diameter of 20 mm and a measurement accuracy of ±4%. The spall strength was 1.75 ± 0.11 GPa (Kanel [1982a]).

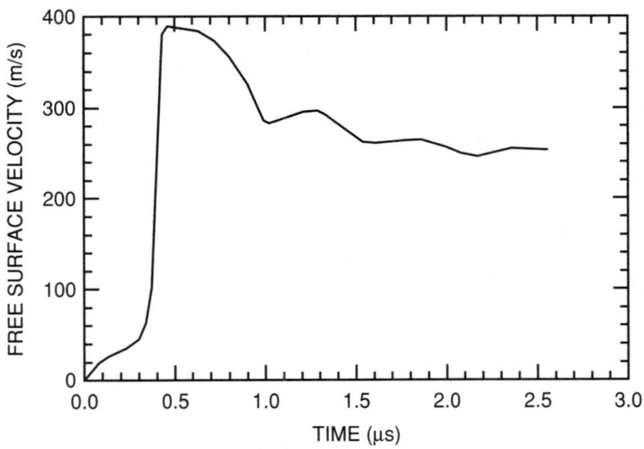

Figure A.20. Velocity history at the free surface of a stainless steel (Kh18N10T) target impacted by an aluminum flyer plate at an impact velocity of 700 ± 30 m/s. The experiment was performed using configuration A (see Table A.2) with a_1 = 2.0 mm and a_2 = 10.0 mm. The data were recorded using a capacitor gauge with an electrode diameter of 20 mm and a measurement accuracy of ±4%. The spall strength was 1.93 ± 0.09 GPa, and the spall thickness was 1.61 (±10%) mm (Kanel [1982a]).

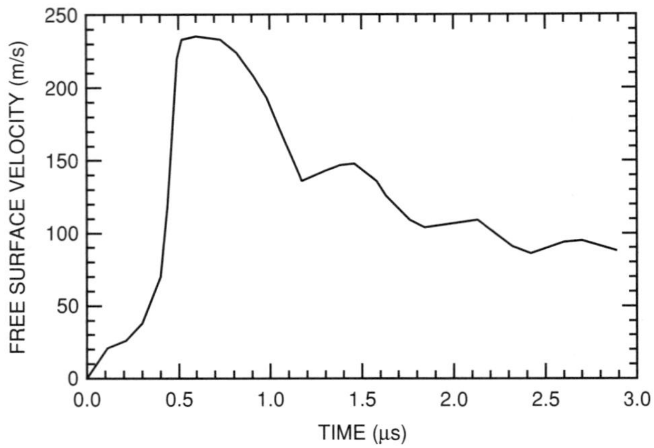

Figure A.21. Velocity history at the free surface of a stainless steel (Kh18N10T) target impacted by an aluminum flyer plate at an impact velocity of 445 ± 15 m/s. The experiment was performed using configuration A (see Table A.2) with a_1 = 2.0 mm and a_2 = 10.0 mm. The data were recorded using a capacitor gauge with an electrode diameter of 20 mm and a measurement accuracy of ±4%. The spall strength was 1.93 ± 0.13 GPa, and the spall thickness was 1.78 (±10%) mm (Kanel [1982a]).

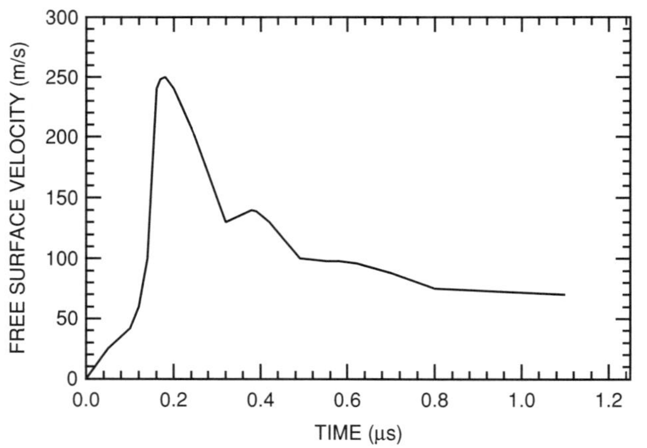

Figure A.22. Velocity history at the free surface of a stainless steel (Kh18N10T) target impacted by an aluminum flyer plate at an impact velocity of 700 ± 30 m/s. The experiment was performed using configuration A (see Table A.2) with a_1 = 0.4 mm and a_2 = 4.0 mm. The data were recorded using a capacitor gauge with an electrode diameter of 5 mm and a measurement accuracy of ±4%. The spall strength was 2.28 ± 0.1 GPa, and the spall thickness was 0.65 (±10%) mm (Kanel and Razorenov [1989]).

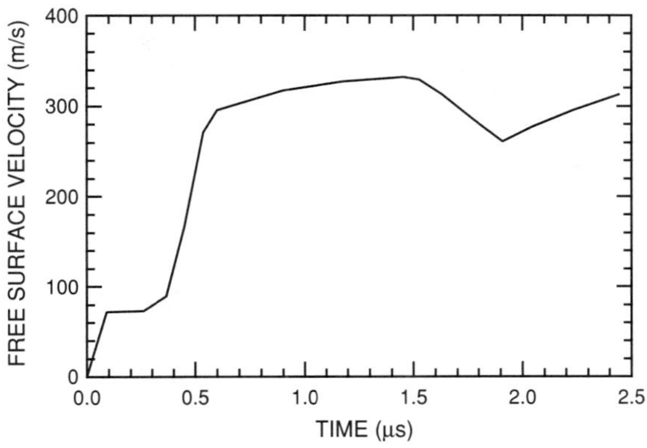

Figure A.23. Velocity history at the free surface of a KhVG steel target impacted by an aluminum flyer plate at an impact velocity of 600 ± 20 m/s. The experiment was performed using configuration A (see Table A.2) with a_1 = 5.0 mm and a_2 = 10.0 mm. The data were recorded using a capacitor gauge with an electrode diameter of 20 mm and a measurement accuracy of ±4%. The spall strength† was calculated to be 1.45 GPa using Eq. (4.2) and 1.59 GPa using Eqs. (4.4) to (4.6) (Kanel and Razorenov [1989]).

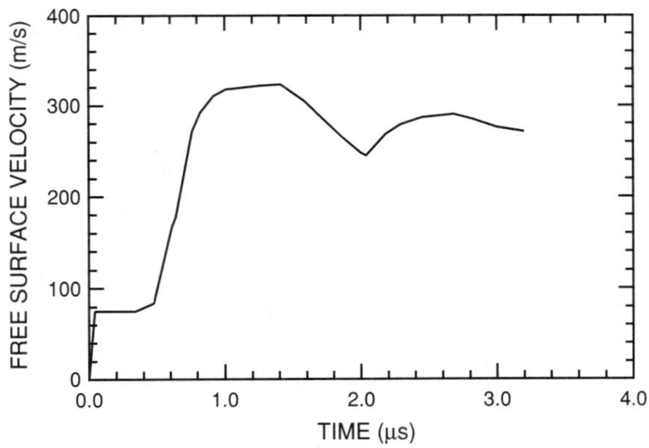

Figure A.24. Velocity history at the free surface of a KhVG steel target impacted by an aluminum flyer plate at an impact velocity of 600 ± 20 m/s. The experiment was performed using configuration A (see Table A.2) with a_1 = 5.0 mm and a_2 = 15.0 mm. The data were recorded using a capacitor gauge with an electrode diameter of 20 mm and a measurement accuracy of ±4%. The spall strength† was calculated to be 1.57 GPa using Eq. (4.2) and 1.68 GPa using Eqs. (4.4) to (4.6) (Kanel and Razorenov [1989]).

† Both reported values underestimate the spall strength due to relatively large spall signal distortion at this ratio of target to impactor thickness.

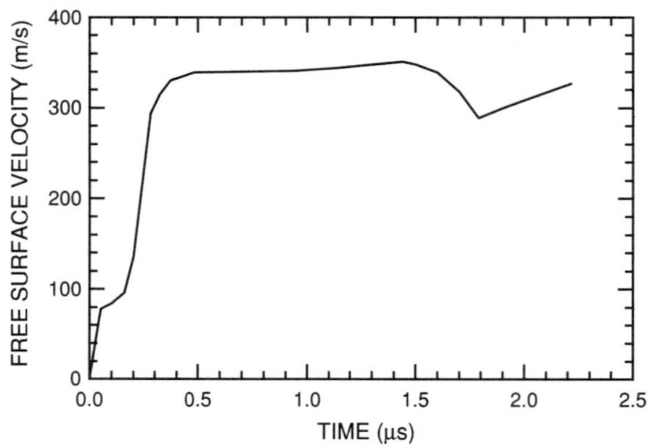

Figure A.25. Velocity history at the free surface of a KhVG steel target impacted by an aluminum flyer plate at an impact velocity of 600 ± 20 m/s. The experiment was performed using configuration A (see Table A.2) with a_1 = 5.0 mm and a_2 = 5.0 mm. The data were recorded using a capacitor gauge with an electrode diameter of 20 mm and a measurement accuracy of ±4%. The spall strength† was calculated to be 1.27 GPa using Eq. (4.2) and 1.49 GPa using Eqs. (4.4) to (4.6) (Kanel and Razorenov [1989]).

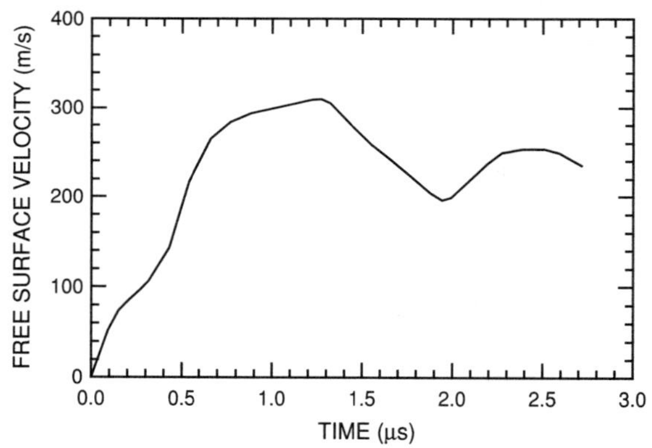

Figure A.26. Velocity history at the free surface of a quenched KhVG steel target impacted by an aluminum plate at an impact velocity of 600 ± 20 m/s. The experiment was performed using configuration A (see Table A.2) with a_1 = 5.0 mm and a_2 = 20.0 mm. The data were recorded using a capacitor gauge with an electrode diameter of 20 mm and a measurement accuracy of ±4%. The spall strength† was calculated to be 2.31 GPa using Eq. (4.2) and 2.41 GPa using Eqs. (4.4) to (4.6) (Kanel and Razorenov [1989]).

† Both reported values underestimate the spall strength due to relatively large spall signal distortion at this ratio of target to impactor thickness.

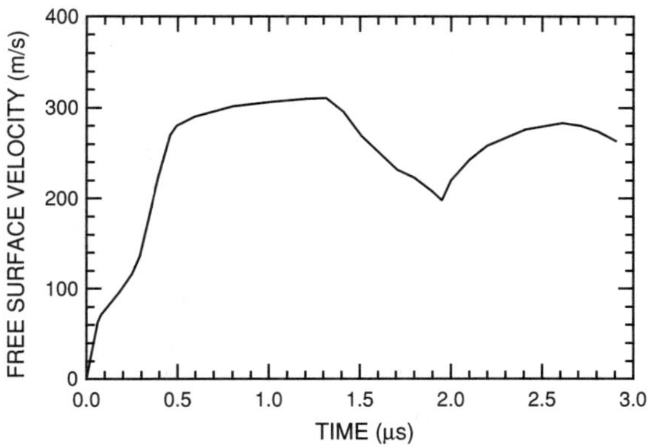

Figure A.27. Velocity history at the free surface of a quenched KhVG steel target impacted by an aluminum plate at an impact velocity of 600 ± 20 m/s. The experiment was performed using configuration A (see Table A.2) with a_1 = 5.0 mm and a_2 = 15.0 mm. The data were recorded using a capacitor gauge with an electrode diameter of 20 mm and a measurement accuracy of ±4%. The spall strength† was calculated to be 2.29 GPa using Eq. (4.2) and 2.51 GPa using Eqs. (4.4) to (4.6) (Kanel and Razorenov [1989]).

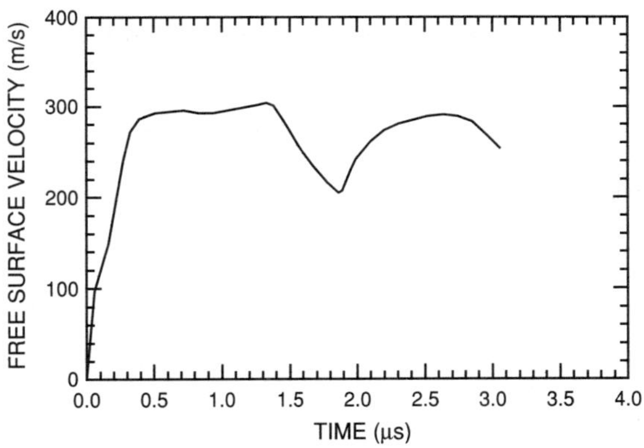

Figure A.28. Velocity history at the free surface of a quenched KhVG steel target impacted by an aluminum plate at an impact velocity of 600 ± 20 m/s. The experiment was performed using configuration A (see Table A.2) with a_1 = 5.0 mm and a_2 = 10.0 mm. The data were recorded using a capacitor gauge with an electrode diameter of 20 mm and a measurement accuracy of ±4%. The spall strength† was calculated to be 2.01 GPa using Eq. (4.2) and 2.32 GPa using Eqs. (4.4) to (4.6) (Kanel and Razorenov [1989]).

† Both reported values underestimate the spall strength due to relatively large spall signal distortion at this ratio of target to impactor thickness.

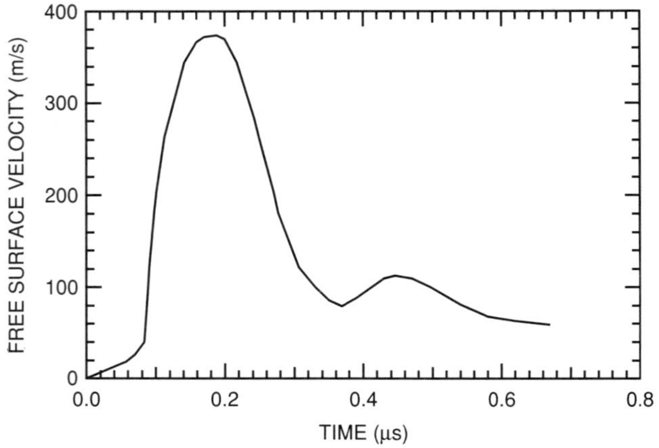

Figure A.29. Velocity history at the free surface of a high-purity titanium target impacted by an aluminum flyer plate at an impact velocity of 660 ± 20 m/s. The experiment was performed using configuration C (see Table A.2) with c_1 = 0.4 mm, c_2 = 0.78 mm (aluminum), and c_3 = 2.06 mm. The data were recorded using a VISAR with a measurement accuracy of 3 to 5 m/s. The spall strength was 3.2 ± 0.2 GPa (Razorenov et al. [1995b]).

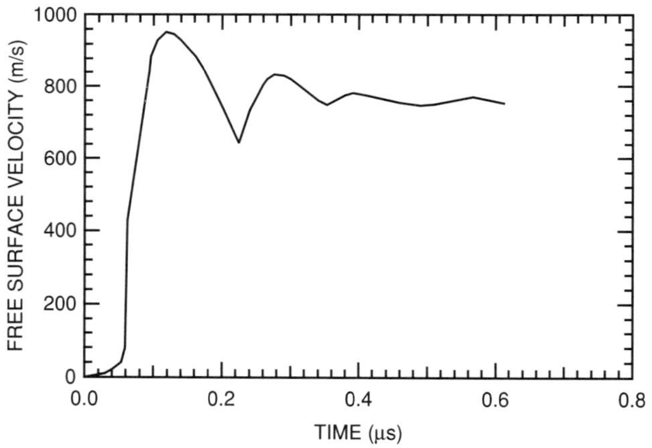

Figure A.30. Velocity history at the free surface of a high-purity titanium target impacted by an aluminum flyer plate at an impact velocity of 1250 ± 50 m/s. The experiment was performed using configuration C (see Table A.2) with c_1 = 0.4 mm, c_2 = 0.77 mm (aluminum), and c_3 = 2.29 mm. The data were recorded using a VISAR with a measurement accuracy of 3 to 5 m/s. The spall strength was 3.34 ± 0.2 GPa, and the spall thickness was 0.37 (±10%) mm (Razorenov et al. [1995b]).

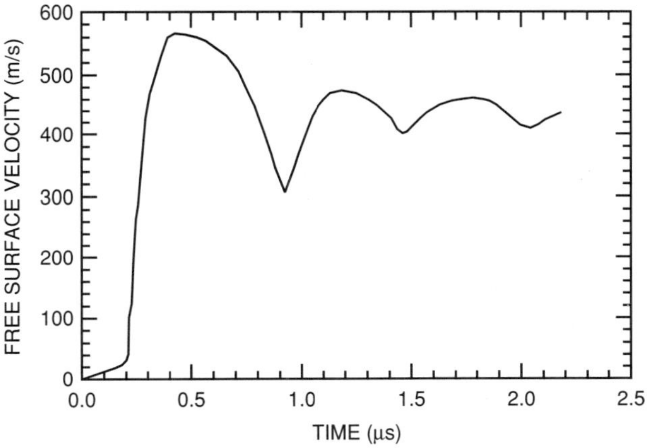

Figure A.31. Velocity history at the free surface of a high-purity titanium target impacted by an aluminum flyer plate at an impact velocity of 660 ± 20 m/s. The experiment was performed using configuration C (see Table A.2) with c_1 = 2.0 mm, c_2 = 2.0 mm (aluminum), and c_3 = 4.4 mm. The data were recorded using a VISAR with a measurement accuracy of 3 to 5 m/s. The spall strength was 2.81 ± 0.1 GPa, and the spall thickness was 1.54 (±10%) mm (Razorenov et al. [1995b]).

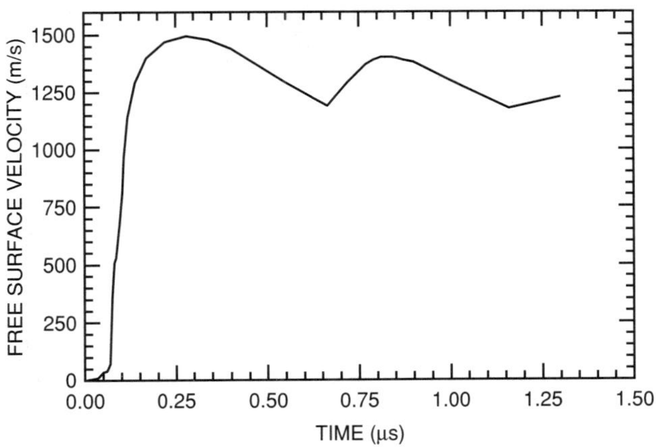

Figure A.32. Velocity history at the free surface of a high-purity titanium target impacted by an aluminum flyer plate at an impact velocity of 1900 ± 70 m/s. The experiment was performed using configuration C (see Table A.2) with c_1 = 2.0 mm, c_2 = 2.0 mm (aluminum), and c_3 = 4.3 mm. The data were recorded using a VISAR with a measurement accuracy of 3 to 5 m/s. The spall strength was 3.47 ± 0.1 GPa, and the spall thickness was 1.43 (±10%) mm (Razorenov et al. [1995b]).

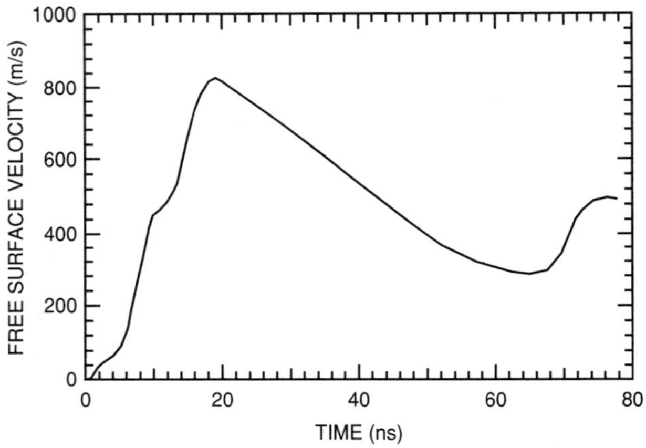

Figure A.33. Velocity history at the free surface of a high-purity titanium target subjected on its front surface to radiation from an ion beam with an 8-mm spot size and an energy density of 0.2 TW/cm^2. The experiment was performed using configuration K (see Table A.2) with k_1 = 0.78 mm. The data were recorded using ORVIS with a measurement accuracy of ±20 m/s. The spall strength was 6.33 ± 0.1 GPa, and the spall thickness was 0.149 (±10%) mm (Razorenov et al. [1995b]).

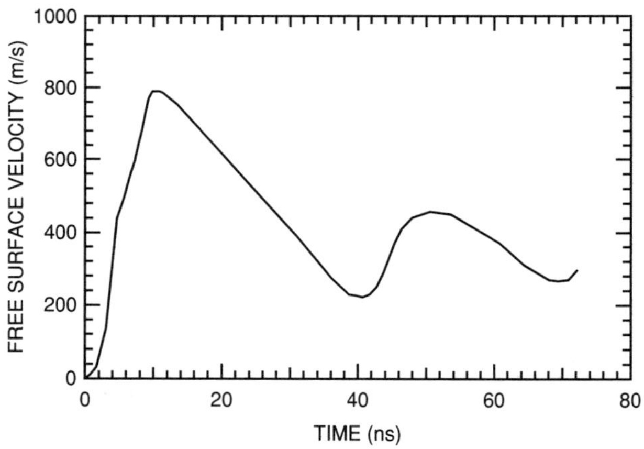

Figure A.34. Velocity history at the free surface of a high-purity titanium target subjected on its front surface to radiation from an ion beam with an 8-mm spot size and an energy density of 0.2 TW/cm^2. The experiment was performed using configuration K (see Table A.2) with k_1 = 0.483 mm. The data were recorded using ORVIS with a measurement accuracy of ±20 m/s. The spall strength was 6.28 ± 0.1 GPa, and the spall thickness was 0.054 (±10%) mm (Razorenov et al. [1995b]).

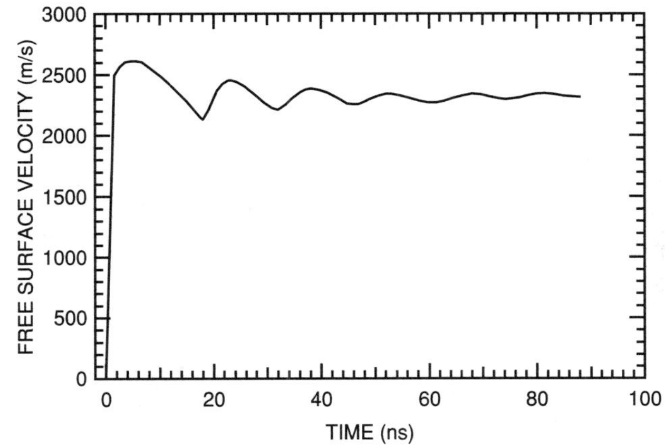

Figure A.35. Velocity history at the free surface of a high-purity titanium target impacted by an ion beam-launched aluminum flyer plate at an impact velocity of 4100 ± 150 m/s. The beam had an 8-mm spot size and an energy density of 0.2 TW/cm^2. The experiment was performed using configuration L (see Table A.2) with l_1 = 0.05 mm and l_2 = 0.45 mm. The data were recorded using ORVIS with a measurement accuracy of ±20 m/s. The spall strength was 6.33 ± 0.1 GPa, and the spall thickness was 0.041 (±10%) mm (Razorenov et al. [1995b]).

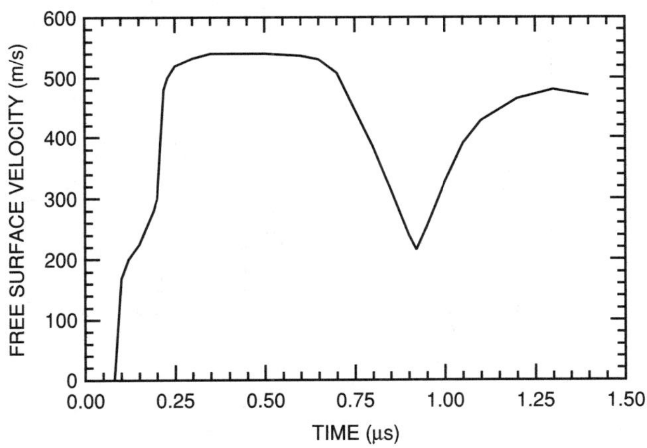

Figure A.36. Velocity history at the free surface of a titanium VT5-1 target impacted by an aluminum flyer plate at an impact velocity of 660 ± 20 m/s. The experiment was performed using configuration B (see Table A.2) with b_1 = 2 mm and b_2 = 4.0 mm. The data were recorded using a VISAR with a measurement accuracy of 3 to 5 m/s. The spall strength was 4.24 ± 0.1 GPa (Kanel et al. [1986]).

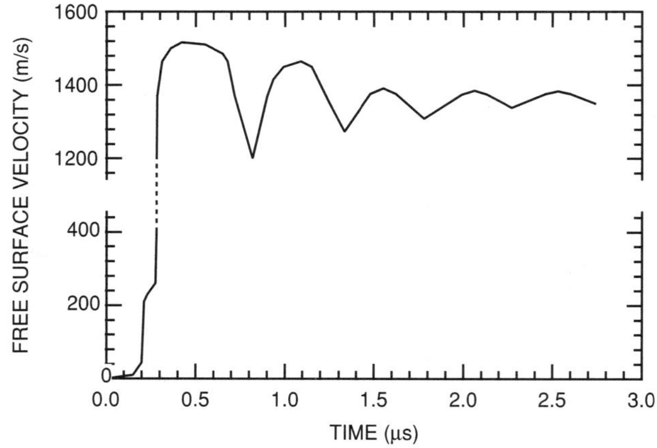

Figure A.37. Velocity history at the free surface of a titanium VT5-1 target impacted by an aluminum flyer plate at an impact velocity of 1900 ± 100 m/s. The experiment was performed using configuration B (see Table A.2) with b_1 = 2 mm and b_2 = 4.0 mm. The data were recorded using a VISAR with a measurement accuracy of 3 to 5 m/s. The spall strength was 4.13 ± 0.1 GPa (Kanel et al. [1986]).

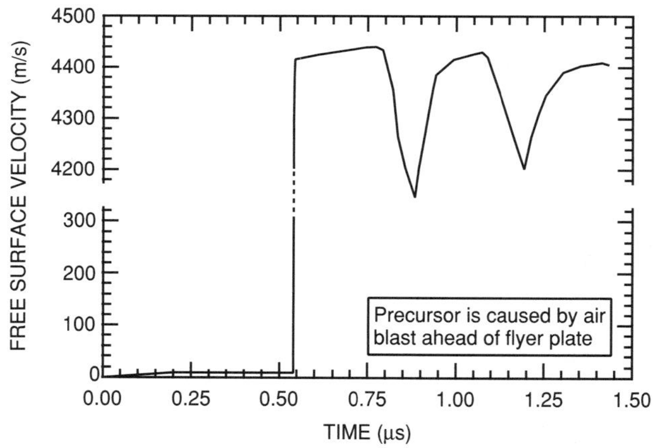

Figure A.38. Velocity history at the free surface of a titanium VT5-1 target impacted by an aluminum flyer plate at an impact velocity of 5300 ± 100 m/s. The experiment was performed using configuration B (see Table A.2) with b_1 = 2 mm and b_2 = 4.0 mm. The data were recorded using a VISAR with a measurement accuracy of 3 to 5 m/s. The spall strength was 3.98 ± 0.1 GPa (Kanel et al. [1986]).

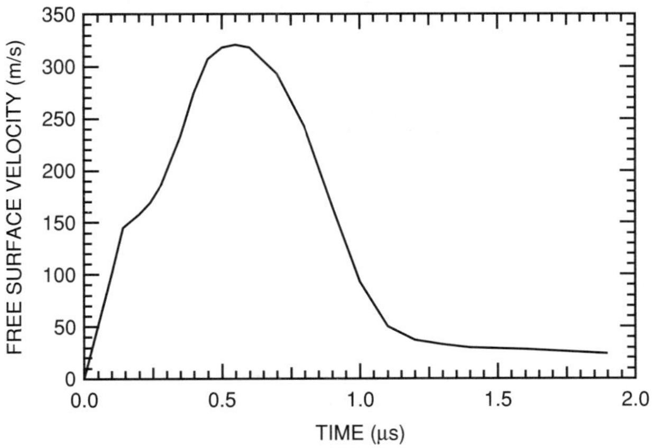

Figure A.39. Velocity history at the free surface of a titanium VT6 target impacted by an aluminum flyer plate at an impact velocity of 450 ± 20 m/s. The experiment was performed using configuration A (see Table A.2) with a_1 = 2.0 mm and a_2 = 9.8 mm. The data were recorded using a capacitor gauge with an electrode diameter of 20 mm and a measurement accuracy of ±4%. No spall was observed (Kanel and Petrova [1981]).

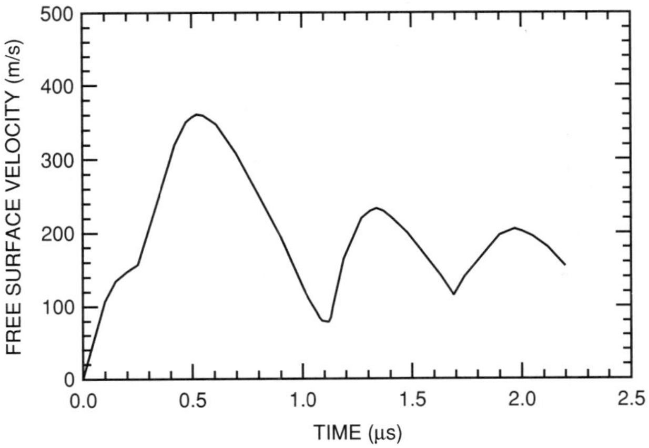

Figure A.40. Velocity history at the free surface of a titanium VT6 target impacted by an aluminum flyer plate at an impact velocity of 675 ± 20 m/s. The experiment was performed using configuration A (see Table A.2) with a_1 = 2.0 mm and a_2 = 11.7 mm. The data were recorded using a capacitor gauge with an electrode diameter of 20 mm and a measurement accuracy of ±4%. The spall strength was 3.5 ± 0.2 GPa, and the spall thickness was 1.75 (±10%) mm (Kanel and Petrova [1981]).

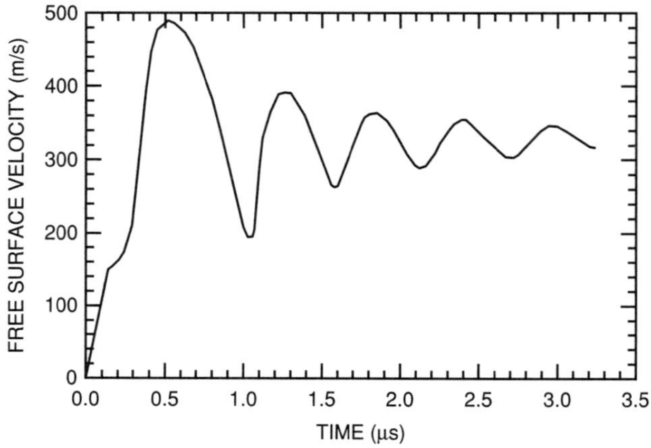

Figure A.41. Velocity history at the free surface of a titanium VT6 target impacted by an aluminum flyer plate at an impact velocity of 700 ± 20 m/s. The experiment was performed using configuration A (see Table A.2) with a_1 = 2.0 mm and a_2 = 9.8 mm. The data were recorded using a capacitor gauge with an electrode diameter of 20 mm and a measurement accuracy of ±4%. The spall strength was 3.87 ± 0.17 GPa, and the spall thickness was 1.66 (±10%) mm (Kanel and Petrova [1981]).

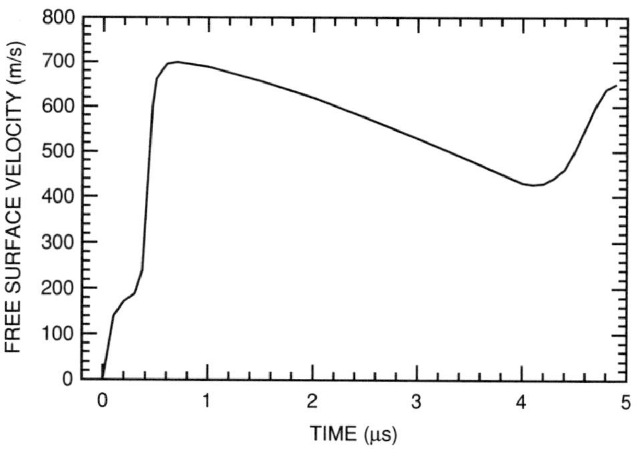

Figure A.42. Velocity history at the free surface of an explosively loaded titanium VT6 target. The experiment was performed using configuration H (see Table A.2) with h_1 = 18.4 mm. The data were recorded using a capacitor gauge with an electrode diameter of 20 mm and a measurement accuracy of ±4%. The spall strength was 3.43 ± 0.2 GPa, and the spall thickness was 9.8 (±10%) mm (Kanel and Petrova [1981]).

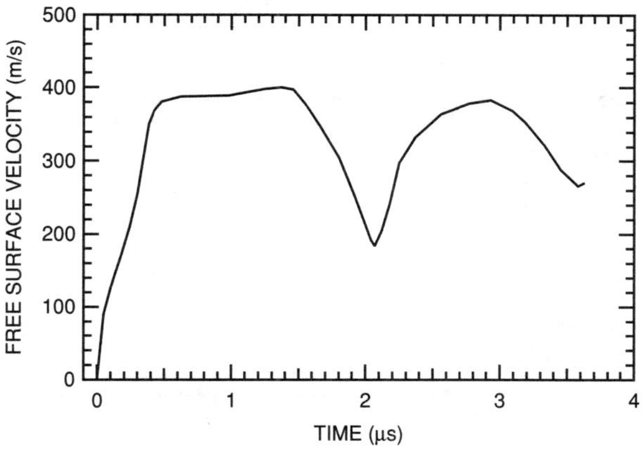

Figure A.43. Velocity history at the free surface of a titanium VT6 target impacted by an aluminum flyer plate at an impact velocity of 600 ± 20 m/s. The experiment was performed using configuration A (see Table A.2) with a_1 = 5.0 mm and a_2 = 9.0 mm. The data were recorded using a capacitor gauge with an electrode diameter of 20 mm and a measurement accuracy of ±4%. The spall strength† was calculated to be 2.60 GPa using Eq. (4.2) and 3.01 GPa using Eqs. (4.4) to (4.6) (Kanel and Petrova [1981]).

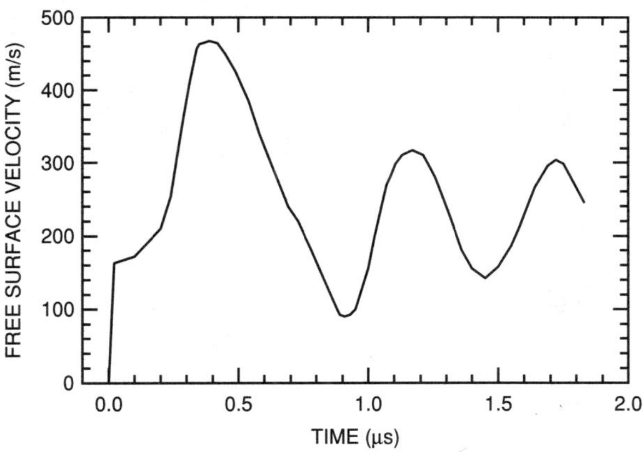

Figure A.44. Velocity history at the free surface of a titanium VT8 target impacted by an aluminum plate at an impact velocity of 660 ± 20 m/s. The experiment was performed using configuration B (see Table A.2) with b_1 = 2 mm and b_2 = 10 mm. The data were recorded using a VISAR with a measurement accuracy of 3 to 5 m/s. The spall strength was 4.65 ± 0.3 GPa, and the spall thickness was 1.6 (±10%) mm (Kanel et al. [1987b]).

† Both reported values underestimate the spall strength due to relatively large spall signal distortion at this ratio of target to impactor thickness.

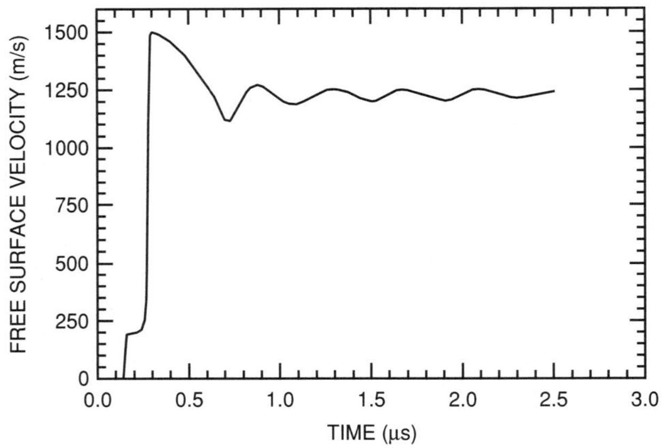

Figure A.45. Velocity history at the free surface of a titanium VT8 target impacted by an aluminum plate at an impact velocity of 1900 ± 70 m/s. The experiment was performed using configuration B (see Table A.2) with b_1 = 2 mm and b_2 = 10 mm. The data were recorded using a VISAR with a measurement accuracy of 3 to 5 m/s. The spall strength was 4.63 ± 0.3 GPa, and the spall thickness was 1.15 (±10%) mm (Kanel et al. [1987b]).

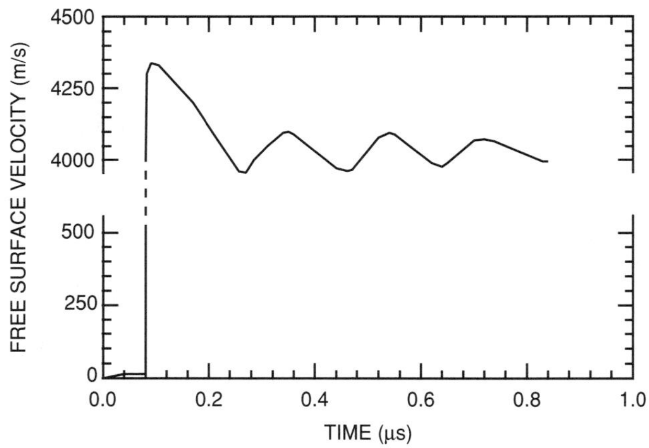

Figure A.46. Velocity history at the free surface of a titanium VT8 target impacted by an aluminum plate at an impact velocity of 5300 ± 150 m/s. The experiment was performed using configuration B (see Table A.2) with b_1 = 2 mm and b_2 = 10 mm. The data were recorded using a VISAR with a measurement accuracy of 3 to 5 m/s. The spall strength was 4.63 ± 0.3 GPa, and the spall thickness was 0.52 (±10%) mm (Kanel et al. [1987b]).

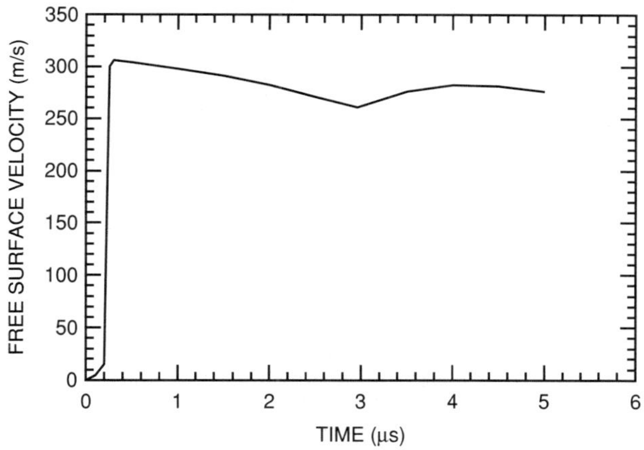

Figure A.47. Velocity history at the free surface of an explosively loaded copper M2 target. The experiment was performed using configuration H (see Table A.2) with h_1 = 12.0 mm. The data were recorded using a capacitor gauge with an electrode diameter of 20 mm and a measurement accuracy of ±4%. The spall strength was 0.8 ± 0.1 GPa, and the spall thickness was 6.0 (±10%) mm (Razorenov and Kanel [1992]).

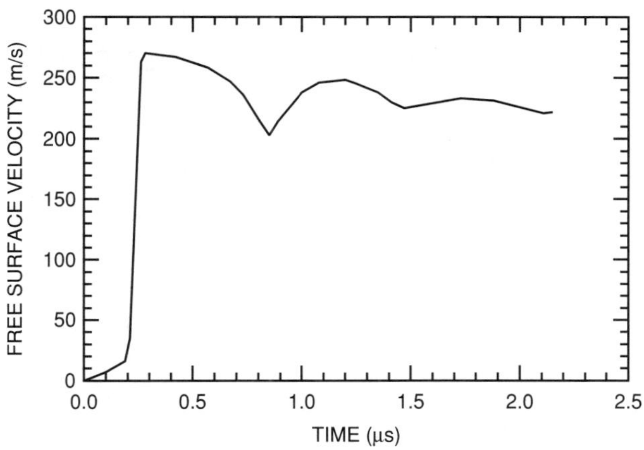

Figure A.48. Velocity history at the free surface of a copper M2 target impacted by an aluminum flyer plate at an impact velocity of 450 ± 20 m/s. The experiment was performed using configuration A (see Table A.2) with a_1 = 2.0 mm and a_2 = 15.0 mm. The data were recorded using a capacitor gauge with an electrode diameter of 20 mm and a measurement accuracy of ±4%. The spall strength was 1.1 ± 0.1 GPa, and the spall thickness was 1.24 (±10%) mm (Razorenov and Kanel [1992]).

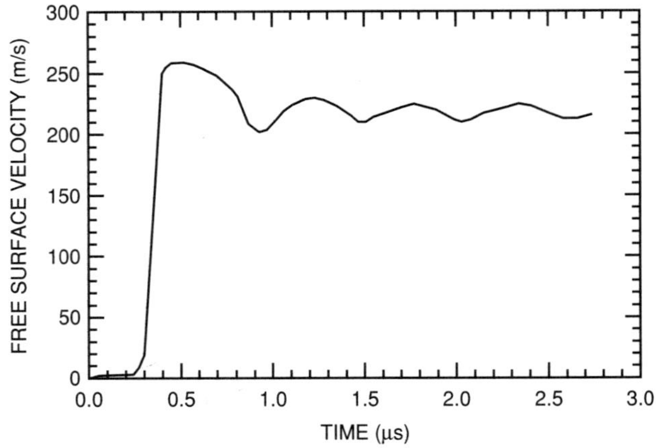

Figure A.49. Velocity history at the free surface of a copper M2 target impacted by an aluminum flyer plate at an impact velocity of 450 ± 20 m/s. The experiment was performed using configuration A (see Table A.2) with a_1 = 2.0 mm and a_2 = 12.0 mm. The data were recorded using a capacitor gauge with an electrode diameter of 20 mm and a measurement accuracy of ±4%. The spall strength was 1.12 ± 0.1 GPa, and the spall thickness was 1.15 (±10%) mm (Razorenov and Kanel [1992]).

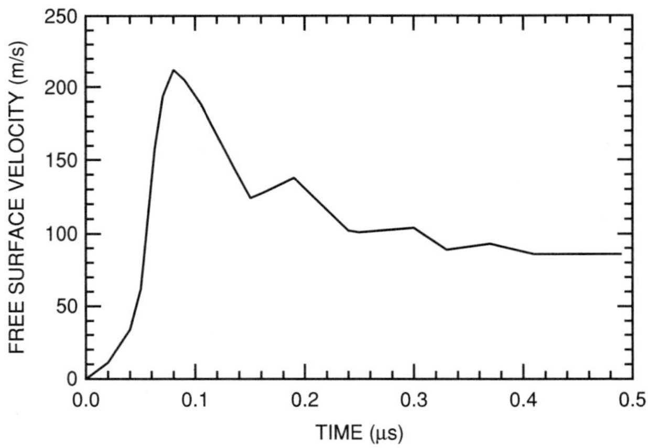

Figure A.50. Velocity history at the free surface of a copper M2 target impacted by an aluminum flyer plate at an impact velocity of 660 ± 20 m/s. The experiment was performed using configuration A (see Table A.2) with a_1 = 0.2 mm and a_2 = 3.9 mm. The data were recorded using a capacitor gauge with an electrode diameter of 5 mm and a measurement accuracy of ±4%. The spall strength was 1.64 ± 0.1 GPa, and the spall thickness was 0.18 (±10%) mm (Razorenov and Kanel [1992]).

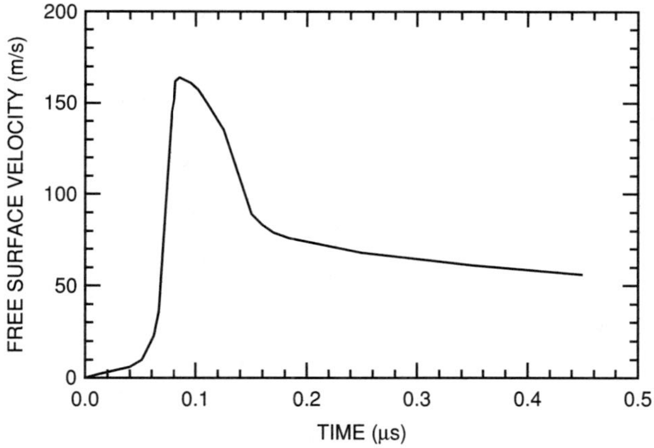

Figure A.51. Velocity history at the free surface of a copper M2 target impacted by an aluminum flyer plate at an impact velocity of 660 ± 20 m/s. The experiment was performed using configuration B (see Table A.2) with b_1 = 0.2 mm and b_2 = 3.9 mm. The data were recorded using a VISAR with a measurement accuracy of 3 to 5 m/s. The spall strength was 1.5 ± 0.1 GPa, and the spall thickness was 0.18 (±10%) mm (Razorenov and Kanel [1992]).

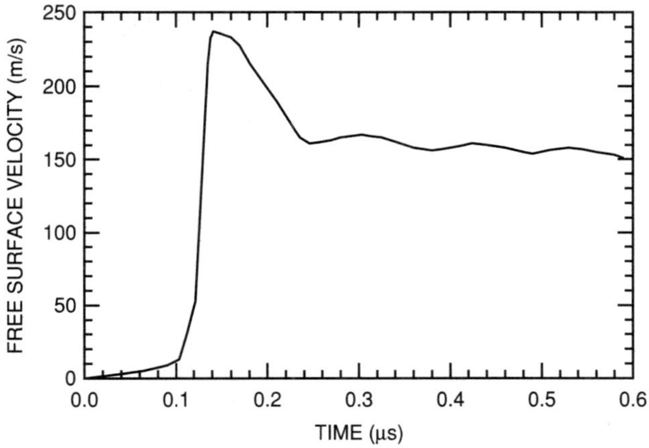

Figure A.52. Velocity history at the free surface of a copper M2 target impacted by an aluminum flyer plate at an impact velocity of 660 ± 20 m/s. The experiment was performed using configuration B (see Table A.2) with b_1 = 0.4 mm and b_2 = 2.7 mm. The data were recorded using a VISAR with a measurement accuracy of 3 to 5 m/s. The spall strength was 1.35 ± 0.1 GPa, and the spall thickness was 0.23 (±10%) mm (Razorenov and Kanel [1992]).

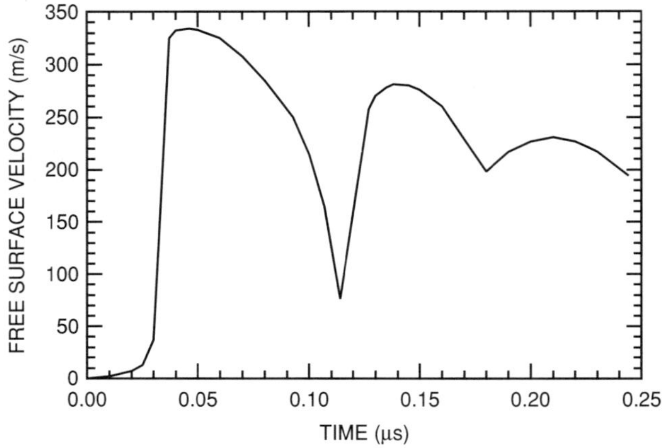

Figure A.53. Velocity history at the free surface of a copper single crystal target impacted in the <111> direction by an aluminum flyer plate at an impact velocity of 660 ± 20 m/s. The experiment was performed using configuration B (see Table A.2) with b_1 = 0.2 mm and b_2 = 0.7 mm. The data were recorded using a VISAR with a measurement accuracy of 3 to 5 m/s. The spall strength was 4.5 ± 0.1 GPa, and the spall thickness was 0.14 (±10%) mm (Razorenov and Kanel [1992]).

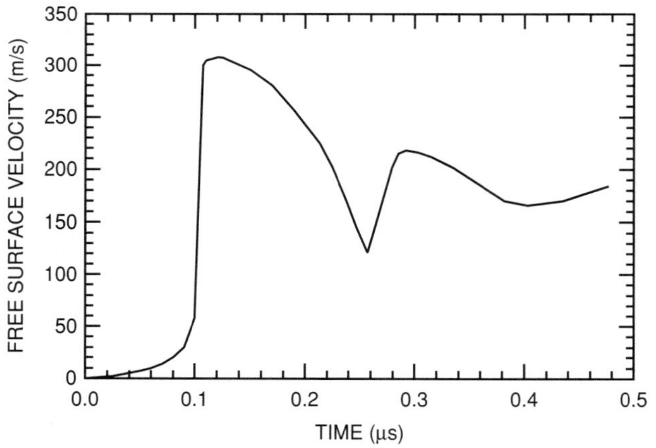

Figure A.54. Velocity history at the free surface of a copper single crystal target impacted in the <111> direction by an aluminum flyer plate at an impact velocity of 660 ± 20 m/s. The experiment was performed using configuration B (see Table A.2) with b_1 = 0.4 mm and b_2 = 1.9 mm. The data were recorded using a VISAR with a measurement accuracy of 3 to 5 m/s. The spall strength was 3.45 ± 0.2 GPa, and the spall thickness was 0.33 (±10%) mm (Razorenov and Kanel [1992]).

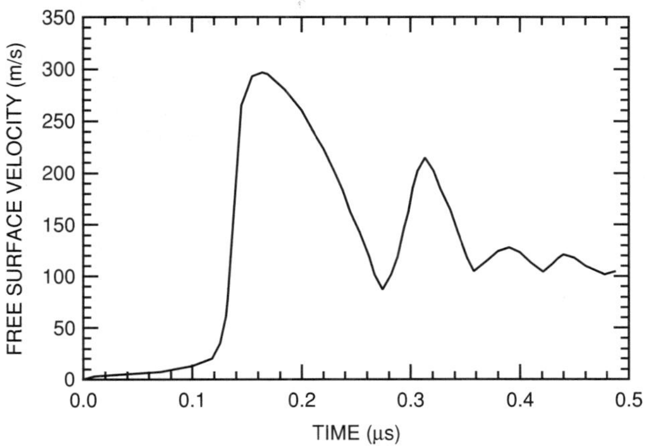

Figure A.55. Velocity history at the free surface of a copper single crystal target impacted in the <111> direction by an aluminum flyer plate at an impact velocity of 660 ± 20 m/s. The experiment was performed using configuration B (see Table A.2) with b_1 = 0.4 mm and b_2 = 1.95 mm. The data were recorded using a VISAR with a measurement accuracy of 3 to 5 m/s. The spall strength was 3.75 ± 0.2 GPa, and the spall thickness was 0.25 (±10%) mm (Razorenov and Kanel [1992]).

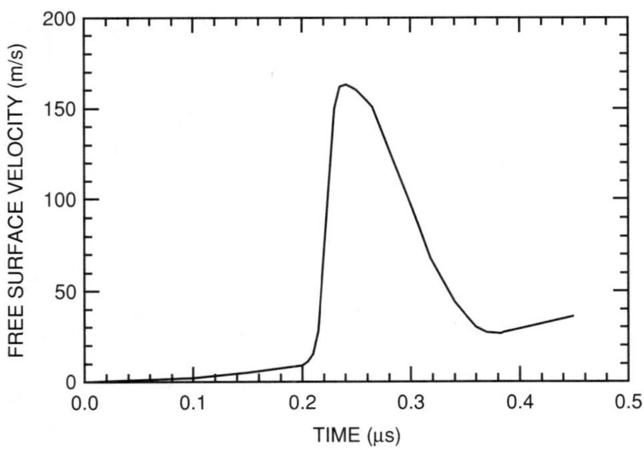

Figure A.56. Velocity history at the free surface of a copper single crystal target impacted in the <111> direction by an aluminum flyer plate at an impact velocity of 660 ± 20 m/s. The experiment was performed using configuration B (see Table A.2) with b_1 = 0.2 mm and b_2 = 4.35 mm. The data were recorded using a VISAR with a measurement accuracy of 3 to 5 m/s. A peak tensile stress of 2.5 ± 0.1 GPa was reached during the test, but no spall was observed (Razorenov and Kanel [1992]).

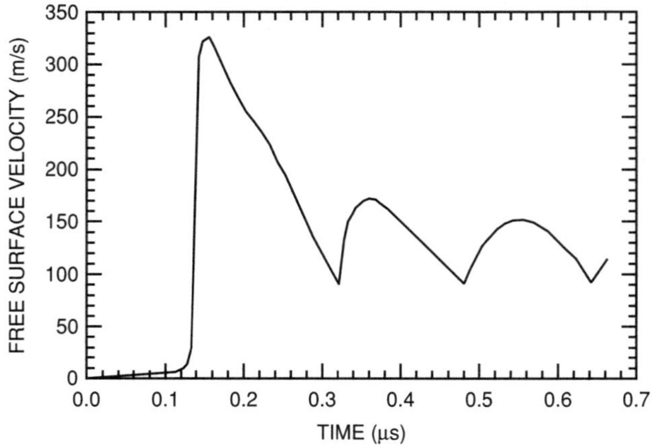

Figure A.57. Velocity history at the free surface of a copper single crystal target impacted in the <100> direction by an aluminum flyer plate at an impact velocity of 660 ± 20 m/s. The experiment was performed using configuration B (see Table A.2) with b_1 = 0.4 mm and b_2 = 4.3 mm. The data were recorded using a VISAR with a measurement accuracy of 3 to 5 m/s. The spall strength was 4.5 ± 0.1 GPa, and the spall thickness was 0.33 (±10%) mm (Razorenov and Kanel [1992]).

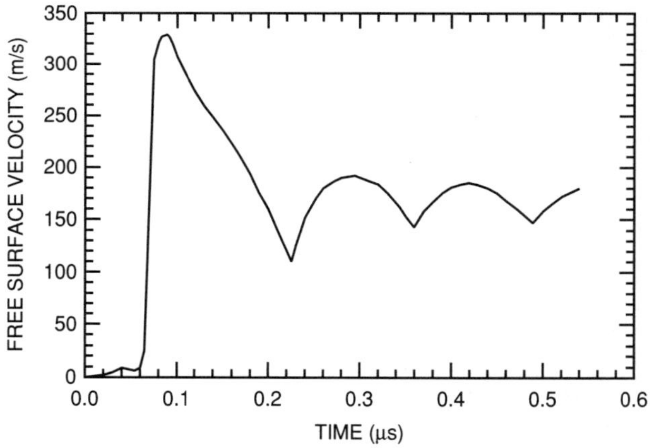

Figure A.58. Velocity history at the free surface of a copper single crystal target annealed for 2 hours at 900°C, then impacted in the <100> direction by an aluminum flyer plate at an impact velocity of 660 ± 20 m/s. The experiment was performed using configuration B (see Table A.2) with b_1 = 0.4 mm and b_2 = 4.5 mm. The data were recorded using a VISAR with a measurement accuracy of 3 to 5 m/s. The spall strength was 3.95 ± 0.1 GPa, and the spall thickness was 0.3 (±10%) mm (Razorenov and Kanel [1992]).

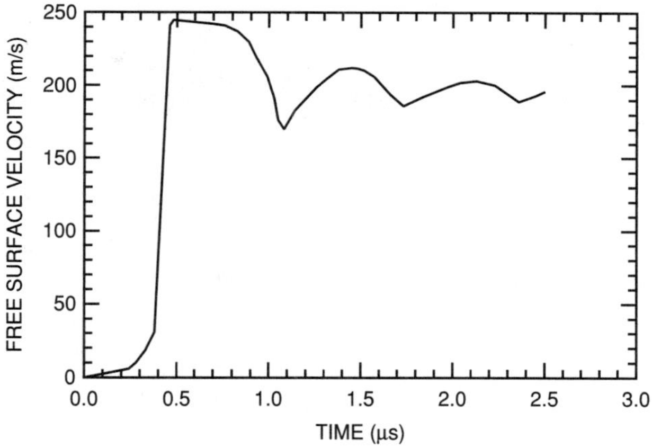

Figure A.59. Velocity history at the free surface of a nickel NP-2 target impacted by an aluminum flyer plate at an impact velocity of 445 ± 15 m/s. The experiment was performed using configuration A (see Table A.2) with a_1 = 2.0 mm and a_2 = 9.5 mm. The data were recorded using a capacitor gauge with an electrode diameter of 20 mm and a measurement accuracy of ±4%. The spall strength was 1.49 ± 0.08 GPa, and the spall thickness was 1.46 (±10%) mm (Kanel [1982a]).

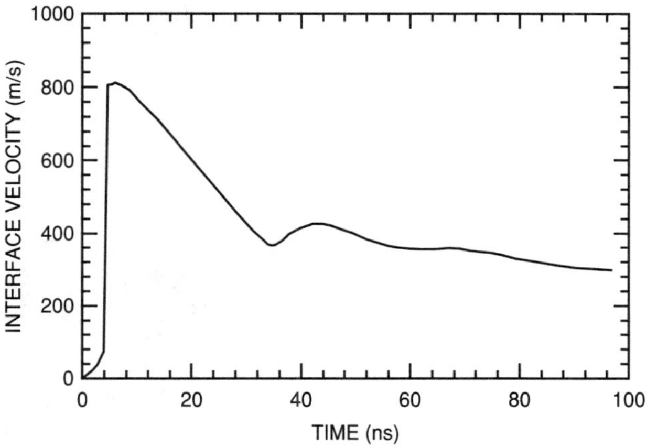

Figure A.60. Velocity history at the interface between a 99.99% pure nickel target and a water window. The front surface of the specimen was irradiated using an ion beam with an 8-mm spot size and an energy density of 0.2 TW/cm^2. The experiment was performed using configuration M (see Table A.2) with m_1 = 0.4 mm. The data were recorded using ORVIS with a measurement accuracy of ±20 m/s. The spall strength was 5.9 ± 0.3 GPa, and the spall thickness was 0.08 (±10%) mm.

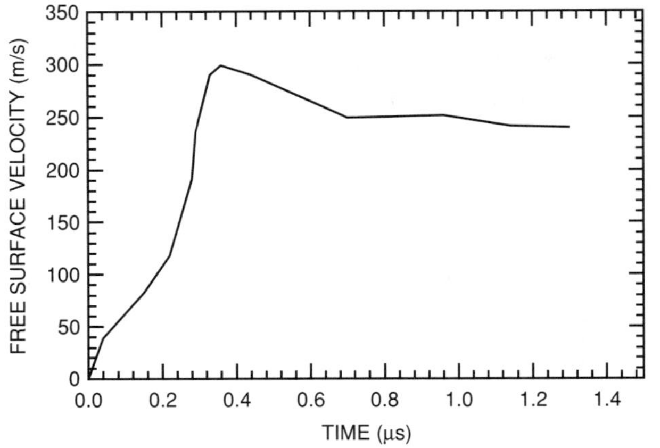

Figure A.61. Velocity history at the free surface of a sintered molybdenum polycrystal target impacted by an aluminum flyer plate at an impact velocity of 660 ± 20 m/s. The experiment was performed using configuration C (see Table A.2) with c_1 = 2 mm, c_2 = 4.9 mm (aluminum), and c_3 = 5.9 mm. The data were recorded using a VISAR with a measurement accuracy of 3 to 5 m/s. The spall strength was 1.32 ± 0.05 GPa, and the spall thickness was 1.13 (±10%) mm (Kanel et al. [1993]).

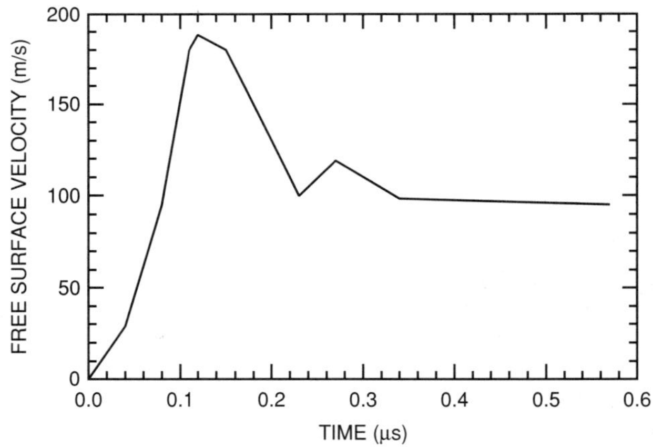

Figure A.62. Velocity history at the free surface of a sintered molybdenum polycrystal target impacted by an aluminum flyer plate at an impact velocity of 660 ± 20 m/s. The experiment was performed using configuration C (see Table A.2) with c_1 = 0.4 mm, c_2 = 2.0 mm (aluminum), and c_3 = 2.07 mm. The data were recorded using a VISAR with a measurement accuracy of 3 to 5 m/s. The spall strength was 2.4 ± 0.05 GPa, and the spall thickness was 0.4 (±10%) mm (Kanel et al. [1993]).

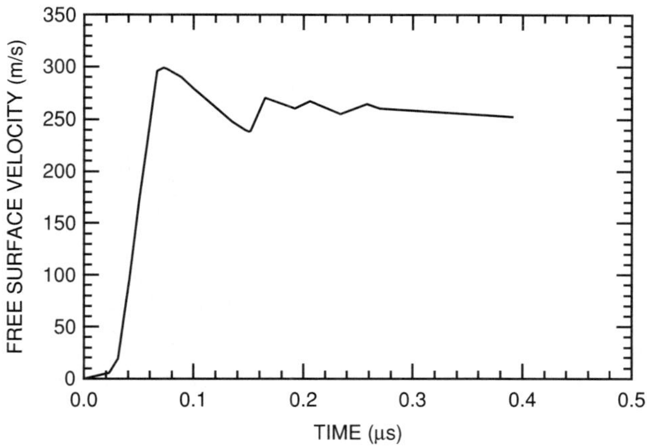

Figure A.63. Velocity history at the free surface of a sintered molybdenum polycrystal target impacted by an aluminum flyer plate at an impact velocity of 660 ± 20 m/s. The experiment was performed using configuration C (see Table A.2) with c_1 = 0.4 mm, c_2 = 2.0 mm (aluminum), and c_3 = 0.75 mm. The data were recorded using a VISAR with a measurement accuracy of 3 to 5 m/s. The spall strength was 1.8 ± 0.05 GPa, and the spall thickness was 0.26 (±10%) mm (Kanel et al. [1993]).

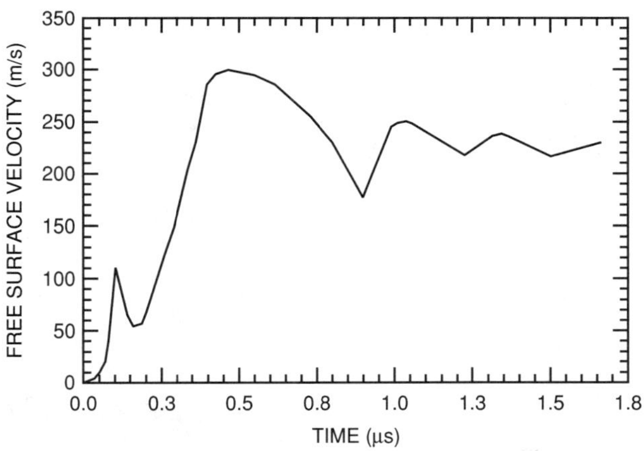

Figure A.64. Velocity history at the free surface of a molybdenum single crystal target impacted in the <100> direction by an aluminum flyer plate at an impact velocity of 660 ± 20 m/s. The experiment was performed using configuration C (see Table A.2) with c_1 = 2.0 mm, c_2 = 4.0 mm (aluminum), and c_3 = 3.88 mm. The data were recorded using a VISAR with a measurement accuracy of 3 to 5 m/s. The spall strength was 3.3 ± 0.2 GPa, and the spall thickness was 1.05 (±10%) mm (Kanel et al. [1993]).

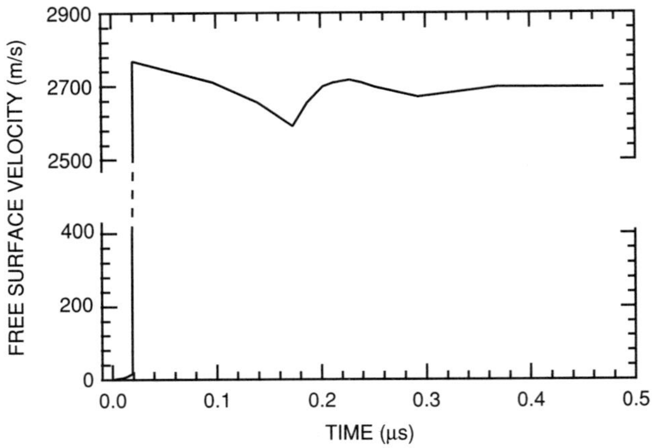

Figure A.65. Velocity history at the free surface of a molybdenum single crystal target impacted in the <100> direction by an aluminum flyer plate at an impact velocity of 5300 ± 150 m/s. The experiment was performed using configuration C (see Table A.2) with c_1 = 2 mm, c_2 = 5.0 mm (aluminum), and c_3 = 4.6 mm. The data were recorded using a VISAR with a measurement accuracy of 3 to 5 m/s. The spall strength was 4.85 ± 0.2 GPa, and the spall thickness was 1.3 (±10%) mm (Kanel et al. [1993]).

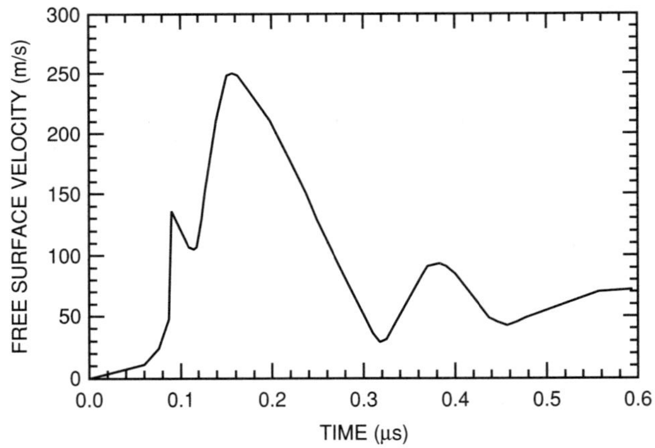

Figure A.66. Velocity history at the free surface of a molybdenum single crystal target impacted in the <100> direction by an aluminum flyer plate at an impact velocity of 660 ± 20 m/s. The experiment was performed using configuration C (see Table A.2) with c_1 = 0.4 mm, c_2 = 2.0 mm (aluminum), and c_3 = 1.4 mm. The data were recorded using a VISAR with a measurement accuracy of 3 to 5 m/s. The spall strength was 5.6 ± 0.05 GPa, and the spall thickness was 0.43 (±10%) mm (Kanel et al. [1993]).

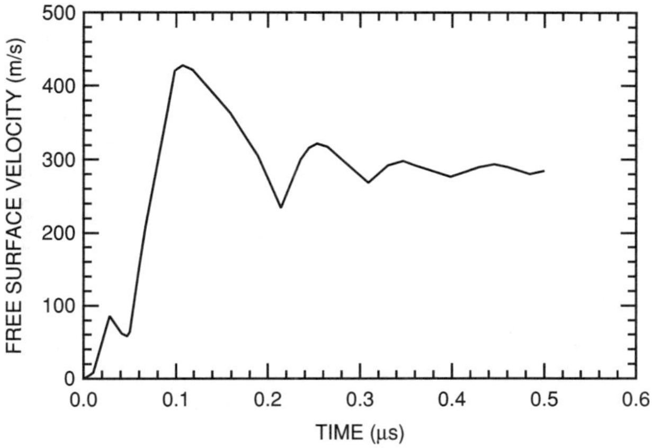

Figure A.67. Velocity history at the free surface of a molybdenum single crystal target impacted in the <100> direction by an aluminum flyer plate at an impact velocity of 1250 ± 50 m/s. The experiment was performed using configuration C (see Table A.2) with c_1 = 0.4 mm, c_2 = 2.0 mm (aluminum), and c_3 = 1.38 mm. The data were recorded using a VISAR with a measurement accuracy of 3 to 5 m/s. The spall strength was 5.4 ± 0.2 GPa, and the spall thickness was 0.29 (±10%) mm (Kanel et al. [1993]).

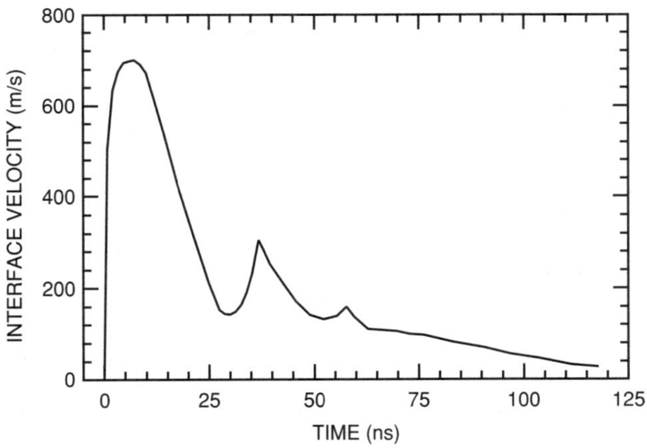

Figure A.68. Velocity history at the interface between a molybdenum single crystal target and a water window. The front surface of the specimen was irradiated in the <100> direction using an ion beam with an 8-mm spot size and an energy density of 0.2 TW/cm^2. The experiment was performed using configuration M (see Table A.2) with m_1 = 0.32 mm. The data were recorded using ORVIS with a measurement accuracy of ±20 m/s. The spall strength was 13.5 ± 0.1 GPa, and the spall thickness was 0.07 (±10%) mm (Kanel et al. [1993]).

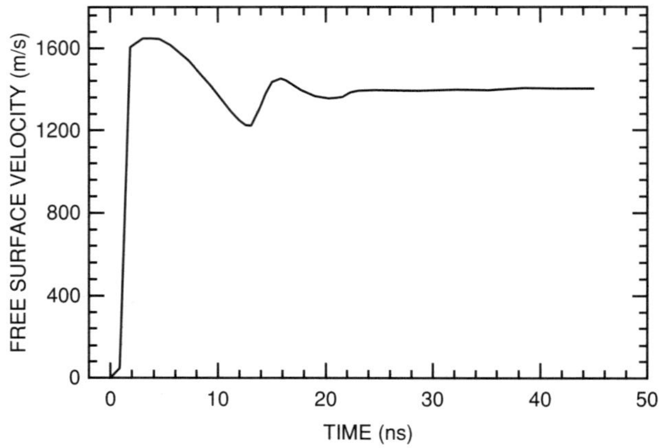

Figure A.69. Velocity history at the free surface of a deformed (90% to 95% prestrain) molybdenum single crystal target impacted in the <100> direction by an ion beam-launched aluminum flyer plate at an impact velocity of 4100 ± 150 m/s. The beam had an 8-mm spot size and an energy density of 0.2 TW/cm^2. The experiment was performed using configuration L (see Table A.2) with l_1 = 0.05 mm and l_2 = 0.275 mm. The data were recorded using ORVIS with a measurement accuracy of ±20 m/s. The spall strength was 11.5 ± 0.2 GPa, and the spall thickness was 0.022 (±10%) mm (Kanel et al. [1993]).

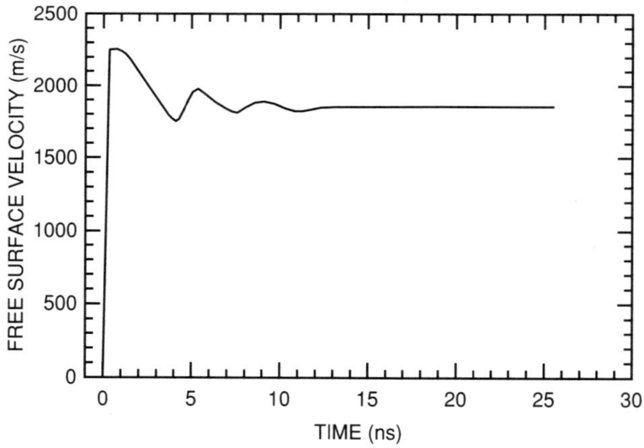

Figure A.70. Velocity history at the free surface of a deformed (90% to 95% prestrain) molybdenum single crystal target impacted in the <100> direction by an ion beam-launched aluminum flyer plate at an impact velocity of 4100 ± 150 m/s. The beam had an 8-mm spot size and an energy density of 0.2 TW/cm^2. The experiment was performed using configuration L (see Table A.2) with l_1 = 0.05 mm and l_2 = 0.1 mm. The data were recorded using ORVIS with a measurement accuracy of ±20 m/s. The spall strength was 13.4 ± 0.2 GPa, and the spall thickness was 0.01 (±10%) mm (Kanel et al. [1993]).

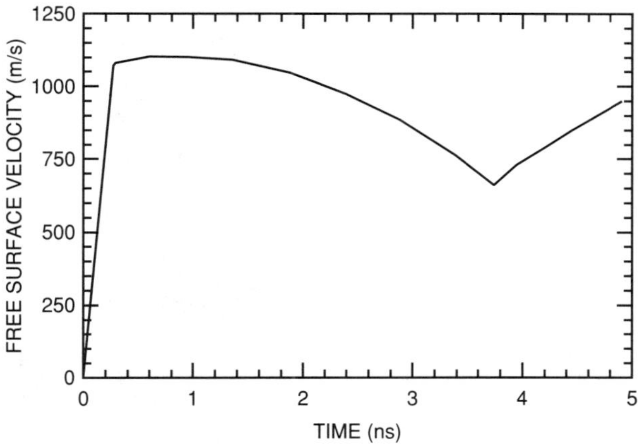

Figure A.71. Velocity history at the free surface of a deformed (90% to 95% prestrain) molybdenum single crystal target impacted in the <100> direction by an ion beam-launched aluminum flyer plate at an impact velocity of 4100 ± 150 m/s. The beam had an 8-mm spot size and an energy density of 0.2 TW/cm^2. The experiment was performed using configuration L (see Table A.2) with l_1 = 0.01 mm and l_2 = 0.275 mm. The data were recorded using ORVIS with a measurement accuracy of ±20 m/s. The spall strength was 16.5 ± 0.2 GPa, and the spall thickness was 0.009 (±10%) mm (Kanel et al. [1993]).

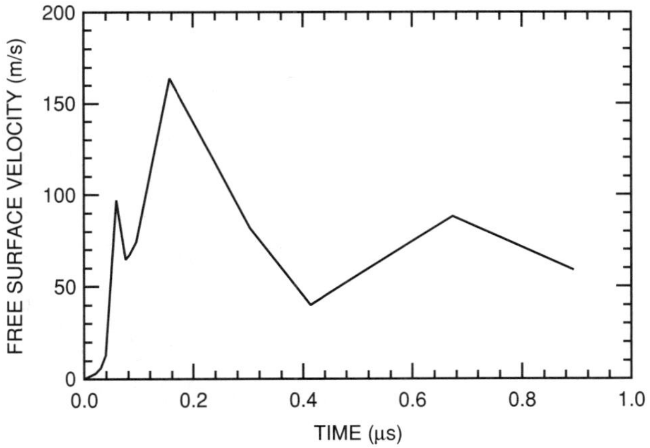

Figure A.72. Velocity history at the free surface of a molybdenum single crystal target impacted in the <110> direction by an aluminum flyer plate at an impact velocity of 660 ± 20 m/s. The experiment was performed using configuration C (see Table A.2) with c_1 = 0.4 mm, c_2 = 2.0 mm (aluminum), and c_3 = 1.37 mm. The data were recorded using a VISAR with a measurement accuracy of 3 to 5 m/s. The spall strength was 3.2 ± 0.2 GPa, and the spall thickness was 1.8 (±10%) mm (Kanel et al. [1993]).

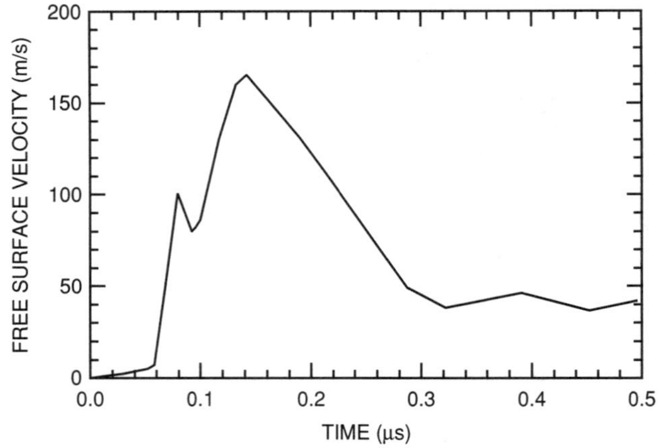

Figure A.73. Velocity history at the free surface of a molybdenum single crystal target impacted in the <110> direction by an aluminum flyer plate at an impact velocity of 660 ± 20 m/s. The experiment was performed using configuration C (see Table A.2) with c_1 = 0.4 mm, c_2 = 1.45 mm (aluminum), and c_3 = 1.45 mm. The data were recorded using a VISAR with a measurement accuracy of 3 to 5 m/s. No spall was observed (Kanel et al. [1993]).

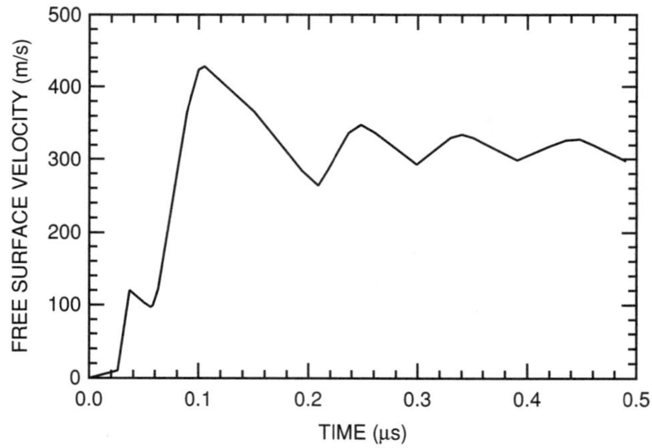

Figure A.74. Velocity history at the free surface of a molybdenum single crystal target impacted in the <110> direction by an aluminum flyer plate at an impact velocity of 1250 ± 70 m/s. The experiment was performed using configuration C (see Table A.2) with c_1 = 0.4 mm, c_2 = 1.96 mm (aluminum), and c_3 = 1.47 mm. The data were recorded using a VISAR with a measurement accuracy of 3 to 5 m/s. The spall strength was 4.6 ± 0.1 GPa, and the spall thickness was 0.29 (±10%) mm (Kanel et al. [1993]).

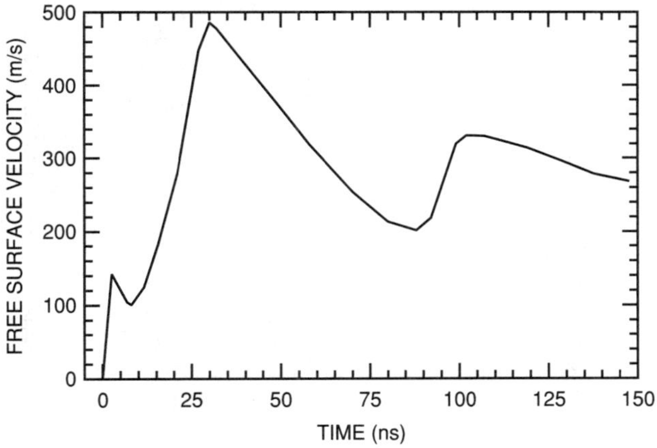

Figure A.75. Velocity history at the free surface of a molybdenum single crystal target subjected on its front surface to radiation (in the <110> direction) from an ion beam with an 8-mm spot size and an energy density of 0.2 TW/cm^2. The experiment was performed using configuration K (see Table A.2) with k_1 = 0.9 mm. The data were recorded using ORVIS with a measurement accuracy of ±20 m/s. The spall strength was 8.0 ± 0.1 GPa, and the spall thickness was 0.19 (±10%) mm (Kanel et al. [1993]).

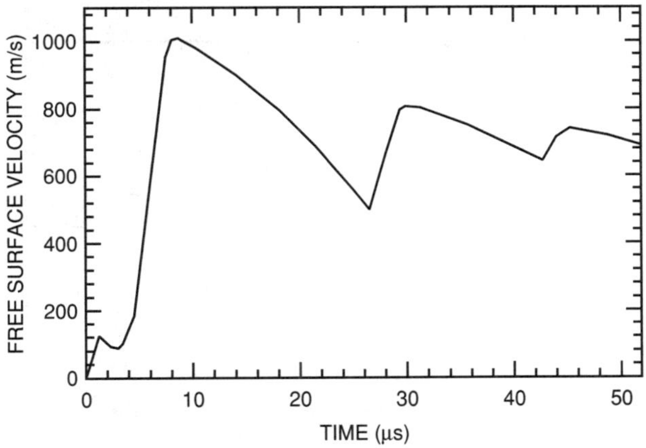

Figure A.76. Velocity history at the free surface of a molybdenum single crystal target impacted in the <110> direction by an ion beam-launched aluminum flyer plate at an impact velocity of 4100 ± 150 m/s. The beam had an 8-mm spot size and an energy density of 0.2 TW/cm^2. The experiment was performed using configuration L (see Table A.2) with l_1 = 0.05 mm and l_2 = 0.416 mm. The data were recorded using ORVIS with a measurement accuracy of ±20 m/s. The spall strength was 13.93 ± 0.1 GPa, and the spall thickness was 0.048 (±10%) mm (Kanel et al. [1993]).

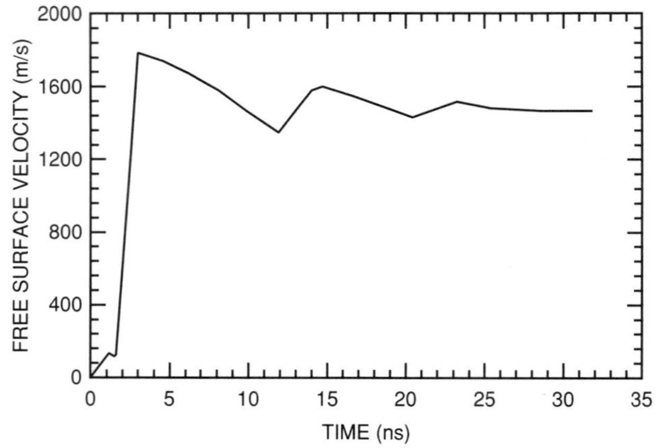

Figure A.77. Velocity history at the free surface of a molybdenum single crystal target impacted in the <110> direction by an ion beam-launched aluminum flyer plate at an impact velocity of 4100 ± 150 m/s. The beam had an 8-mm spot size and an energy density of 0.2 TW/cm^2. The experiment was performed using configuration L (see Table A.2) with l_1 = 0.05 mm and l_2 = 0.286 mm. The data were recorded using ORVIS with a measurement accuracy of ±20 m/s. The spall strength was 11.82 ± 0.1 GPa, and the spall thickness was 0.027 (±10%) mm (Kanel et al. [1993]).

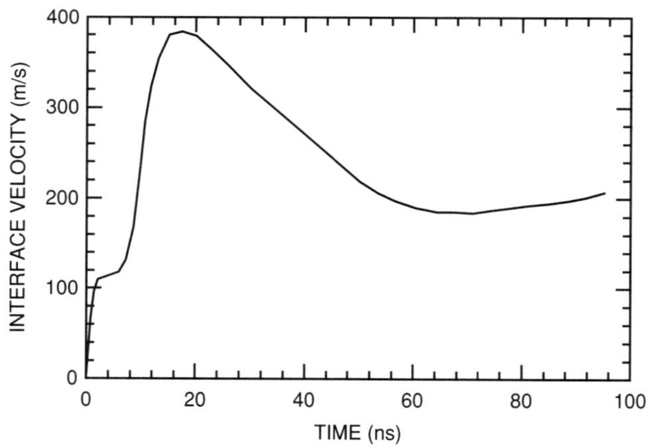

Figure A.78. Velocity history at the interface between a molybdenum single crystal target and a water window. The front surface of the specimen was irradiated in the <110> direction using an ion beam with an 8-mm spot size and an energy density of 0.2 TW/cm^2. The experiment was performed using configuration M (see Table A.2) with m_1 = 0.37 mm. The data were recorded using ORVIS with a measurement accuracy of ±20 m/s. No spall was observed (Kanel et al. [1993]).

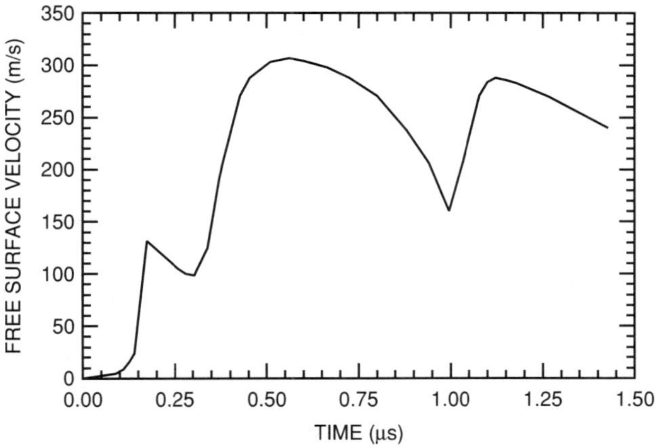

Figure A.79. Velocity history at the free surface of a molybdenum single crystal target impacted in the <111> direction by an aluminum flyer plate at an impact velocity of 660 ± 20 m/s. The experiment was performed using configuration C (see Table A.2) with c_1 = 2.0 mm, c_2 = 4.7 mm (aluminum), and c_3 = 3.91 mm. The data were recorded using a VISAR with a measurement accuracy of 3 to 5 m/s. The spall strength was 4.2 ± 0.1 GPa, and the spall thickness was 1.1 (±10%) mm (Kanel et al. [1993]).

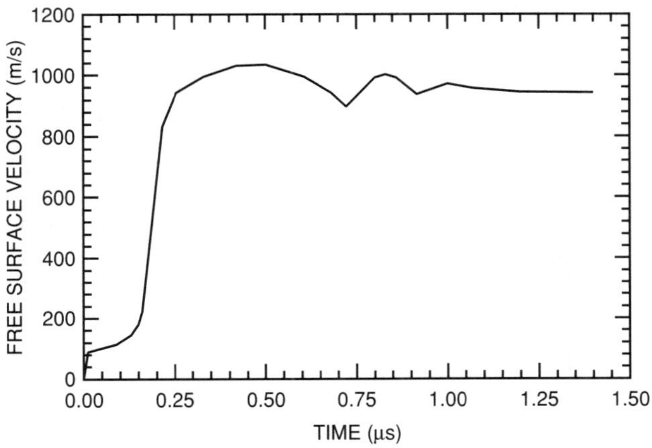

Figure A.80. Velocity history at the free surface of a molybdenum single crystal target impacted in the <111> direction by an aluminum flyer plate at an impact velocity of 1900 ± 70 m/s. The experiment was performed using configuration C (see Table A.2) with c_1 = 2.0 mm, c_2 = 4.7 mm (aluminum), and c_3 = 3.7 mm. The data were recorded using a VISAR with a measurement accuracy of 3 to 5 m/s. The spall strength was 3.5 ± 0.1 GPa, and the spall thickness was 0.6 (±10%) mm (Kanel et al. [1993]).

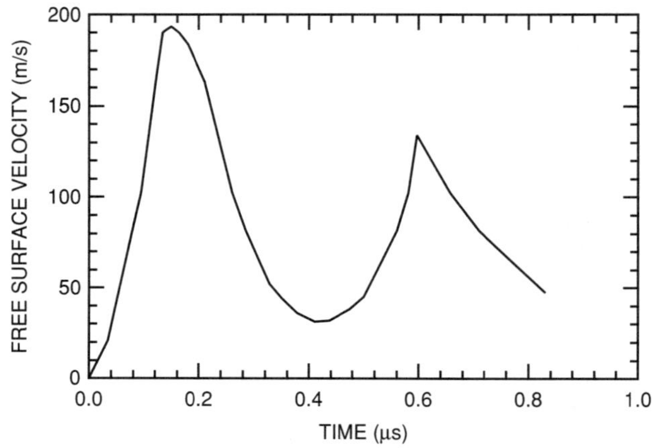

Figure A.81. Velocity history at the free surface of a molybdenum single crystal target impacted in the <111> direction by an aluminum flyer plate at an impact velocity of 660 ± 20 m/s. The experiment was performed using configuration C (see Table A.2) with c_1 = 0.4 mm, c_2 = 2.0 mm (aluminum), and c_3 = 1.25 mm. The data were recorded using a VISAR with a measurement accuracy of 3 to 5 m/s. The spall strength was 3.7 ± 0.2 GPa, and the spall thickness was 0.7 (±10%) mm (Kanel et al. [1993]).

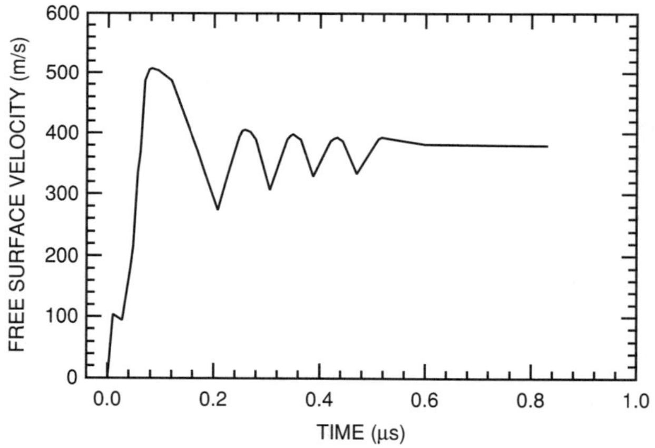

Figure A.82. Velocity history at the free surface of a molybdenum single crystal target impacted in the <111> direction by an aluminum flyer plate at an impact velocity of 1250 ± 50 m/s. The experiment was performed using configuration C (see Table A.2) with c_1 = 0.4 mm, c_2 = 2.0 mm (aluminum), and c_3 = 1.34 mm. The data were recorded using a VISAR with a measurement accuracy of 3 to 5 m/s. The spall strength was 6.3 ± 0.2 GPa, and the spall thickness was 0.27 (±10%) mm (Kanel et al. [1993]).

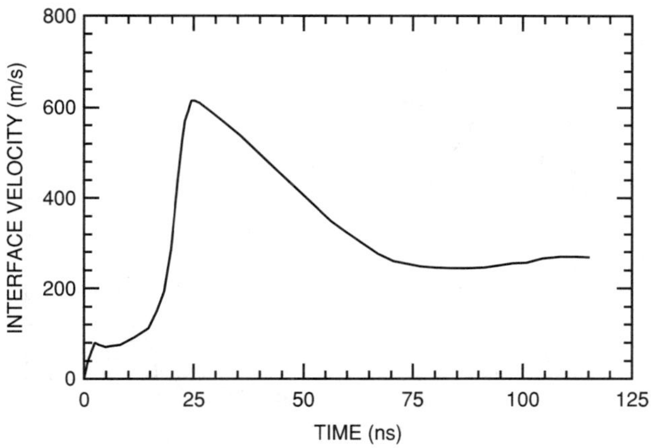

Figure A.83. Velocity history at the interface between a molybdenum single crystal target and a water window. The front surface of the specimen was irradiated in the <111> direction using an ion beam with an 8-mm spot size and an energy density of 0.2 TW/cm^2. The experiment was performed using configuration M (see Table A.2) with m_1 = 0.66 mm. The data were recorded using ORVIS with a measurement accuracy of ±20 m/s. No spall was observed (Kanel et al. [1993]).

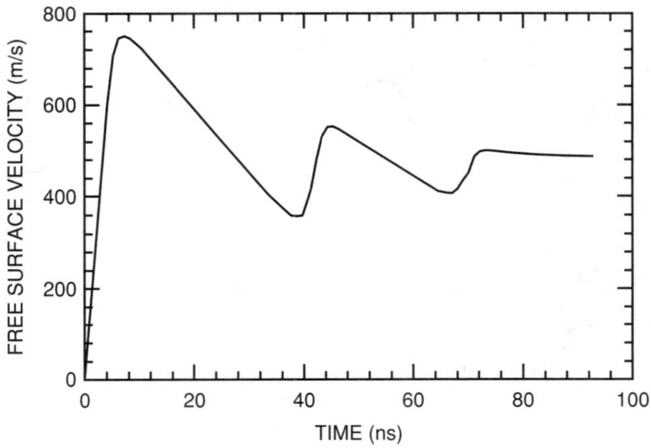

Figure A.84. Velocity history at the free surface of a niobium single crystal target irradiated in the <100> direction using an ion beam with an 8-mm spot size and an energy density of 0.2 TW/cm^2. The experiment was performed using configuration K (see Table A.2) with k_1 = 0.53 mm. The data were recorded using ORVIS with a measurement accuracy of ±20 m/s. The spall strength was 7.41 ± 0.2 GPa, and the spall thickness was 0.07 (±10%) mm (Kanel et al. [1994a]).

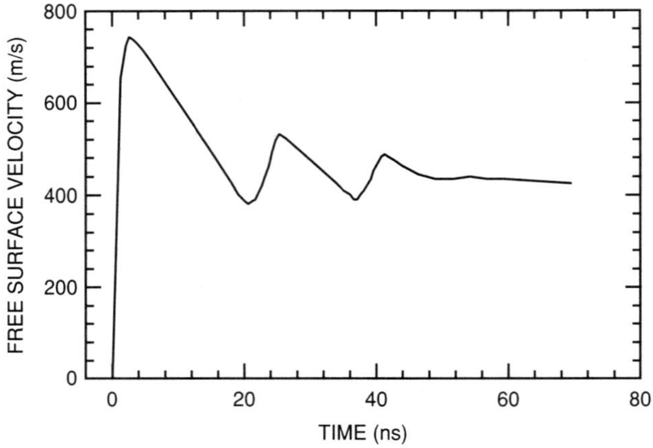

Figure A.85. Velocity history at the free surface of a niobium single crystal target irradiated in the <110> direction using an ion beam with an 8-mm spot size and an energy density of 0.2 TW/cm^2. The experiment was performed using configuration K (see Table A.2) with k_1 = 0.455 mm. The data were recorded using ORVIS with a measurement accuracy of ±20 m/s. The spall strength was 6.86 ± 0.2 GPa, and the spall thickness was 0.042 (±10%) mm (Kanel et al. [1994a]).

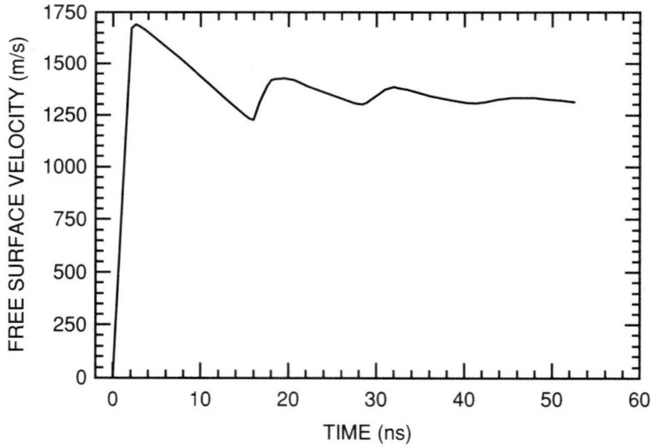

Figure A.86. Velocity history at the free surface of a niobium single crystal target impacted in the <100> direction by an ion beam-launched aluminum flyer plate at an impact velocity of 4100 ± 150 m/s. The beam had an 8-mm spot size and an energy density of 0.2 TW/cm^2. The experiment was performed using configuration L (see Table A.2) with l_1 = 0.05 mm and l_2 = 0.49 mm. The data were recorded using ORVIS with a measurement accuracy of ±20 m/s. The spall strength was 8.66 ± 0.2 GPa, and the spall thickness was 0.029 (±10%) mm (Kanel et al. [1994a]).

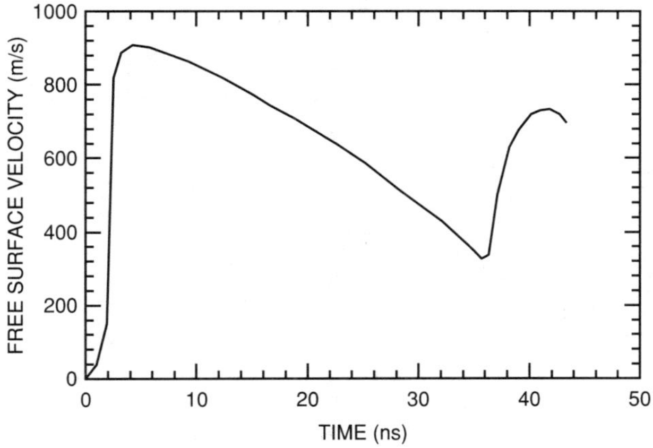

Figure A.87. Velocity history at the free surface of a deformed (85% to 90% prestrain) niobium single crystal target irradiated in the <110> direction using an ion beam with an 8-mm spot size and an energy density of 0.2 TW/cm^2. The experiment was performed using configuration K (see Table A.2) with k_1 = 0.4 mm. The data were recorded using ORVIS with a measurement accuracy of ±20 m/s. The spall strength was 11.0 ± 0.2 GPa, and the spall thickness was 0.07 (±10%) mm (Kanel et al. [1994a]).

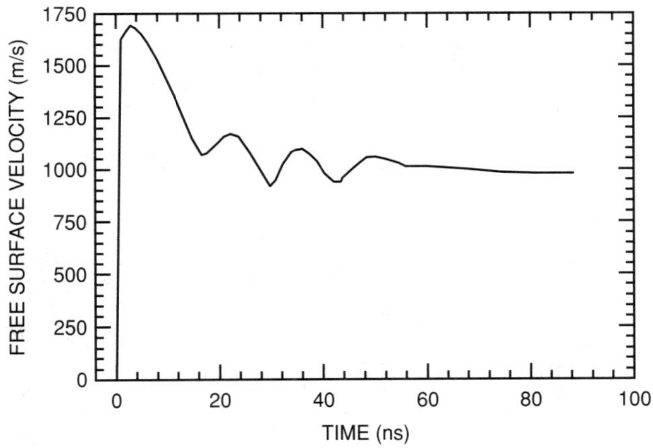

Figure A.88. Velocity history at the free surface of a deformed (85% to 90% prestrain) niobium single crystal target impacted in the <110> direction by an ion beam-launched aluminum flyer plate at an impact velocity of 4100 ± 150 m/s. The beam had an 8-mm spot size and an energy density of 0.2 TW/cm^2. The experiment was performed using configuration L (see Table A.2) with l_1 = 0.05 mm and l_2 = 0.44 mm. The data were recorded using ORVIS with a measurement accuracy of ±20 m/s. The spall strength was 11.15 ± 0.2 GPa, and the spall thickness was 0.029 (±10%) mm (Kanel et al. [1994a]).

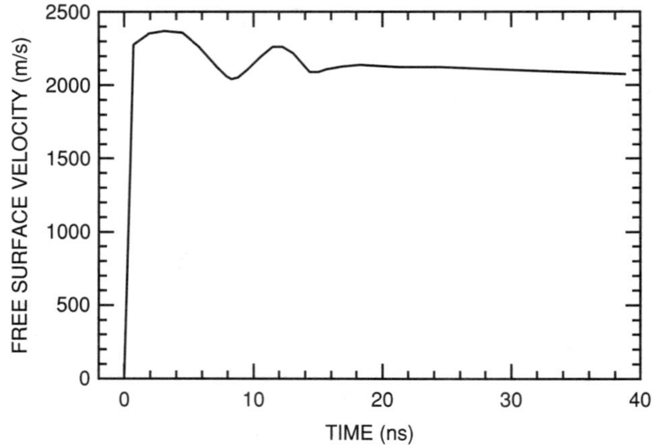

Figure A.89. Velocity history at the free surface of a deformed (85% to 90% prestrain) niobium single crystal target impacted in the <110> direction by an ion beam-launched aluminum flyer plate at an impact velocity of 4100 ± 150 m/s. The beam had an 8-mm spot size and an energy density of 0.2 TW/cm^2. The experiment was performed using configuration L (see Table A.2) with l_1 = 0.05 mm and l_2 = 0.09 mm. The data were recorded using ORVIS with a measurement accuracy of ±20 m/s. The spall strength was 6.45 ± 0.2 GPa, and the spall thickness was 0.014 (±10%) mm (Kanel et al. [1994a]).

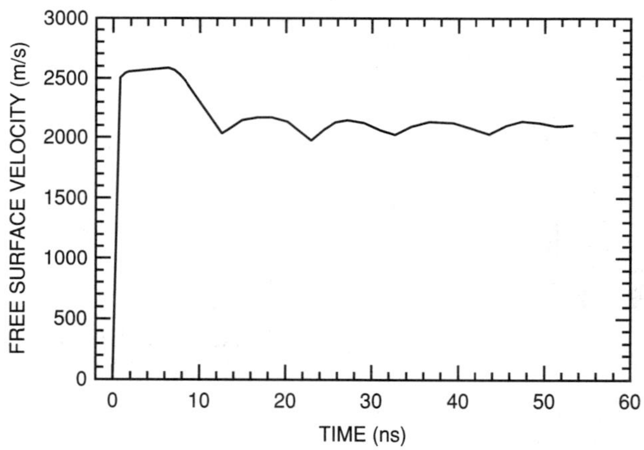

Figure A.90. Velocity history at the free surface of a deformed (85% to 90% prestrain) niobium single crystal target impacted in the <110> direction by an ion beam-launched aluminum flyer plate at an impact velocity of 4100 ± 150 m/s. The beam had an 8-mm spot size and an energy density of 0.2 TW/cm^2. The experiment was performed using configuration L (see Table A.2) with l_1 = 0.05 mm and l_2 = 0.068 mm. The data were recorded using ORVIS with a measurement accuracy of ±20 m/s. The spall strength was 10.1 ± 0.2 GPa, and the spall thickness was 0.023 (±10%) mm (Kanel et al. [1994a]).

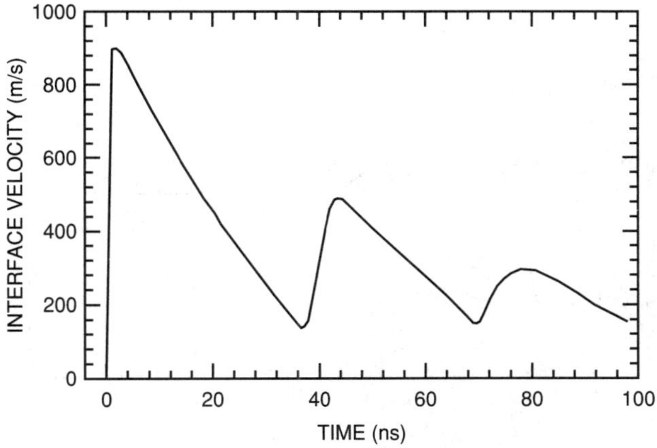

Figure A.91. Velocity history at the interface between a deformed (85% to 90% prestrain) niobium single crystal target and a water window. The front surface of the specimen was irradiated in the <110> direction using an ion beam with an 8-mm spot size and an energy density of 0.2 TW/cm^2. The experiment was performed using configuration M (see Table A.2) with m_1 = 0.41 mm. The data were recorded using ORVIS with a measurement accuracy of ±20 m/s. The spall strength was 10.1 ± 0.2 GPa, and the spall thickness was 0.07 (±10%) mm (Kanel et al. [1993]).

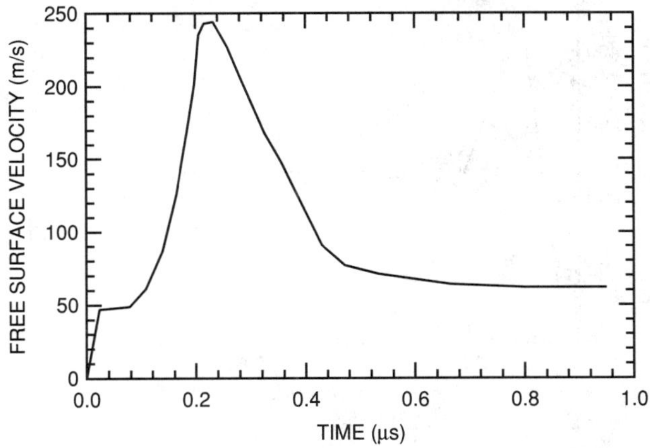

Figure A.92. Velocity history at the free surface of a magnesium Ma1 target impacted by an aluminum flyer plate at an impact velocity of 660 ± 20 m/s. The experiment was performed using configuration A (see Table A.2) with a_1 = 0.2 mm and a_2 = 4.9 mm. The data were recorded using a capacitor gauge with an electrode diameter of 10 mm and a measurement accuracy of ±4%. The spall strength was 0.8 ± 0.05 GPa, and the spall thickness was 0.62 (±10%) mm (Kanel [1984b]).

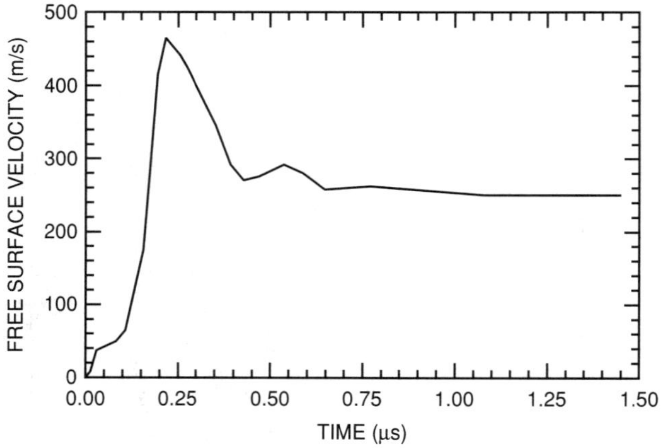

Figure A.93. Velocity history at the free surface of a magnesium Ma1 target impacted by an aluminum flyer plate at an impact velocity of 660 ± 20 m/s. The experiment was performed using configuration A (see Table A.2) with a_1 = 0.4 mm and a_2 = 4.9 mm. The data were recorded using a capacitor gauge with an electrode diameter of 5 mm and a measurement accuracy of ±4%. The spall strength was 0.88 ± 0.05 GPa, and the spall thickness was 0.3 (±10%) mm (Kanel [1984b]).

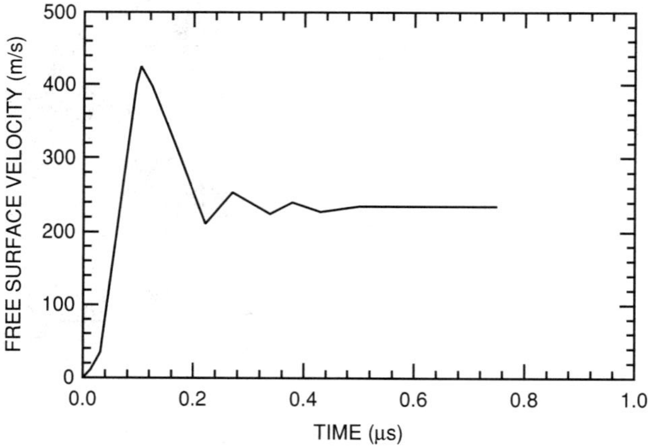

Figure A.94. Velocity history at the free surface of a magnesium Ma1 target impacted by an aluminum flyer plate at an impact velocity of 660 ± 20 m/s. The experiment was performed using configuration A (see Table A.2) with a_1 = 0.2 mm and a_2 = 2.0 mm. The data were recorded using a capacitor gauge with an electrode diameter of 10 mm and a measurement accuracy of ±4%. No spall was observed (Kanel [1984b]).

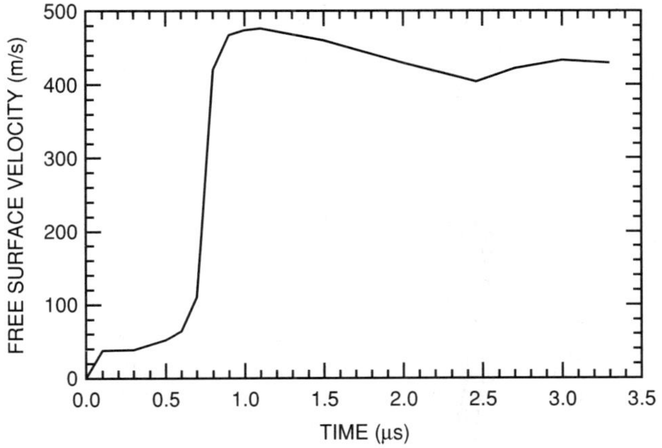

Figure A.95. Velocity history at the free surface of an explosively loaded Armco iron target. The experiment was performed using configuration H (see Table A.2) with $h_1 = 20$ mm. The data were recorded using a capacitor gauge with an electrode diameter of 20 mm and a measurement accuracy of ±4%. The spall strength was 1.31 ± 0.09 GPa, and the spall thickness was 4.2 (±10%) mm (Kanel and Shcherban [1980]).

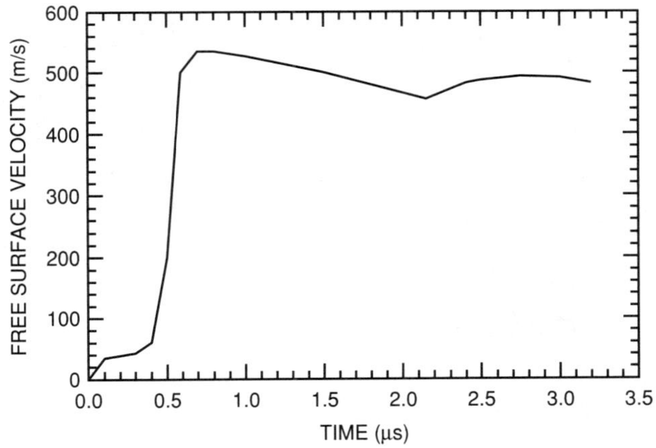

Figure A.96. Velocity history at the free surface of an explosively loaded Armco iron target. The experiment was performed using configuration H (see Table A.2) with $h_1 = 12$ mm. The data were recorded using a capacitor gauge with an electrode diameter of 20 mm and a measurement accuracy of ±4%. The spall strength was 1.41 ± 0.09 GPa, and the spall thickness was 3.9 (±10%) mm (Kanel and Shcherban [1980]).

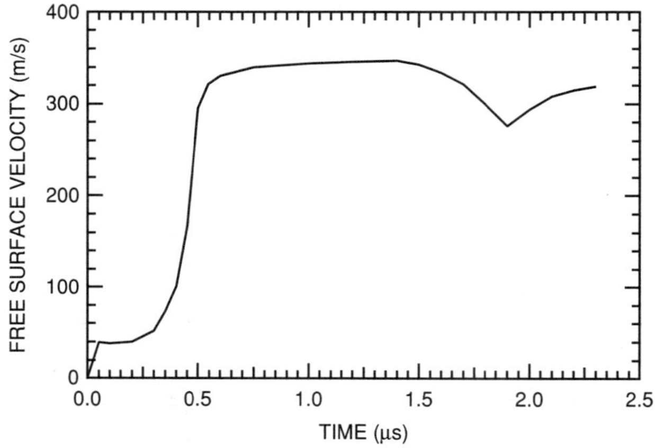

Figure A.97. Velocity history at the free surface of an Armco iron target impacted by an aluminum flyer plate at an impact velocity of 590 ± 20 m/s. The experiment was performed using configuration A (see Table A.2) with a_1 = 5.0 mm and a_2 = 10.0 mm. The data were recorded using a capacitor gauge with an electrode diameter of 20 mm and a measurement accuracy of ±4%. The spall strength was 1.31 ± 0.07 GPa, and the spall thickness was 3.5 (±10%) mm (Kanel and Shcherban [1980]).

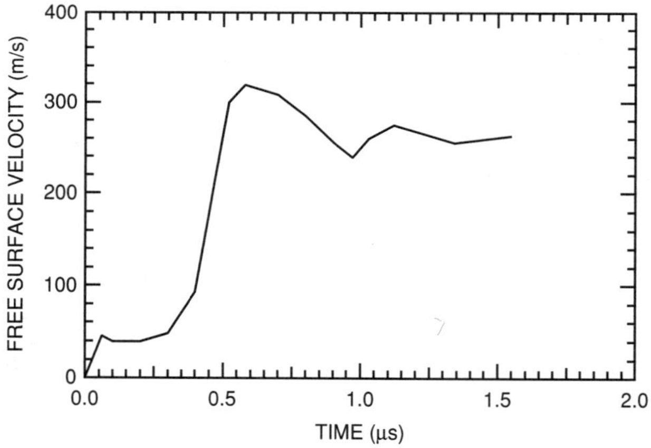

Figure A.98. Velocity history at the free surface of an Armco iron target impacted by an aluminum flyer plate at an impact velocity of 590 ± 20 m/s. The experiment was performed using configuration A (see Table A.2) with a_1 = 2.0 mm and a_2 = 10.0 mm. The data were recorded using a capacitor gauge with an electrode diameter of 20 mm and a measurement accuracy of ±4%. The spall strength was 1.54 ± 0.09 GPa, and the spall thickness was 1.1 (±10%) mm (Kanel and Shcherban [1980]).

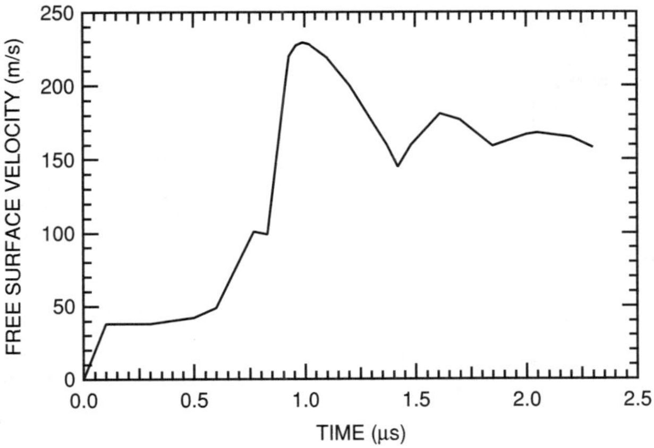

Figure A.99. Velocity history at the free surface of an Armco iron target impacted by an aluminum flyer plate at an impact velocity of 590 ± 20 m/s. The experiment was performed using configuration A (see Table A.2) with a_1 = 2.0 mm and a_2 = 20.0 mm. The data were recorded using a capacitor gauge with an electrode diameter of 20 mm and a measurement accuracy of ±4%. The spall strength was 1.49 ± 0.09 GPa, and the spall thickness was 1.2 (±10%) mm (Kanel and Shcherban [1980]).

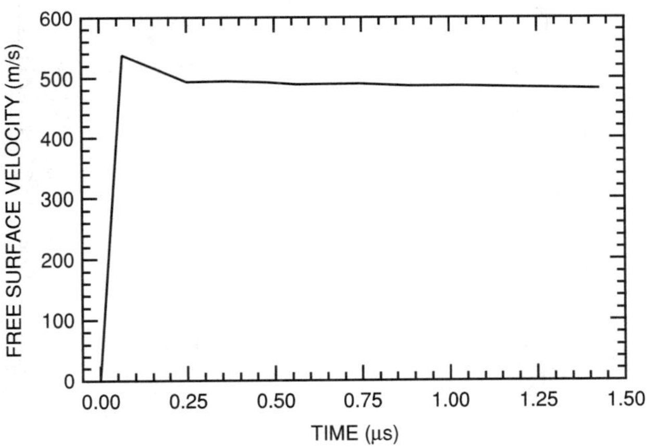

Figure A.100. Velocity history at the free surface of a lead target impacted by an aluminum flyer plate at an impact velocity of 700 ± 20 m/s. The experiment was performed using configuration C (see Table A.2) with c_1 = 2.0 mm, c_2 = 4.53 mm (aluminum), and c_3 = 3.75 mm. The data were recorded using a VISAR with a measurement accuracy of 3 to 5 m/s. The spall strength was 0.5 (± 6%) GPa (Kanel et al. [1996b]).

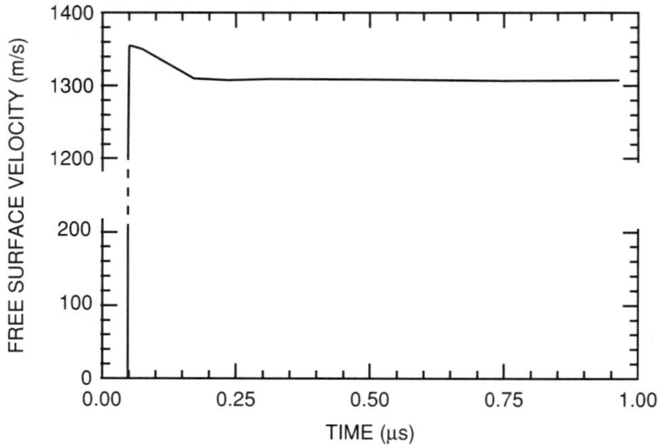

Figure A.101. Velocity history at the free surface of a lead target impacted by an aluminum flyer plate at an impact velocity of 2000 ± 70 m/s. The experiment was performed using configuration C (see Table A.2) with c_1 = 1.9 mm, c_2 = 4.0 mm (aluminum), and c_3 = 3.75 mm. The data were recorded using a VISAR with a measurement accuracy of 3 to 5 m/s. The spall strength was 0.5 (± 6%) GPa (Kanel et al. [1996b]).

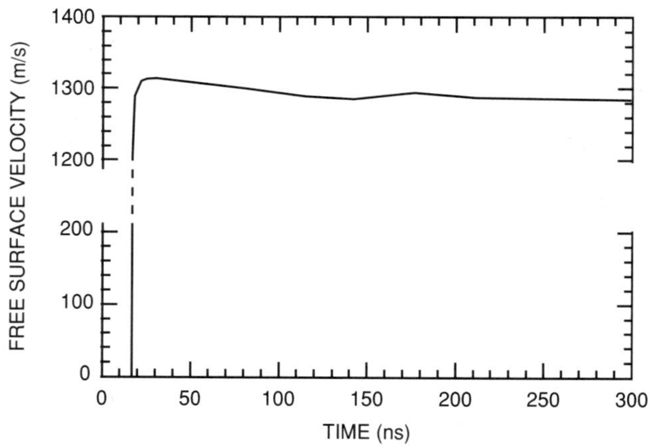

Figure A.102. Velocity history at the free surface of a lead target impacted by an aluminum flyer plate at an impact velocity of 1900 ± 70 m/s. The experiment was performed using configuration C (see Table A.2) with c_1 = 2.0 mm, c_2 = 4.1 mm (aluminum), and c_3 = 4.0 mm. The data were recorded using a VISAR with a measurement accuracy of 3 to 5 m/s. The spall strength was 0.35 (± 6%) GPa (Kanel et al. [1996b]).

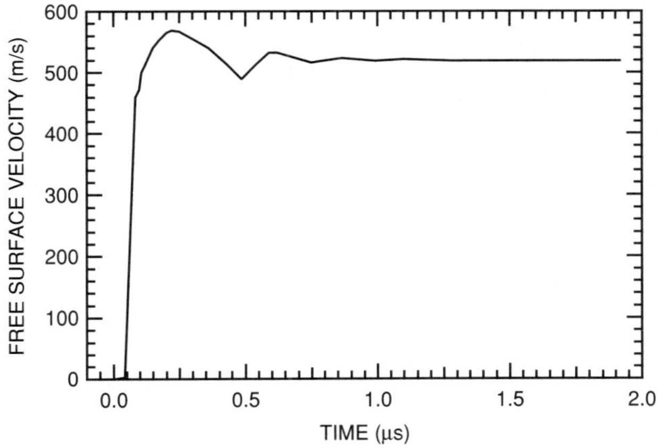

Figure A.103. Velocity history at the free surface of a tin target impacted by an aluminum flyer plate at an impact velocity of 660 ± 20 m/s. The experiment was performed using configuration C (see Table A.2) with c_1 = 2.0 mm, c_2 = 4.0 mm (aluminum), and c_3 = 4.0 mm. The data were recorded using a VISAR with a measurement accuracy of 3 to 5 m/s. The spall strength was 0.7 ± 0.05 GPa, and the spall thickness was 0.44 (±10%) mm (Kanel et al. [1996b]).

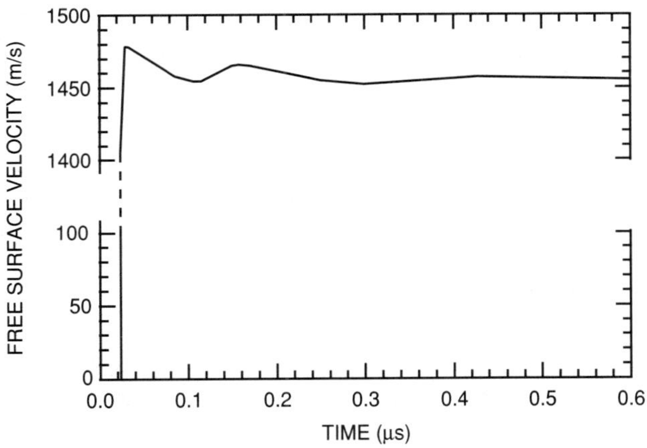

Figure A.104. Velocity history at the free surface of a tin target impacted by an aluminum flyer plate at an impact velocity of 2000 ± 70 m/s. The experiment was performed using configuration C (see Table A.2) with c_1 = 1.8 mm, c_2 = 4.95 mm (aluminum), and c_3 = 5.0 mm. The data were recorded using a VISAR with a measurement accuracy of 3 to 5 m/s. The spall strength was 0.22 ± 0.02 GPa, and the spall thickness was 0.44 (±10%) mm (Kanel et al. [1996b]).

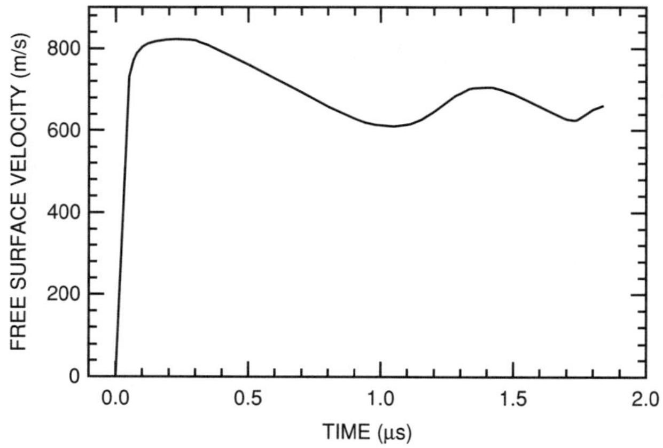

Figure A.105. Velocity history at the free surface of an Epoxy EDT-10 target impacted by a PMMA flyer plate at an impact velocity of 850 ± 30 m/s. The experiment was performed using configuration B (see Table A.2) with b_1 = 1.4 mm and b_2 = 4.5 mm. The data were recorded using a VISAR with a measurement accuracy of 3 to 5 m/s. The spall strength was 0.3 ± 0.05 GPa, and the spall thickness was 0.9 (±10%) mm (Parhomenko and Utkin [1990]).

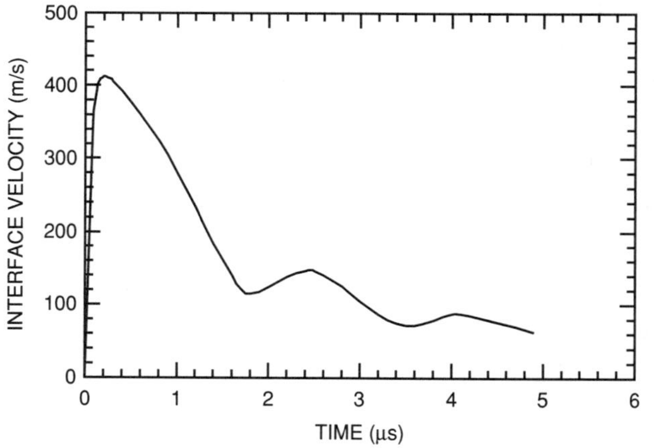

Figure A.106. Velocity history at the interface between a PMMA target and a hexan window. The target was impacted by a PMMA flyer plate at an impact velocity of 850 ± 30 m/s. The experiment was performed using configuration E (see Table A.2) with e_1 = 1.33 mm and e_2 = 11.5 mm. The data were recorded using a VISAR with a measurement accuracy of 3 to 5 m/s. The spall strength was 0.17 ± 0.01 GPa, and the spall thickness was 2.8 (±10%) mm (Parhomenko and Utkin [1990]).

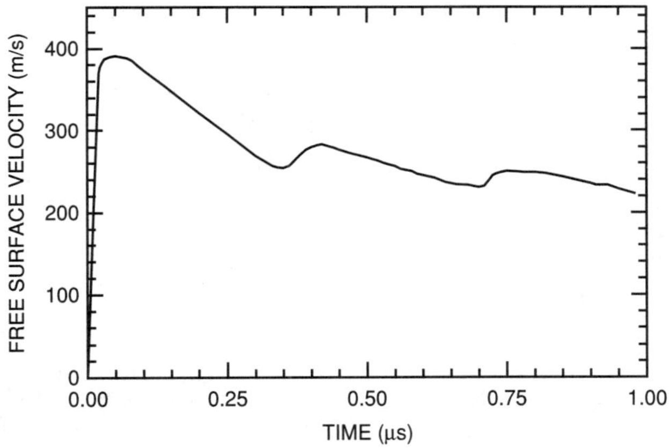

Figure A.107. Velocity history at the free surface of a PMMA target impacted by an aluminum flyer plate at an impact velocity of 660 ± 20 m/s. The experiment was performed using configuration B (see Table A.2) with b_1 = 2.0 mm and b_2 = 8.32 mm. The data were recorded using a VISAR with a measurement accuracy of 3 to 5 m/s. The spall strength was 0.21 ± 0.01 GPa, and the spall thickness was 0.46 (±10%) mm (Parhomenko and Utkin [1990]).

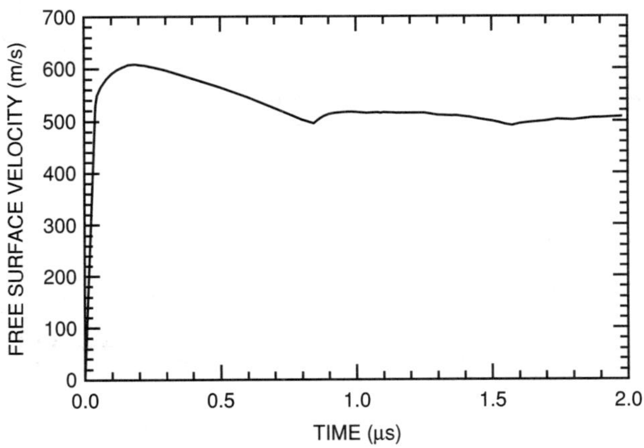

Figure A.108. Velocity history at the free surface of a PMMA target impacted by a PMMA flyer plate at an impact velocity of 850 ± 30 m/s. The experiment was performed using configuration B (see Table A.2) with b_1 = 1.36 mm and b_2 = 11.68 mm. The data were recorded using a VISAR with a measurement accuracy of 3 to 5 m/s. The spall strength was 0.13 ± 0.01 GPa, and the spall thickness was 0.86 (±10%) mm (Parhomenko and Utkin [1990]).

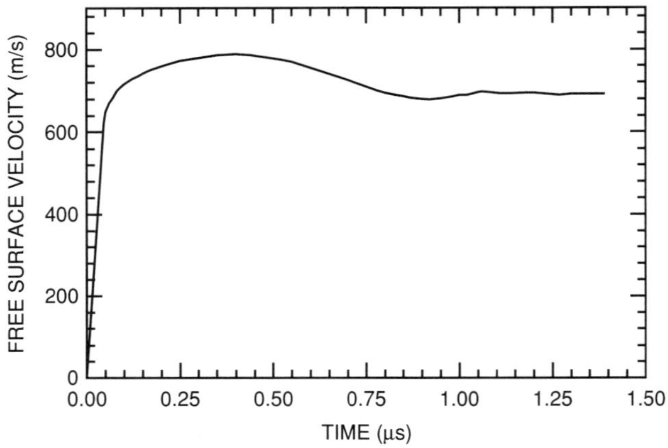

Figure A.109. Velocity history at the free surface of a PMMA target impacted by a PMMA flyer plate at an impact velocity of 850 ± 30 m/s. The experiment was performed using configuration B (see Table A.2) with b_1 = 2.16 mm and b_2 = 8.30 mm. The data were recorded using a VISAR with a measurement accuracy of 3 to 5 m/s. The spall strength was 0.18 ± 0.02 GPa (Parhomenko and Utkin [1990]).

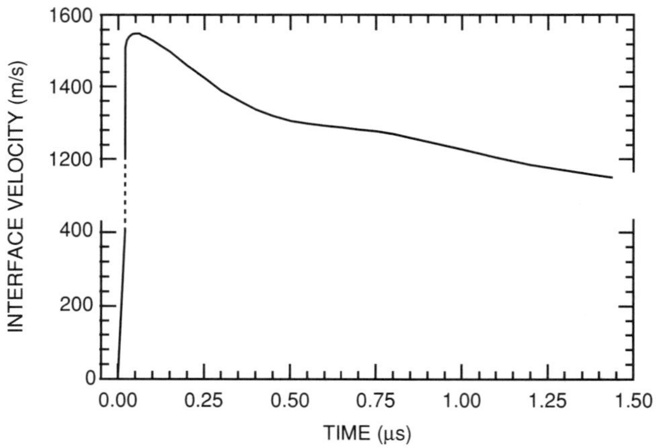

Figure A.110. Velocity history at the interface between a PMMA target and a hexan window. The target was impacted by a PMMA flyer plate at an impact velocity of 1900 ± 70 m/s. The experiment was performed using configuration E (see Table A.2) with e_1 = 2.0 mm and e_2 = 8.2 mm. The data were recorded using a VISAR with a measurement accuracy of 3 to 5 m/s (Parhomenko and Utkin [1990]).

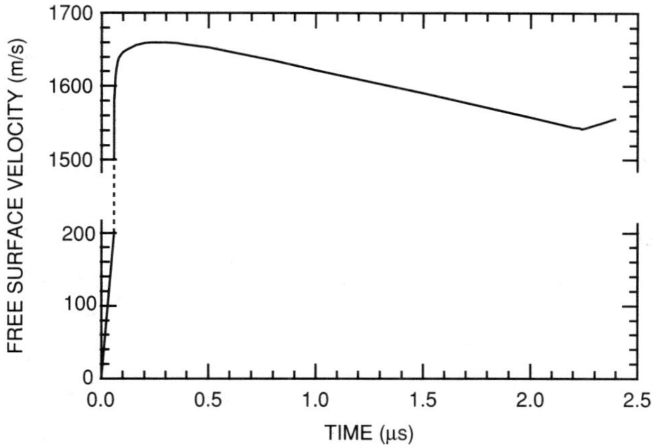

Figure A.111. Velocity history at the free surface of an explosively loaded PMMA target. The experiment was performed using configuration J (see Table A.2) with j_1 = 18.21 mm. The data were recorded using a VISAR with a measurement accuracy of 3 to 5 m/s. The spall strength was 0.17 ± 0.01 GPa, and the spall thickness was 2.85 (±10%) mm (Parhomenko and Utkin [1990]).

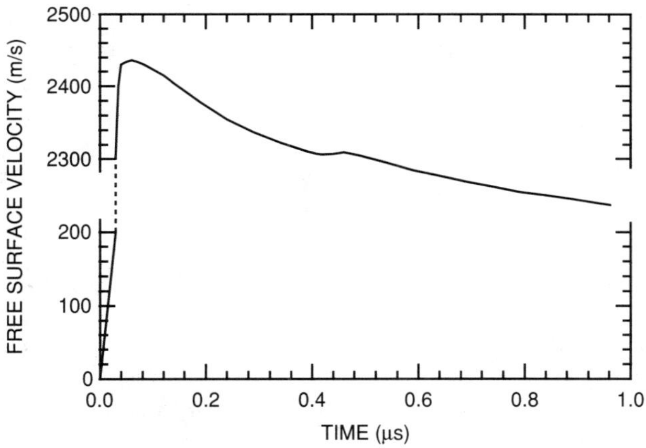

Figure A.112. Velocity history at the free surface of a PMMA target impacted by an aluminum flyer plate at an impact velocity of 1900 ± 70 m/s. The experiment was performed using configuration B (see Table A.2) with b_1 = 2.0 mm and b_2 = 8.28 mm. The data were recorded using a VISAR with a measurement accuracy of 3 to 5 m/s. The spall strength was 0.21 ± 0.015 GPa (Parhomenko and Utkin [1990]).

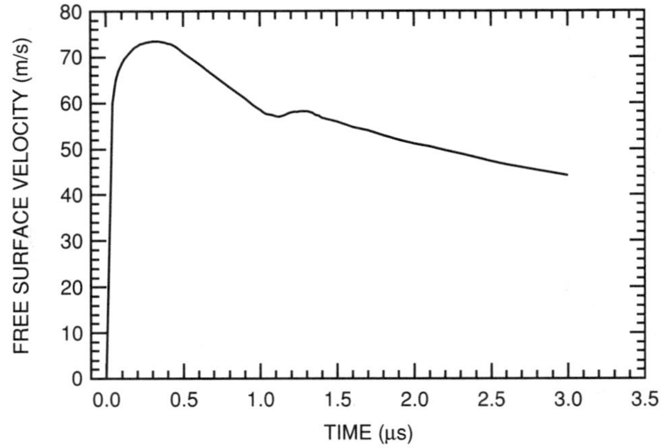

Figure A.113. Velocity history at the free surface of a white rubber (Grade 7889) target impacted by a PMMA flyer plate at an impact velocity of 380 ± 20 m/s. The experiment was performed using configuration F (see Table A.2) with f_1 = 1.5 mm, f_2 = 5.0 mm (copper), and f_3 = 10.4 mm. The data were recorded using a VISAR with a measurement accuracy of 3 to 5 m/s. The spall strength was 0.016 GPa. The spall plane could not be identified at 100X magnification (Kalmykov et al. [1990]).

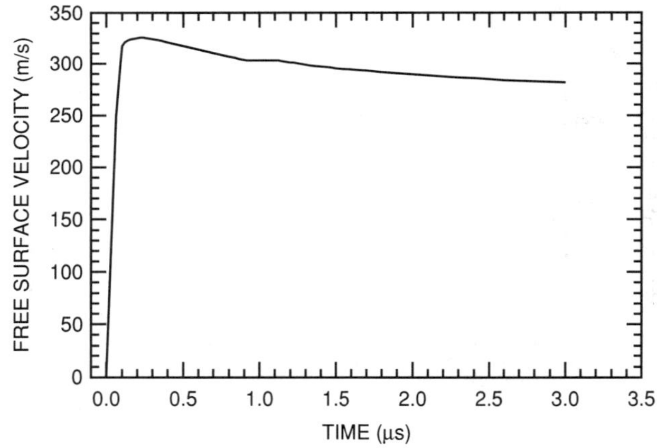

Figure A.114. Velocity history at the free surface of a a white rubber (Grade 7889) target impacted by a PMMA flyer plate at an impact velocity of 380 ± 20 m/s. The experiment was performed using configuration D (see Table A.2) with d_1 = 1.5 mm and d_2 = 9.95 mm. The data were recorded using a VISAR with a measurement accuracy of 3 to 5 m/s. The spall strength was 0.022 GPa. The spall plane could not be identified at 100X magnification (Kalmykov et al. [1990]).

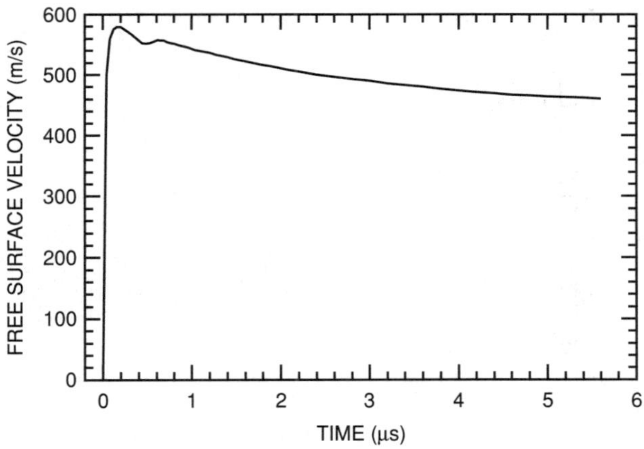

Figure A.115. Velocity history at the free surface of a white rubber (Grade 7889) target impacted by a PMMA flyer plate at an impact velocity of 850 ± 30 m/s. The experiment was performed using configuration F (see Table A.2) with f_1 = 2.3 mm, f_2 = 1.9 mm (PMMA), and f_3 = 10.0 mm. The data were recorded using a VISAR with a measurement accuracy of 3 to 5 m/s. The spall strength was 0.027 GPa. The spall plane could not be identified at 100X magnification (Kalmykov et al. [1990]).

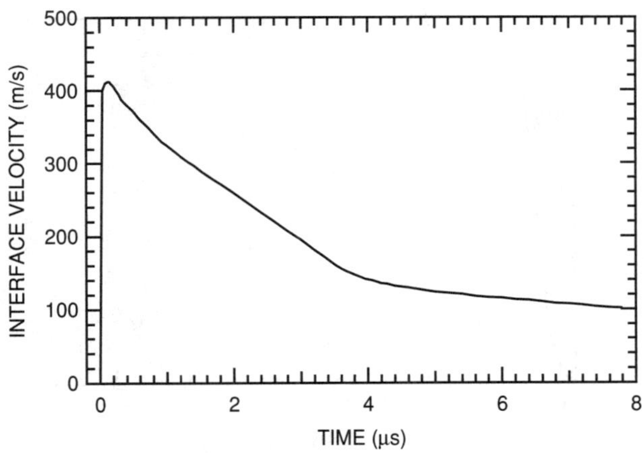

Figure A.116. Velocity history at the interface between a white rubber (Grade 7889) target and a hexan window. The target was impacted by a PMMA flyer plate at an impact velocity of 850 ± 30 m/s. The experiment was performed using configuration G (see Table A.2) with g_1 = 2.40 mm, g_2 = 2.2 mm (PMMA), and g_3 = 9.68 mm. The data were recorded using a VISAR with a measurement accuracy of 3 to 5 m/s (Kalmykov et al. [1990]).

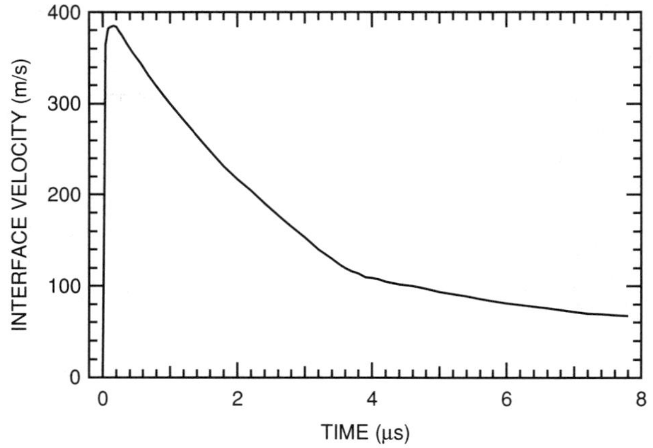

Figure A.117. Velocity history at the interface between a white rubber (Grade 7889) target and an ethanol window. The target was impacted by a PMMA flyer plate at an impact velocity of 850 ± 30 m/s. The experiment was performed using configuration G (see Table A.2) with g_1 = 2.15 mm, g_2 = 2.25 mm (PMMA), and g_3 = 10.68 mm. The data were recorded using a VISAR with a measurement accuracy of 3 to 5 m/s (Kalmykov et al. [1990]).

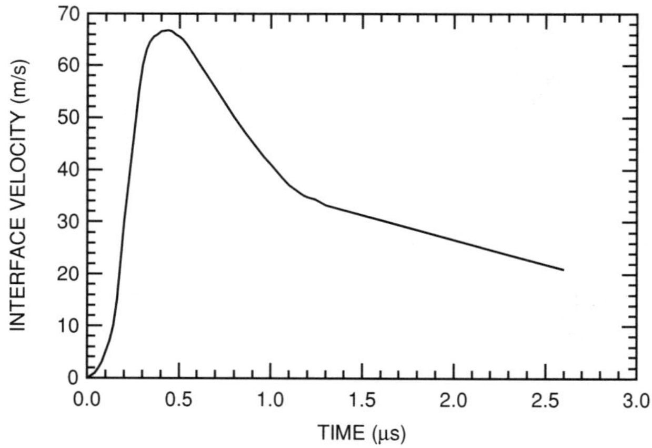

Figure A.118. Velocity history at the interface between a propellant simulant (filled rubber) target and a water window. The target was impacted by a PMMA flyer plate at an impact velocity of 380 ± 20 m/s. The experiment was performed using configuration G (see Table A.2) with g_1 = 1.60 mm, g_2 = 5.03 mm (copper), and g_3 = 4.5 mm. The data were recorded using a VISAR with a measurement accuracy of 3 to 5 m/s. The spall strength was 0.01 (± 6%) GPa (Kanel et al. [1993b]).

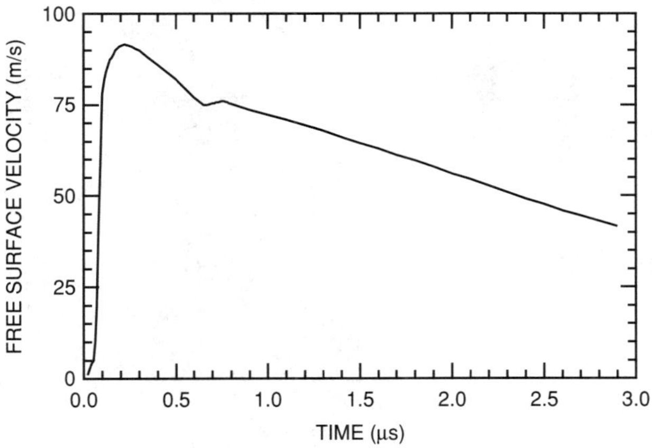

Figure A.119. Velocity history at the free surface of a propellant simulant (filled rubber) target impacted by a PMMA flyer plate at an impact velocity of 380 ± 20 m/s. The experiment was performed using configuration F (see Table A.2) with f_1 = 1.7 mm, f_2 = 5.0 mm (copper), and f_3 = 4.8 mm. The data were recorded using a VISAR with a measurement accuracy of 3 to 5 m/s. The spall strength was 0.025 (± 6%) GPa (Kanel et al. [1993b]).

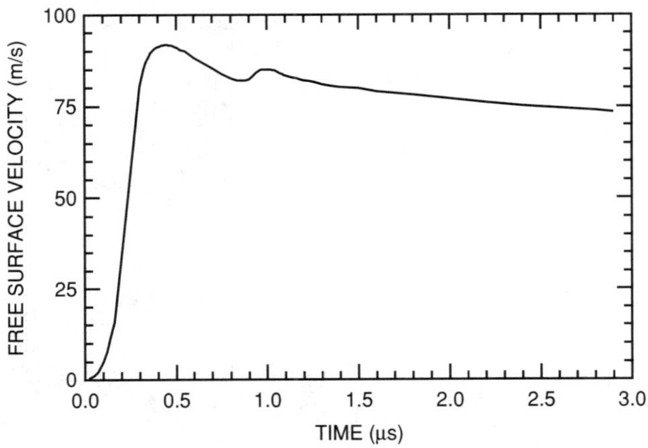

Figure A.120. Velocity history at the free surface of a propellant simulant (filled rubber) target impacted by a PMMA flyer plate at an impact velocity of 380 ± 20 m/s. The experiment was performed using configuration F (see Table A.2) with f_1 = 1.72 mm, f_2 = 5.0 mm (copper), and f_3 = 4.63 mm. The data were recorded using a VISAR with a measurement accuracy of 3 to 5 m/s. The spall strength was 0.015 (± 6%) GPa (Kanel et al. [1993b]).

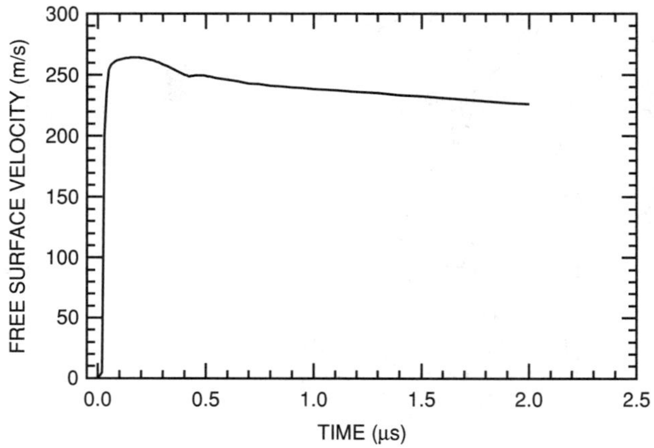

Figure A.121. Velocity history at the free surface of a propellant simulant (filled rubber) target impacted by a PMMA flyer plate at an impact velocity of 850 ± 30 m/s. The experiment was performed using configuration F (see Table A.2) with f_1 = 1.4 mm, f_2 = 5.0 mm (copper), and f_3 = 5.0 mm. The data were recorded using a VISAR with a measurement accuracy of 3 to 5 m/s. The spall strength was 0.024 (± 6%) GPa (Kanel et al. [1993b]).

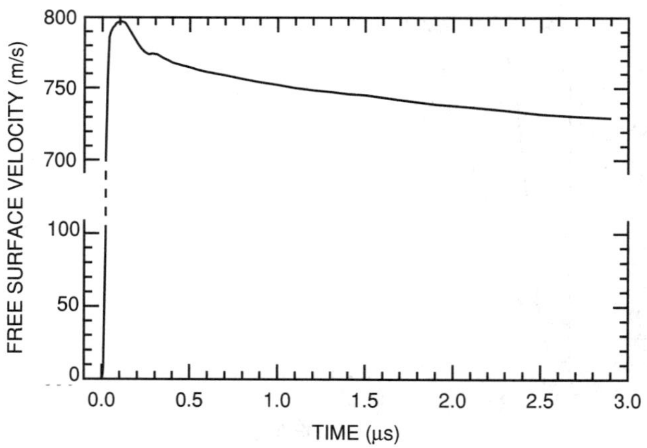

Figure A.122. Velocity history at the free surface of a propellant simulant (filled rubber) target impacted by a PMMA flyer plate at an impact velocity of 850 ± 30 m/s. The experiment was performed using configuration F (see Table A.2) with f_1 = 1.4 mm, f_2 = 1.2 mm (PMMA), and f_3 = 5.0 mm. The data were recorded using a VISAR with a measurement accuracy of 3 to 5 m/s. The spall strength was 0.03 (± 6%) GPa (Kanel et al. [1993b]).

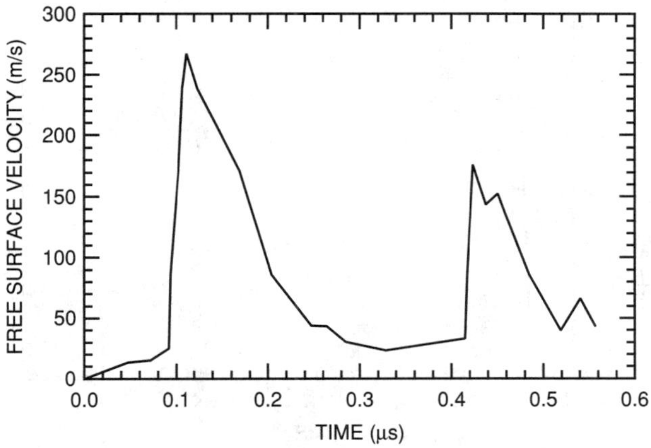

Figure A.123. Velocity history at the free surface of an alumina (ruby) target impacted normal to {1100} by an aluminum flyer plate at an impact velocity of 660 ± 20 m/s. The experiment was performed using configuration C (see Table A.2) with c_1 = 0.2 mm, c_2 = 3.0 mm (aluminum), and c_3 = 1.93 mm. The data were recorded using a VISAR with a measurement accuracy of 3 to 5 m/s. The peak stress was 6.0 GPa, and the peak tensile stress was 5.9 GPa. No spall was observed (Razorenov et al. [1993]).

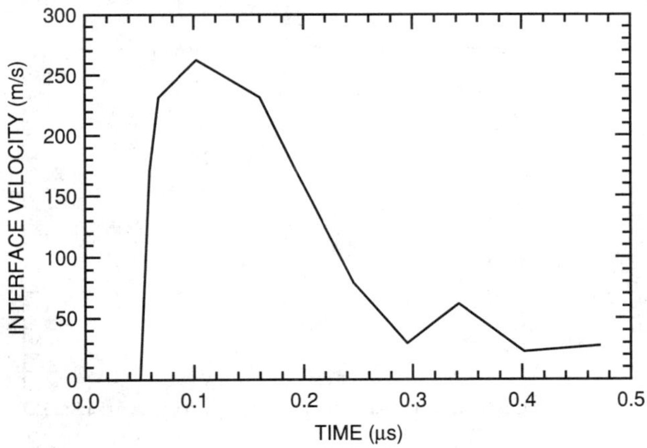

Figure A.124. Velocity history at the interface between an alumina (ruby) target and a water window. The target was impacted normal to {1100} by an aluminum flyer plate at an impact velocity of 660 ± 20 m/s. The experiment was performed using configuration G (see Table A.2) with g_1 = 0.4 mm, g_2 = 2.0 mm (aluminum), and g_3 = 4.98 mm. The data were recorded using a VISAR with a measurement accuracy of 3 to 5 m/s. The peak stress was 6.2 GPa, and the peak tensile stress was 5.5 GPa. No spall was observed (Razorenov et al. [1993]).

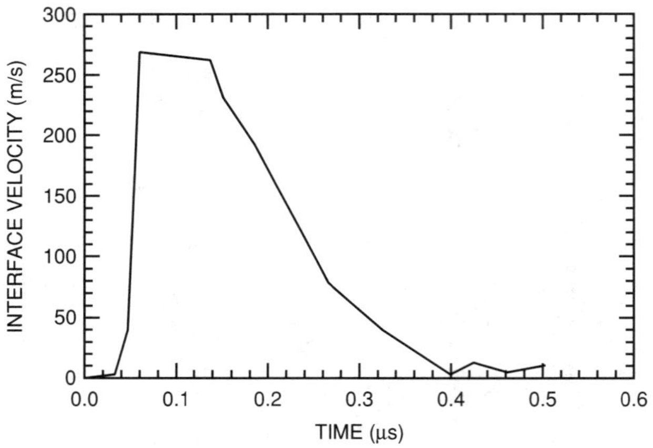

Figure A.125. Velocity history at the interface between an alumina (ruby) target and a water window. The target was impacted normal to {1120} by an aluminum flyer plate at an impact velocity of 660 ± 20 m/s. The experiment was performed using configuration G (see Table A.2) with g_1 = 0.4 mm, g_2 = 1.98 mm (aluminum), and g_3 = 3.63 mm. The data were recorded using a VISAR with a measurement accuracy of 3 to 5 m/s. The peak stress was 6.3 GPa, and the peak tensile stress was 5.6 GPa. No spall was observed (Razorenov et al. [1993]).

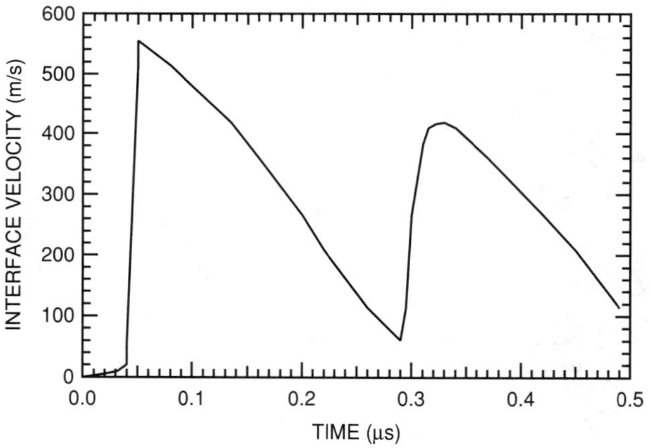

Figure A.126. Velocity history at the interface between an alumina (ruby) target and a water window. The target was impacted normal to {1120} by an aluminum flyer plate at an impact velocity of 1250 ± 50 m/s. The experiment was performed using configuration G (see Table A.2) with g_1 = 0.4 mm, g_2 = 4.37 mm (aluminum), and g_3 = 2.21 mm. The data were recorded using a VISAR with a measurement accuracy of 3 to 5 m/s. The peak stress was 13.4 GPa, and the spall strength was 10.0 ± 0.1 GPa (Razorenov et al. [1993]).

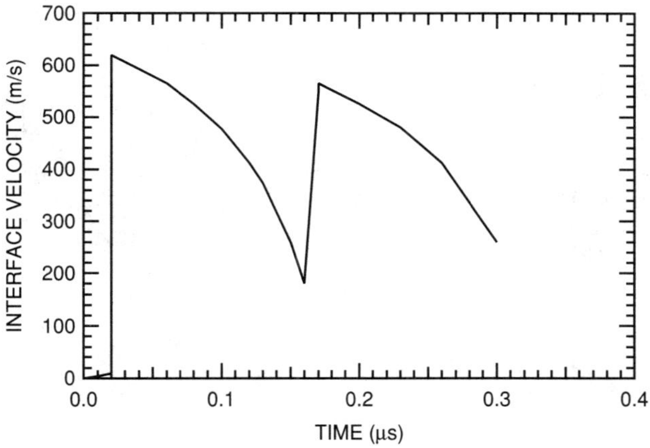

Figure A.127. Velocity history at the interface between an alumina (ruby) target and a water window. The target was impacted normal to {1120} by an aluminum flyer plate at an impact velocity of 1250 ± 50 m/s. The experiment was performed using configuration G (see Table A.2) with g_1 = 0.4 mm, g_2 = 2.0 mm (aluminum), and g_3 = 3.65 mm. The data were recorded using a VISAR with a measurement accuracy of 3 to 5 m/s. The peak stress was 15.0 GPa, and the spall strength was 8.6 ± 0.1 GPa (Razorenov et al. [1993]).

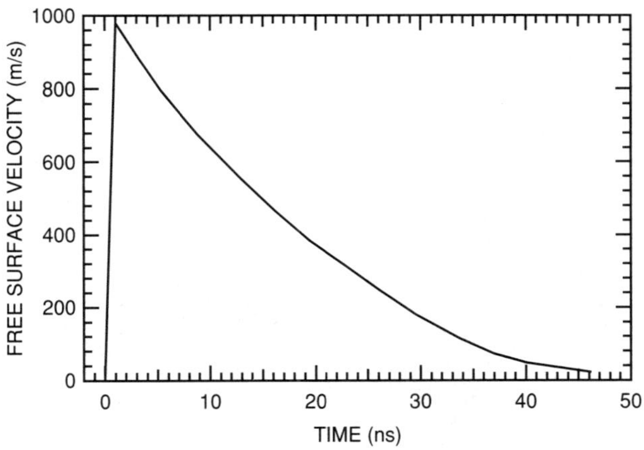

Figure A.128. Velocity history at the free surface of a z-cut sapphire target irradiated using an ion beam with an 5-mm spot size and an energy density of 0.2 TW/cm^2. The experiment was performed using configuration K (see Table A.2) with k_1 = 2.3 mm. The data were recorded using ORVIS with a measurement accuracy of ±20 m/s. The peak stress was 22.5 GPa, and the peak tensile stress was 20.5 GPa. No spall was observed (Kanel et al. [1994a]).

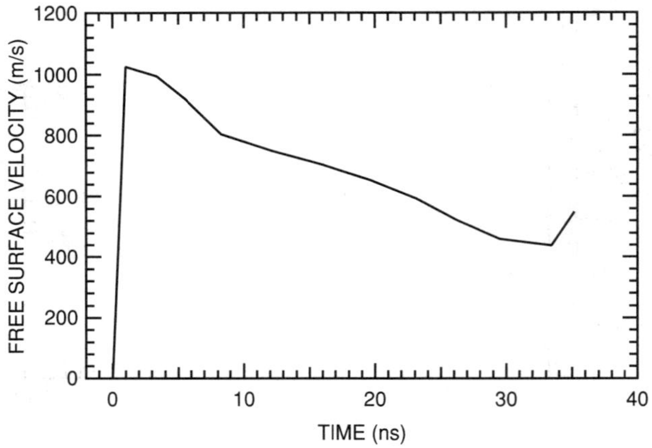

Figure A.129. Velocity history at the free surface of a z-cut sapphire target irradiated using an ion beam with an 5-mm spot size and an energy density of 0.2 TW/cm^2. The experiment was performed using configuration K (see Table A.2) with k_1 = 2.3 mm. The data were recorded using ORVIS with a measurement accuracy of ±20 m/s. The peak stress was 24.0 GPa, and the spall strength was 12.8 (± 6%) GPa (Kanel et al. [1994a]).

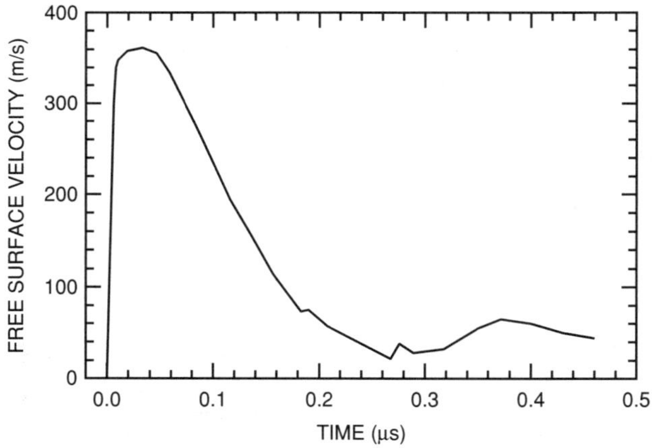

Figure A.130. Velocity history at the free surface of a x-cut quartz target impacted by an aluminum flyer plate at an impact velocity of 660 ± 20 m/s. The experiment was performed using configuration C (see Table A.2) with c_1 = 0.2 mm, c_2 = 2.9 mm (aluminum), and c_3 = 1.98 mm. The data were recorded using a VISAR with a measurement accuracy of 3 to 5 m/s. No spall was observed (Kanel et al. [1992b]).

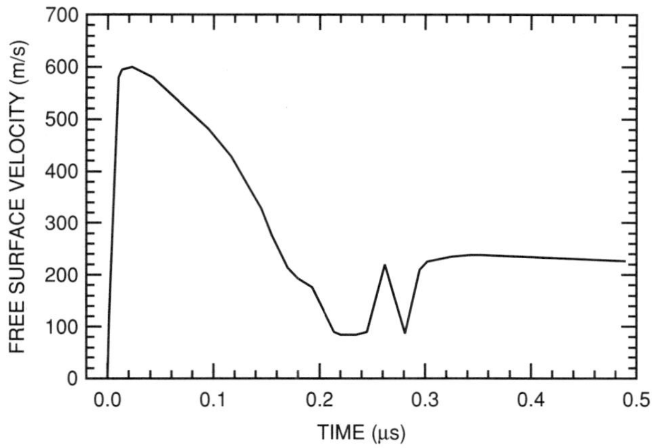

Figure A.131. Velocity history at the free surface of a x-cut quartz target impacted by an aluminum flyer plate at an impact velocity of 660 ± 20 m/s. The experiment was performed using configuration C (see Table A.2) with c_1 = 0.4 mm, c_2 = 2.0 mm (aluminum), and c_3 = 1.98 mm. The data were recorded using a VISAR with a measurement accuracy of 3 to 5 m/s. The spall strength was 4.0 ± 0.03 GPa (Kanel et al. [1992b]).

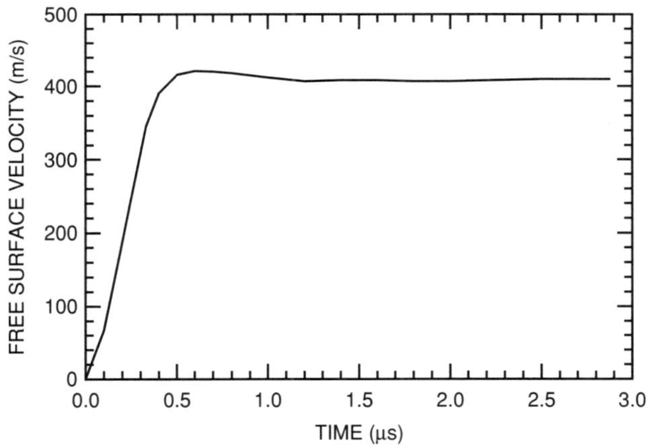

Figure A.132. Velocity history at the free surface of an explosively loaded titanium carbide (84.5% by weight) with nickel binder target. The experiment was performed using configuration H (see Table A.2) with h_1 = 10.3 mm. The data were recorded using a capacitor gauge with an electrode diameter of 20 mm and a measurement accuracy of ±4%. The spall strength was 0.35 ± 0.05 GPa (Kanel and Pityulin [1984]).

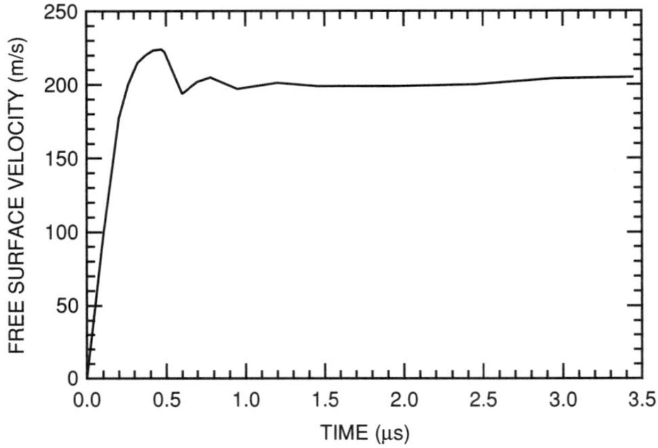

Figure A.133. Velocity history at the free surface of a titanium carbide (84.5% by weight) with nickel binder target impacted by an aluminum flyer plate at an impact velocity of 450 ± 20 m/s. The experiment was performed using configuration A (see Table A.2) with a_1 = 2.0 mm and a_2 = 10.0 mm. The data were recorded using a capacitor gauge with an electrode diameter of 20 mm and a measurement accuracy of ±4%. The spall strength was 0.5 ± 0.05 GPa (Kanel and Pityulin [1984]).

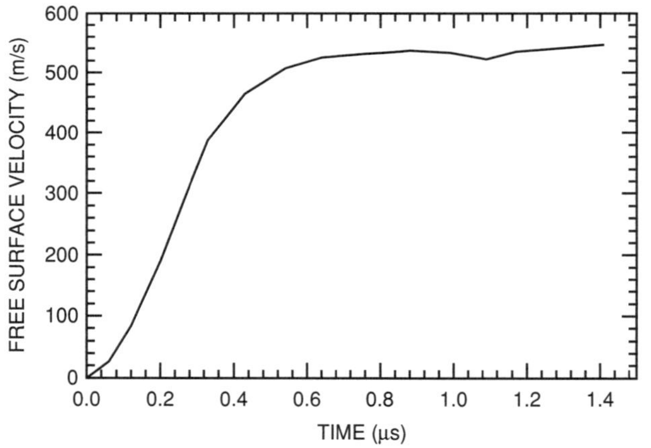

Figure A.134. Velocity history at the free surface of a titanium carbide (84.5% by weight) with nickel binder target impacted by an aluminum flyer plate at an impact velocity of 1050 ± 50 m/s. The experiment was performed using configuration A (see Table A.2) with a_1 = 7.0 mm and a_2 = 11.9 mm. The data were recorded using a capacitor gauge with an electrode diameter of 20 mm and a measurement accuracy of ±4%. No spall was observed (Kanel and Pityulin [1984]).

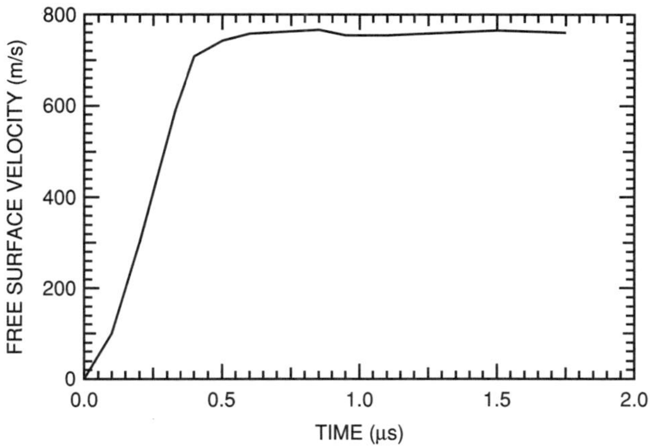

Figure A.135. Velocity history at the free surface of a titanium carbide (84.5% by weight) with nickel binder target impacted by an aluminum flyer plate at an impact velocity of 1500 ± 50 m/s. The experiment was performed using configuration A (see Table A.2) with a_1 = 4.0 mm and a_2 = 10.0 mm. The data were recorded using a capacitor gauge with an electrode diameter of 20 mm and a measurement accuracy of ±4%. No spall was observed (Kanel and Pityulin [1984]).

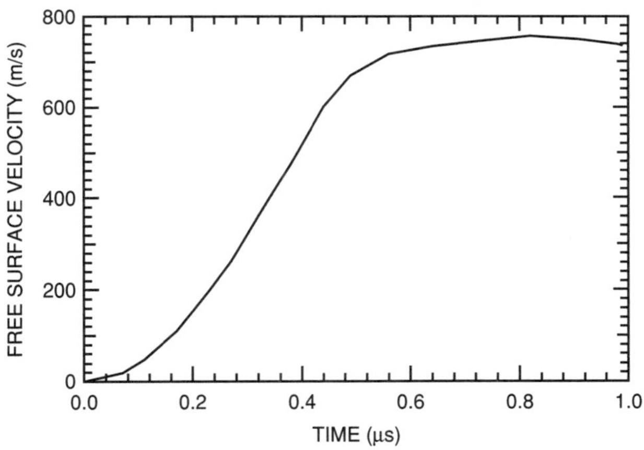

Figure A.136. Velocity history at the free surface of a titanium carbide (84.5% by weight) with nickel binder target impacted by an aluminum flyer plate at an impact velocity of 1500 ± 50 m/s. The experiment was performed using configuration A (see Table A.2) with a_1 = 4.0 mm and a_2 = 12.0 mm. The data were recorded using a capacitor gauge with an electrode diameter of 20 mm and a measurement accuracy of ±4%. No spall was observed (Kanel and Pityulin [1984]).

References

Ahrens, T.J., Anderson, D.L., and Ringwood, A.E., "Equation of State and Crystal Structures of High-Pressure Phases of Shocked Silicates and Oxides," *Rev. Geophys.*, **7**, 667–707 (1969).

Al'tshuler, L.V., "Application of Shock Waves in the Physics of High Pressures," *Usp. Fiz. Nauk.,* **85**(2), 197–258 (1965), (in Russian).

Al'tshuler, L.V., Novikov, S.A., and Divnov, I.I., "Relationship Between Critical Fracture Stresses and the Fracture Time Under Explosive Loading of Metals," *Dokl. Akad. Nauk., (USSR)* **166**(1), 67–70 (1966), (in Russian).

Anderson, C.E., Jr., O'Donoghue, P.E., and Bodner, S.R., "Tensile Failure in Tungsten Alloy Fragments," in *Shock Waves in Condensed Matter—1987*, edited by N.H. Holmes, and S.C. Schmidt, North Holland, Amsterdam, 371–374 (1988).

Antoun, T.H., "Continuum Damage Mechanics Based Constitutive Model for the Impact Behavior of Concrete," Ph.D. Dissertation, University of Dayton, Dayton, OH (December 1991).

Antoun, T.H., Seaman, L., Curran, D.R., and Glinsky, M., "Damage and Failure Mechanisms Associated with Photoablation of Biological Tissues," in *Shock Compression of Condensed Matter—1995*, edited by S.C. Schmidt and W.C. Tao, AIP Press, Woodbury, NY, 503–506 (1996).

Arad, B., Moshe, E., Eliezer, S., Dekel, E., Ludmirsky, A., Henis, Z., Goldberg, I.B., Eliaz, N., and Eliezer, D., "Spall Strength Measurements in Aluminum, Copper and Metallic Glass at Strain Rates of $\sim 10^7$ s^{-1}," in *Shock Compression of Condensed Matter—1997*, edited by S.C. Schmidt, D.P. Dandekar, and J.W. Forbes, AIP Press, Woodbury, NY, 459–462 (1998).

Arrhenius, S., "On the Reaction Velocity of the Inversion of Cane Sugar by Acids," *Z. Phys. Chem.*, **4**, 226 (1889).

Asay, J.R., and Barker, L.M., "Interferometric Measurement of Shock-Induced Internal Particle Velocity and Spatial Variations of Particle Velocity," *J. Appl. Phys.,* **45**(6), 2540–2544 (1974).

Asay, J.R., Chhabildas, L.C., and Dandecar, D.P., "Shear-Strength of Shock-Loaded Polycrystalline Tungsten,"*J. Appl. Phys.,* **51**(9), 4774–4783 (1980).

Auroux, E., "Study of the Adhesion of Thin Layers above a Substrate Subjected to Laser Shocks," Ph.D. Dissertation, University of Poitiers, submitted to the Ecole Nationale Supérieure de Mécanique et d'Aérotechnique et Faculté des Sciences Fondamentales et Appliquées, France (October 19, 1999).

Bancroft, D., Peterson, E.L., and Minshall, S., "Polymorphism of Iron at High Pressure," *J. Appl. Phys.*, **27**, 291–298 (1956).

Barbee, T.W., Jr., Seaman, L., and Crewdson, R.C., "Dynamic Fracture Criteria of Homogeneous Materials," Technical Report No. AFWL-TR-70-99, Air Force Weapons Laboratory, Kirtland Air Force Base, NM (1970).

Barbee, T.W., Jr., Seaman, L., Crewdson, R., and Curran, D.R., "Dynamic Fracture Criteria for Ductile and Brittle Metals," *J. of Materials,* **7**, 393–401 (1972).

Barker, L.M., in *Shock Compression of Condensed Matter—1997,* edited by S.C. Schmidt, D.P. Dandekar, and J.W. Forbes, AIP Press, Woodbury, NY, 833 (1998).

Barker, L.M., and Hollenbach, R.E., "Shock-Wave Studies of PMMA, Fused Silica, and Sapphire," *J. Appl. Phys.,* **41**(10), 4208–4226 (1970).

Barker, L.M., and Hollenbach, R.E., "Laser Interferometry for Measuring High Velocities of Any Reflecting Surface," *J. Appl. Phys.,* **43**(11), 4669 (1972).

Barker L.M., and Hollenbach, R.E., "Shock Wave Study of the α–ε Phase transition in Iron," *J. Appl. Phys.*, **45**(11), 4872–4887 (1974).

Barker, L.M., and Schuler, K.W., "Correction to Velocity-Per Fringe Relationship for VISAR Interferometer," *J. Appl. Phys.,* **45**, 3692–3693 (1974).

Baumung, K., Bluhm, H.J., Hoppe, P., Karow, H.U., Rush, D., Stoltz, O., Singer, J., Kanel, G.I., Utkin, A.V., and Razorenov, S.V., "Hydrodynamic Beam-Target Interaction Experiments on the Karlsruhe Light Ion Facility KALIF," to be published in Proceedings of the BEAMS—94 Conference (1994).

Bazant, Z.P., "Imbricate Continuum and Its Variational Derivation," *J. Eng. Mech., ASCE,* **110**(12), 1693–1712 (1984).

Bazant, Z.P., "Softening Instability: Part I—Localization into a Planar Band," *J. Appl. Mech,* **55**, 517–522 (1988a).

Bazant, Z.P., "Softening Instability: Part II—Localization into Ellipsoidal Regions," *J. Appl. Mech,* **55**, 523–529 (1988b).

Bazant, Z.P., Belytschko, T.B., and Chang, T.-P., "Continuum Theory for Strain-Softening," *J. Eng. Mech., ASCE,* **110**(12), 1666–1692 (1984).

Bazant, Z.P., and Chang, T.-P., "Nonlocal Finite Element Analysis of Strain-Softening Solids," *J. Eng. Mech, ASCE,* **113**(1), 89–105 (1987).

Bazant, Z.P., and Pijaudier-Cabot, G., "Nonlocal Continuum Damage, Localization Instability and Convergence," *J. Appl. Mech,* **55**, 287–293 (1988).

Belak, J., private communication (June 1999).

Belytschko, T., Lu, Y.Y., and Gu, L., "Element-Free Galerkin Methods," *Int. J. Numerical Methods Eng.*, **37**, 229–256 (1994).

Bernstein, D., and Keough, D.D., "Piezoresistivity of Manganin," *J. Appl. Phys.,* **35**, 1471 (1964).

Bless, S.J., and Paisley, D.L., in *Shock Waves in Condensed Matter—1983,* edited by J.R. Asay, R.A. Graham, and G.K. Straub, Elsevier Science Publishers B.V., Amsterdam, 163 (1984).

Blinkow, D.W., and Keller, D.W., *ASTM Spec. Tech. Publ.*, 252 (1962).

Bloomquist, D.D., and Sheffield, S.A., "Optically Recording Interferometer for Velocity Measurements with Subnanosecond Resolution," *J. Appl. Phys.,* **54**, 1717 (1983).

Bourne, N.K., Rozenberg, Z., Crouch, I.G., and Field, J.E., "Microstructural Variations in Seven Aluminas and Their Effect on Impact Response," in *High-Pressure Science and Technology—1993*, edited by S.C. Schmidt, J.W. Shaner, G.A. Samara, and M. Ross, AIP Press, Woodbury, NY, 769–772 (1994).

Boustie, M., Cottet, F., and Romain, J.P., "Spalling Due to a Strong Shock Wave Decay Process in Solid Targets Irradiated by a Pulsed Laser," in *Shock Compression of*

Condensed Matter—1991, edited by S.C. Schmidt, R.D. Dick, J.W. Forbes, and D.G. Tasker, Elsevier Science Publishers B.V., Amsterdam, 805–808 (1992a).

Boustie, M., Cottet, F., and Chauveau, Y., "Spalling of Aluminum and Copper Targets by Laser Shocks," in *Shock-Wave and High-Strain-Rate Phenomena in Materials*, edited by M.A. Meyers, L.E. Murr, and K.P. Staudham, Marcel Dekker, New York, 741–750 (1992b).

Brannon, R.M., Drugan, W.J., and Shen, Y., "Requirements of Thermodynamics in the Analysis of Elastic-Plastic Shock Waves," *J. Mech. Phys. Solids*, **43**(6), 973–1001 (1995).

Brannon, R., and Wong, M., "MIG Version 0.0 Model Interface Guidelines: Rules to Accelerate Installation of Numerical Models into any Compliant Parent Code," Sandia National Laboratories Report No. SAND96-2000 (1996).

Brar, N.S., and Bless, S.J., "Dynamic Fracture and Failure Mechanisms of Ceramic Bars," in *Shock-Wave and High-Strain-Rate Phenomena in Materials*, edited by M.A. Meyers, L.E. Murr, and K.P. Staudham, Marcel Dekker, New York, 1041–1049 (1992).

Brar, N.S., Rosenberg, Z., and Bless, S.J., "Spall Strength and Failure Waves in Glass," *J. de Physique IV*, 1, C3–639 (1991).

Breed, B.R., Mader, C.L., and Venable, D., "Technique for the Determination of Dynamic-Tensile-Strength Characteristics," *J. Appl. Phys.*, **38**(8), 3271–3275 (1967).

Bridgman, P., *Proc. Am. Acad. Arts and Sci.*, **47**, 321 (1911).

Bridgman, P., "The Measurements of Hydrostatic Pressure to 30000 kg/cm^2," *Proc. Am. Acad. Arts and Sci.*, **74**, 1 (1940).

Bridgman, P., *Studies in Large Plastic Flow and Fracture*, Harvard University Press, Cambridge, MA (1964).

Buchar, J., Rolc, S., and Hrebicek, J., "Dynamic Fracture (Spalling) of Some Structural Steels," in *Shock-Wave and High-Strain-Rate Phenomena in Materials*, edited by M.A. Meyers, L.E. Murr, and K.P. Staudham, Marcel Dekker, New York, 769–774 (1992).

Budiansky, B., Hutchinson, J.W., and Slutsky, S., "Void Growth and Collapse in Viscous Solids," *Mechanics of Solids*, edited by H.G. Hopkins, and M.J. Sewell, Pergamon Press, Oxford, 13–45 (1982).

Bushman, A.V., Kanel, G.I., Ni, A.L., Fortov, V.E., *Intense Dynamic Loading of Condensed Matter*, Taylor and Francis, Washington, DC (1993).

Butcher, B.M., "Spallation in 4340 Steel," *J. Appl. Mech. (Trans. ASME, series E)*, **34**, 209–210 (1967).

Caldirola, P., and Knoepfel, H. (editors), *Physics of High Energy Density*, Academic Press, New York (1971).

Carroll, M.M., and Holt, A.C., "Static and Dynamic Pore-Collapse Relations for Ductile Porous Materials," *J. Appl. Phys.*, **43**(4), 1626–1636 (1972a).

Carroll, M.M., and Holt, A.C., "Suggested Modification of the P–α Model for Porous Materials," *J. Appl. Phys.*, **43**(2), 759–761 (1972b).

Castañeda, P., and Zaidman, M., "Constitutive Models for Porous Materials with Evolving Microstructure," *J. Mech. Phys. Solids*, **42**(9), 1459–1497 (1994).

Chang, S.N., and Choi, J.H., "High Strain Rate Response of a Tungsten Heavy Alloy," in *Shock Compression of Condensed Matter—1997*, edited by S.C. Schmidt, D.P. Dandekar, and J.W. Forbes, AIP Press, Woodbury, NY, 415–418 (1998).

Chang, S.N., and Nemat-Nasser, S., "Compression-Induced High-Strain-Rate Void Collapse," in *Shock-Wave and High-Strain-Rate Phenomena in Materials*, edited by M.A. Meyers, L.E. Murr, and K.P. Staudham, Marcel Dekker, New York, 783–791 (1992).

Chen, D.Y., Gupta, Y.M., and Miles, M.H., "Quasistatic Experiments to Determine Material Constants for the Piezoresistance Foils Used in Shock Wave Experiments," *J. Appl. Phys.,* **55**(11), 3984–3993 (1984).

Chhabildas, L.C., and Asay, J.R., in *Shock-Wave and High-Strain-Rate Phenomena in Materials*, edited by M.A. Meyers, L.E. Murr, and K.P. Staudham, Marcel Dekker, New York, 947 (1992).

Chhabildas, L.C., Barker, L.M., Asay, J.R., and Trucano, T.G., in *Shock Compression of Condensed Matter—1989*, edited by S.C. Schmidt, J.N. Johnson, and L.W. Davison, Elsevier Science Publishers B.V., Amsterdam, 429 (1990).

Chhabildas, L.C., and Graham, R.A., "Development in Measurement Techniques for Shock Loaded Solids," in *Techniques and Theory of Stress Measurements for Shock Wave Applications*, edited by R.R. Stout, E.R. Norwood, and M.E. Fourney, American Society of Mechanical Engineers, New York, AMD **83**, 1–18 (1987).

Chou, P.C., and Hopkins, A.K., *Dynamic Response of Materials to Intense Impulsive Loading*, Air Force Materials Laboratory, Wright Patterson Air Force Base, Dayton, OH (1972).

Christman, D.R., Isbell, W.M., Babcock, S.G., McMillan, A.R., and Green, S.J., DNA Report DASA 2501–3, AD735966 (1971).

Christman, D.R., Michaels, T.E., Isbell, W.M., and Babcock, S.G., DNA Report DASA 2510–4, AD738862 (1972).

Christy, S., Pak, H.-R., and Meyers, M.A., "Effect of Metallurgical Parameters on Dynamic Fracture by Spalling of Copper," in *Metallurgical Applications of Shock-Wave and High-Strain-Rate Phenomena*, edited by L.E. Murr, K.P. Staudhammer, and M.A. Meyers, Marcel Dekker, New York, 835–863 (1986).

Cochran, S., and Banner, D., "Spall Studies in Uranium," *J. Appl. Phys.,* **48**(7), 2729–2737 (1977).

Coleman, B.D., and Gurtin, M.E., "Thermodynamics with Internal State Variables," *J. Chem. Phys.,* **47**(2), 597–613 (1967).

Couch, R., Sharp R., Otero, I., Tipton, R., and McCallen R., "Application of ALE Techniques to Metal Forming Simulations," in *Advanced Computational Methods for Material Modeling*, edited by D.J. Benson, and R.J. Asaro, ASME AMD, **180**, New York, 133–140 (1993).

Courant, R., and Friedrichs, K.O., *Supersonic Flow and Shock Waves*, Interscience Publishers, New York (1948).

Couzens, D.C.F., and Trevena, D.H., "Critical Tension in a Liquid Under Dynamic Conditions of Stressing," *Nature,* **222**(5192), 473–474 (1969).

Cowperthwaite, M., "Relationships Between Incomplete Equations of State," *J. Franklin Institute*, **287**(5), 379 (1969).

Curran, D.R., and Seaman, L., "Computational Models for Nucleation, Growth, and Coalescence of Adiabatic Shear Bands," *J. Physique*, Colloque C5, Supplement au no. 8, Tome 46 (1985).

Curran, D.R., and Seaman, L., "Computational Models for Nucleation, Growth, and Coalescence of Adiabatic Shear Bands," in *Shock Waves in Condensed Matter—1985*, edited by Y. Gupta, Plenum Press, New York (1986).

Curran, D.R., Seaman, L., and Shockey, D.A., "Dynamic Failure of Solids," *Physics Reports,* **147**(5&6), 253–388 (1987).

Curran, D.R., Seaman, L., and Simons, J., "Equation of State (EOS) Models for Hydrocodes," in *New Models and Numerical Codes for Shock Wave Processes in Condensed Media*, Proceedings of the International Workshop held at Oxford, UK, edited by I .G. Cameron, 155–164 (1997).

Curran, D.R., Shockey, D.A., and Seaman, L., "Dynamic Fracture Criteria for a Polycarbonate," *J. Appl. Phys.,* **44**(9), 4025–4038 (1973).

Dandekar, D.P., "Response of Ceramics Under Shock Wave Loading," in *High-Pressure Science and Technology—1993*, edited by S.C. Schmidt, J.W. Shaner, G.A. Samara, and M. Ross, AIP Press, Woodbury, NY, 729–732 (1994).

Dandekar, D.P., and Bartkowski, P., "Shock Response of AD995 Alumina," in *High-Pressure Science and Technology—1993*, edited by S.C. Schmidt, J.W. Shaner, G.A. Samara, and M. Ross, AIP Press, NY, 733–736 (1994).

Davison, L., Grady, D.E., and Shahinpoor, M., editors, *High-Pressure Shock Compression of Solids II: Dynamic Fracture and Fragmentation*, Springer, New York (1996).

Davison, L., Stevens, A.L., and Kipp, M.E., "Theory of Spall Damage Accumulation in Ductile Metals," *J. Mech. Phys. Solids,* **25**(1), 11–28 (1977).

DeCarli, P.S., "Stress Gage System for the Megabar (100 GPa) Range," Technical Report DNA 4066F, Defense Nuclear Agency, Washington, DC (1976).

DeCarli, P.S., and Jamieson, J.C., "Formation of Diamond by Explosive Shock," *Science,* **133**, 1821 (1961).

Dienes, J.K., "A Statistical Theory of Fragmentation," *19th U.S. Symposium on Rock Mechanics*, Stateline, NV (May 1–3, 1978).

Dienes, J.K., "Statistical Crack Mechanics: Theory and Experiment," *Proceedings of the 22nd U.S. Symposium on Rock Mechanics*, Cambridge, MA (June 29–July 2, 1981).

Dienes, J.K., "A Statistical Theory of Fragmentation Processes," *Mechanics of Materials*, **4**(3), 325–335 (1985).

Dienes, J.K., "A Unified Theory of Flow, Hot Spots, and Fragmentation, with an Application to Explosive Sensitivity," in *High Pressure Shock Compression of Solids II: Dynamic Fracture and Fragmentation*, edited by L. Davison, D.E. Grady, and M. Shahinpoor, Springer, New York, 366–398 (1996).

Dienes, J.K., and Margolin, L.G., "A Computational Approach to Rock Fragmentation," *Proceedings of the 20th U.S. Symp. on Rock Mechanics*, Austin, TX (June 4–6, 1979).

Dienes, J.K., and Margolin, L.G., "A Computational Approach to Rock Fragmentation," *21st U.S. Symposium on Rock Mechanics*, Rolla, MS (May 28–30, 1980).

Dingus, R., Goldman, S.R., Kirkpatrick, R., Montierth, L.M., and and Stover, E.K., "Pulsed Laser Effects Phenomenology," SPIE Vol. 1064, *Thermal and Optical Interactions with Biological and Related Composite Materials*, 66–75 (1989).

Dremin, A.N., Kanel, G.I., and Glusman, V.D., "Experimental Investigation of the Pressure Profiles at Nonregular Reflection of Conical Shock Wave in PMMA Cylinders," *Fiz. Gorenia i Vzryva (in Russian),* **8**(1), 104–109 (1972).

Dremin, A.N., and Molodets, A.M., "Kinetic Characteristics of Spall Fracture," *J. Appl. Mech. Tech. Phys.,* **21**(6), 806 (in translation) (1980).

Dremin, A.N., and Molodets, A.M., in *Shock Compression of Condensed Matter—1989*, edited by S.C. Schmidt, J.N. Johnson, and L.W. Davison, Elsevier Science Publishers B.V., Amsterdam, 415–418 (1990).

Dremin, A.N., Molodets, A.M., Melkumov, A.I., and Kolesnikov, A.V., "On Anomalous Increase of Steel Spall Strength and Its Relationship to Martensitic Transformation," in *Shock-Wave and High-Strain-Rate Phenomena in Materials*, edited by M.A. Meyers, L.E. Murr, and K.P. Staudham, Marcel Dekker, New York, 751–757 (1992).

Dube, E., Neely, R., Nichols, A., Sharp, R., and Couch, R., "Users Manual for ALE3D" (unpublished), Lawrence Livermore National Laboratory, Livermore, CA (2000).

Duffy, T.S., and Ahrens, T.J., "Free Surface Velocity Profiles in Molybdenum Shock Compressed at 1400°C," in *High Pressure Science and Technology—1993*, edited by S.C. Schmidt, J.W. Shaner, G.A. Samara, and M. Ross, AIP Press, Woodbury, NY, 1079–1082 (1994).

Durand, M., Laharrague, P., Lalle, P., Lebihan, A., Morvan, J., and Pujois, H., "Interferometric Laser Technique for Accurate Velocity Measurement in Shock-Wave Physics," *Rev. Sci. Instrum.*, **48**, 275–278 (1977).

Duvall, G.E., and Fowles, G.R., "Shock Waves," in *High Pressure Physics and Chemistry,* **2**, edited by R.S. Bradley, Academic Press, London, 209–291 (1963).

Duvall, G.E., and Graham, R.A., "Phase Transitions Under Shock Wave Loading," *Reviews of Modern Physics*, **49**(3), 523–579 (1977).

Elias, C.N., Rios, P.R., and Romero, A.W., "Spall of Differently Treated High-Strength Low-Alloy Steel," in *Shock-Wave and High-Strain-Rate Phenomena in Materials*, edited by M.A. Meyers, L.E. Murr, and K.P. Staudham, Marcel Dekker, New York, 733–739 (1992).

Englman, R., Jaeger, Z., and Levi, A., "Percolation Theoretical Treatment of Two-Dimensional Fragmentation in Solids," *Phil. Mag.*, **B50**, 307–315 (1984).

Erlich, D.C., and Gran, J.K., "Effect of Static Transverse Prestress on Dynamic Tensile Failure in Metals," in *Shock Waves in Condensed Matter—1987*, edited by N.H. Holmes, and S.C. Schmidt, North Holland, Amsterdam, 355–358 (1988).

Erlich, D.C., Wooten, D.C., and Crewdson, R.C., "Dynamic Tensile Failure of Glycerol," *J. Appl. Phys.*, **42**, 5495–5502 (1971).

Ernsberger, F.M., "Role of Densification in Deformation of Glasses Under Point Loading," *J. Amer. Ceram. Soc.*, **51**(10), 545–547 (1968).

Espinosa, H.D., "On the Dynamic Shear Resistance of Ceramic Composites and Its Dependence on Applied Multiaxial Deformation," *Int. J. Solids Structures*, **32**(21), 3105–3128 (1995).

Espinosa, H.D., and Brar, N.S., "Dynamic Failure Mechanisms of Ceramic Bars: Experiments and Numerical Simulations," *J. Mech. Phys. Solids*, **43**(10), 1615–1638 (1995).

Espinosa, H.D., Zavattieri, P.D., and Emore, G.L., "Adaptive FEM Computation of Geometric and Material Nonlinearities with Application to Brittle Failure," *Mech. Matl.*, **29**, 275-305 (1998).

Evans, A.G., Ito, Y.M., and Rosenblatt, M., "Impact Damage Thresholds in Brittle Materials Impacted by Water Drops," *J. Appl. Phys.*, **51**(5), 2473–2482 (1980).

Fowles, G.R., Duvall, G.E., Asay, J., Bellamy, P., Feistmann, F., Grady, D., Micheals, T., and Mitchell, R., "Gas Gun for Impact Studies," *Rev. Sci. Instr.*, **41**, 984 (1970).

Frantziskonis, G., and Desai, C.S., "Elastoplastic Model with Damage for Strain Softening Geomaterials," *Acta Mechanica*, **68**, 151–170, (1987a).

Frantziskonis, G., and Desai, C.S., "Constitutive Model with Strain Softening," *Int. J. Solids Structures*, **23**(6), 733–750 (1987b).

Frantziskonis, G., and C.S. Desai, "Analysis of a Strain Softening Constitutive Model," *Int. J. Solids Structures*, **23**(6), 751–767 (1987c).

Freund, L.B., "Crack Propagation in an Elastic Solid Subjected to General Loading—III. Stress Wave Propagation," *J. Mech. Phys. Solids.*, **21**, 47–61 (1973).

Freund, L.B. *Dynamic Fracture Mechanics*, Cambridge University Press, Cambridge, MA (1990).

Fuller, J.A., and Price, J.H., "Dynamic Pressure Measurements to 300 Kilobars with a Resistance Transducer," *Brit. J. Appl. Phys.*, **15**, 751 (1964).

Gaeta, P.J., and Dandekar, D.P., "Shock Response of Kennertium Grade W-2," in *Shock Waves in Condensed Matter—1987*, edited by N.H. Holmes, and S.C. Schmidt, North Holland, Amsterdam, 269–272 (1988).

Garrett, R.K., Jr, Rajendran, A.M., and Last, H.R., "Modeling Spall in HY100, HY130, and AF1410 Steels," in *Shock Compression of Condensed Matter—1995*, edited by S.C. Schmidt, and W.C. Tao, AIP press, Woodbury, NY, 323–326 (1996).

Gent, A.N., and Lindley, P.B., "Internal Rupture of Bounded Rubber Cylinders in Tension," *Proc. Roy. Soc. (London), Series A*, **249**(1), (1959).

Gidon, S., and Behar, G., "Instantaneous Velocity-Field Measurements—Application to Shock-Wave Studies," *Appl. Opt.*, **25**, 14291–433 (1986).

Giles, A.R., and Maw, J.R., "Modelling the Temperature and Strain Rate Dependence of Spallation in Metals," in *Shock Compression of Condensed Matter—1997*, edited by S.C. Schmidt, D.P. Dandekar, and J.W. Forbes, AIP Press, Woodbury, NY, 243–246 (1998).

Glenn, L.A., and Chudnovsky, A., "Strain Energy Effects on Dynamic Fragmentation," *J. Appl. Phys.*, **59**, 1379–1380 (1986).

Glusman, V.D., and Kanel, G.I., "Measurement of the Tensile Stresses Behind a Spalling Plane," *J. Appl. Mechanics Tech. Phys. (USSR)*, **24**(4), 146–150 (1983).

Glusman, V.D., Kanel, G.I., Loskutov, V.F., Fortov, V.E., and Khorev, I.E., "The Resistance to Deformation and Fracture of Steel 35X3NM in Shock Loading Conditions," *Problemy Prochnosti (USSR)*, (8), 52 (1985).

Godse, R., Ravichandran, G., and Clifton, R.J., "Micromechanisms of Dynamic Crack Propagation in an AISI 4340 Steel," *Materials Sci. Engng.*, **A112**, 79–88 (1989).

Gologanu, M., Leblond, J.-B., and Devaux, J., "Approximate Models for Ductile Metals Containing Non-Spherical Voids—Case of Axisymmetric Prolate Ellipsoidal Cavities," *J. Mech. Phys. Solids*, **41**(11), 1723–1754 (1993).

Golubev, V.K., Korshunov, A.I., Novikov, S.A., Sobolev, Yu.S., and Yukina, N.A., "Strength and Fracture of Aluminum Alloy Amg6 Under Shock-Wave Loading," *J. Appl. Mech. Tech. Phys. (USSR)*, (2), 121 (1988).

Golubev, V.K., and Novikov, S.A., "On an Influence of the Heat Treatment on the Strength of 40 kH Steel at Impact," *Problems of Strength (USSR)*, (4), 41 (1984).

Golubev, V.K., Novikov, S.A., and Sobolev, Yu.S., "The Spall Fracture of PMMA, Teflon, and Polyethylene at Normal and Elevated Temperatures," *Problems of Strength (USSR)*, No.2, 88–91 (1982a).

Golubev, V.K., Novikov, S.A., and Sobolev, Yu.S., "On the Temperature Influence on the Spall Strength of Polymer materials," *J. Appl. Mech. Tech. Phys. (USSR)*, (1), 143–150 (1982b).

Golubev, V.K., Novikov, S.A., Sobolev, Yu.S., and Yukina, N.A., "On Character of Spall Fracture of Aluminum Alloys D16 and Amg6 Within the Temperature Range 196–600°C," *Problems of Strength (USSR)*, (2), 53–59 (1983).

Golubev, V.K., Novikov, S.A., Sobolev, Yu.S., and Yukina, N.A., "Influence of Temperature Upon the Strength and Fracture of Copper and Nickel at Spalling," *Problems of Strength (USSR)*, (1), 63–65 (1985).

Grady, D.E., "Application of Survival Statistics to the Impulsive Fragmentation of Ductile Rings," in *Shock Waves and High-Strain-Rate Phenomena in Metals*, 181–192, edited by Meyers, M.A., and Murr, L.E., Plenum Press, NY (1981).

Grady, D.E., "Local Inertial Effects in Dynamic Fragmentation," *J. Appl. Phys.*, **53**(1), 322–325 (1982).

Grady, D.E., "Steady-Wave Risetime and Spall Measurements on Uranium (3–15 GPa)," in *Metallurgical Applications of Shock-Wave and High-Strain-Rate Phenomena*, edited by L.E. Murr, K.P. Staudhammer, and M.A. Meyers, Marcel Dekker, New York, 763–780 (1986).

Grady, D.E., "The Spall Strength of Condensed Matter," *J. Mech. Phys. Solids,* **36**(3), 353–384 (1988).

Grady, D.E., "Particle Size Statistics in Dynamic Fragmentation," *J. Appl. Phys.*, **68**(12), 6099–6105 (1990).

Grady, D.E., "Shock-Wave Properties of High-Strength Ceramics," in *Shock Compression of Condensed Matter—1991*, edited by S.C. Schmidt, R.D. Dick, J.W. Forbes, and D.G. Tasker, Elsevier Science Publishers B.V., Amsterdam, 455–458 (1992).

Grady, D.E., "Shock-Wave Properties of Brittle Solids," in *Shock Compression of Condensed Matter—1995*, edited by S.C. Schmidt, and W.C. Tao, AIP press, Woodbury, NY, 9–20 (1996).

Grady, D.E., and Kipp, M.E., "The Micromechanics of Impact Fracture of Rock," *Int J. Rock Mech. Min. Sci. Geomech. Abstr.,* **16**, 293–302 (1979).

Grady, D.E., and Kipp, M.E., "Geometric Statistics and Dynamic Fragmentation," *J. Appl. Phys.*, **58**(3), 1210–1222 (1985).

Grady, D.E., and Kipp, M.E., "Experimental Measurement of Dynamic Failure and Fragmentation Properties of Metals," *Int. J. Solids Structures*, **32**(17/18) 2779–2791 (1995).

Grady, D.E., and Kipp, M.E., in *High Pressure Shock Compression of Solids II: Dynamic Fracture and Fragmentation*, edited by L. Davison, D.E. Grady, and M. Shahinpoor, Springer, New York (1996).

Graham, R.A., "Shock-Wave Compression of X-Cut Quartz as Determined by Electrical Response Measurements," *J. Phys. Chem. Solids,* **35**(3), 355–372 (1974).

Graham, R.A., *Solids Under High-Pressure Shock Compression*, Springer, New York (1993).

Graham, R.A., and Asay, J.R., "Measurements of Wave Profiles in Shock-Loaded Solids," *High Temp. High Press.,* **10**, 355–390 (1978).

Graham, R.A., and Brooks, W.P., "Shock-Wave Compression of Sapphire from 15 to 420 kbar. The Effects of Large Anisotropic Compression," *J. Phys. Chem. Solids,* **32**, 2311–2330, (1971).

Gran, J., and Seaman, L., "Observations and Analyses of Microcracks Produced in Dynamic Tension Tests of Concrete," SRI International Technical Report prepared for AFOSR under contract no. F49620-82-K-0021 (1986).

Griffith, A.A., "The Phenomena of Rupture and Flow in Solids," *Phil. Trans. Roy. Soc. London, Series A,* **221**, 163–179 (1921).

Gu, Z., and Jin, X., "Temperature Dependence on Shock Response of Stainless Steel," in *Shock Compression of Condensed Matter—1997*, edited by S.C. Schmidt, D.P. Dandekar, and J.W. Forbes, AIP Press, Woodbury, NY, 467–470 (1998).

Guéry, J.-F., and Seaman, L., "Simulations of Microfractures in Solid Rocket Propellants," in *Shock Compression of Condensed Matter—1995*, edited by S.C. Schmidt, and W.C. Tao, AIP press, Woodbury, NY, 13–18 (1996).

Gupta, S.C., and Gupta, Y.M., "Experimental Measurements and Analysis of the Loading and Unloading Response of Longitudinal and Lateral Manganin Gauges Shocked to 90 kbar," *J. Appl. Phys.*, **62**(7), 2903–2609 (1987).

Gupta, Y.M., "Stress Measurements Using Piezoresistance Gauges: Modeling the Gauge as an Elastic-Plastic Inclusion," *J. Appl. Phys.*, **54**(11), 6256–6266 (1983).

Gurson, A.L., "Continuum Theory of Ductile Rupture by Void Nucleation and Growth: Part I—Yield Criteria and Flow Rules for Porous Ductile Media," *J. Eng. Materials Technology, Trans of ASME*, Paper No. 76-Mat-CC, 2–15 (1977).

Guy, A.G., *Elements of Physical Metallurgy*, Addison-Wesley, Reading, (1960).

Hemsing, W.F., Mathews, A.R., Warnes, R.H., George, M.J., and Whittemore, G.R., "VISAR: Line Imaging Interferometer," in *Shock Compression of Condensed Matter—1991*, edited by S.C. Schmidt, R.D. Dick, J.W. Forbes, and D.G. Tasker, Elsevier Science Publishers B.V., Amsterdam, 767–770 (1992).

Herrmann, W., "Constitutive Equation for the Dynamic Compaction of Ductile Porous Materials," *J. Appl. Phys.*, **40**(6), 2490–2499 (1969).

Herrmann, W., "Constitutive Equations for Compaction of Porous Materials," *Symp. Appl. Mech. Aspects of Nuclear Effects in Materials*, ASME Winter Annual Meeting, Washington, DC (November 28–December 3, 1971).

Hill, R., *The Mathematical Theory of Plasticity*, Oxford University Press, Oxford (1950).

Hixson, R.S., Vorthman, J.E., Gustavsen, R.L., Zurek, A.K., Thissell, W.R., and Tonks, D.L., "Spall Wave-Profile and Shock-Recovery Experiments on Depleted Uranium," in *Shock Compression of Condensed Matter—1997*, edited by S.C. Schmidt, D.P. Dandekar, and J.W. Forbes, AIP Press, Woodbury, NY, 479–482 (1998).

Holsapple, K.A., and Schmidt, R.M., "On the Scaling of Crater Dimensions 2. Impact Processes, *J. Geophysical Res.*, **87**, 1849–1870 (1982).

Hopkinson,B., *Proc. Roy. Soc.*, **A89**, 411 (1914a).

Horii, H., and Nemat-Nasser, S., "Brittle Failure in Compression: Splitting, Faulting and Brittle–Ductile Transition, *Phil. Trans. Roy. Soc. London*, **A319**, 337–374 (1986).

Ivanov, A.G., "Spallation in the Quasiacoustic Approach," *Phys. Combustion and Explosion (USSR)*, (3), 475–480 (1975).

Johnson, G.R., Stryk, R.A., Holmquist, T.J., and Souka, O.A., "Recent EPIC Code Developments for High Velocity Impact: 3D Element Arrangements and 2D Fragment Distributions," *Int. J. Impact Engng.*, **10**, 281–294 (1990).

Johnson, J.N., "Dynamic Fracture and Spallation in Ductile Solids," *J. Appl. Phys.*, **52**(4), 2812–2825 (1981).

Johnson, J.N., and Addessio, F.L., "Tensile Plasticity and Ductile Fracture," *J. Appl. Phys.*, **64**(12) 6699–6712 (1988).

Johnson, J.N., Gray, G.T., and Bourne, N.K., "The Effect of Pulse Duration and Strain Rate on Incipient Spall Fracture in Copper," *J. Appl Phys.*, **86**(9), 4892–4901 (1999).

Johnson, J.N., Hixson, R.S., Tonks, D.L., and Zurek, A.K., "Rate-Dependent Spallation Properties of Tantalum," in *Shock Compression of Condensed Matter—1995*, edited by S.C. Schmidt, and W.C. Tao, AIP press, Woodbury, NY, 523–526 (1996).

Johnson, P.M., and Burgess, T.J., *Rev. Sci. Instrum.*, **39**, 1100–1103 (1968).

Ju, J.W., "On Energy-Based Coupled Elastoplastic Damage Theories: Constitutive Modeling and Computational Aspects," *Int. J. Solids Structures*, **25**(7), 803–833 (1989).

Kachanov, M., "On the Effective Moduli of Solids with Cavities and Cracks," *Int. J. Fracture*, **59**(1), R17–R21 (1993).

Kalmikov, Y.B., Kanel, G.I., Parchomenko, I.P., Utkin, A.V., and Fortov, V.E., "The Behaviour of Rubber in Shock and Rarefaction Waves, *J. Appl. Mechanics Tech. Phys. (USSR)*, (1), 126–130 (1990).

Kanel, G.I., "Resistance of Metals to Spall Fracture," *Combustions, Explosions Shock Waves,* **18**(3), 329–335 (1982a).

Kanel, G.I., "The Work of Spall Fracture," *Combustions, Explosions Shock Waves,* **18** (4) 461–464 (1982b).

Kanel, G.I., and Chernych, L.G., "The Process of Spall Fracture," *J. Appl. Mechanics Tech. Phys. (USSR)*, (6), 78–84 (1980).

Kanel, G.I., and Molodets, A.M., "Behavior of the K8 Glass at Dynamic Compression and Following Unloading," *Zh. Tehn. Fiz.,* **46**, 398 (1976).

Kanel, G.I., and Petrova, E.N., "The Strength of Titanium VT6 at Shock Wave Loading," in *Workshop on Detonation*, Chernogolovka, Russia, 136–142 (1981).

Kanel, G.I., and Pityulin, A.N., "Shock-Wave Deformation of Titanium Carbide-Based Ceramics," *Combustion, Explosion Shock Waves*, **16**(4), 436–438 (1985).

Kanel, G.I., and Razorenov, S.V., "*Shock Wave Loading of Metals: The Motion of the Free Surface of Sample*," preprint, Institute of Chemical Physics, USSR Academy of Sciences, Chernogolovka, Russia, 46 (1989), (in Russian).

Kanel, G.I., Razorenov, S.V., Bogatch, A.A., Utkin, A.V., Fortov, V.E., and Grady, D.E., "Spall Fracture Properties of Aluminum and Magnesium at High Temperatures," *J. Appl. Phys.*, **79**(11), 8310–8317 (1996a).

Kanel, G.I., Razorenov, S.V., and Fortov, V.E., "Kinetics of Spallation Rupture in the Aluminum Alloy AMg6M," *J. Appl. Mech. Tech. Phys.*, **25**(5), 707–711 (1984a).

Kanel, G.I., Razorenov, S.V., and Fortov, V.E., "The Spall Strength of Metals in a Wide Range of the Load Duration," *Acad. of Sciences USSR—Doklady,* **275**(2), 269–271 (1984b).

Kanel, G.I., Razorenov, S.V., and Fortov, V.E., "Fracture in Titanium Caused by Shock Waves of Various Intensity," *Sov. Phys. Tech. Phys.*, **31**(3), 354–355 (1986).

Kanel, G.I., Razorenov, S.V., and Fortov, V.E., "The Spall Strength of Metals in a Wide Range of Shock Load Amplitudes," *Academy of Sciences USSR—Doklady* **294**(2), 350–352 (1987a).

Kanel, G.I., Razorenov, S.V., and Fortov, V.E., "Cleavage Strength of Metals over a Wide Range of Shock-Load Amplitudes," *Sov. Phys. Dokl.*, **32**(5), 413–414 (1987b).

Kanel, G.I., Rasorenov, S.V., and Fortov, V.E., "The Dynamic Strength of Copper Single Crystals," in *Shock-Wave and High-Strain-Rate Phenomena in Materials*, edited by M.A. Meyers, L.E. Murr, and K.P. Staudham, Marcel Dekker, New York, 775–782 (1992a).

Kanel, G.I., Razorenov, S.V., and Fortov, V.E., "The Failure Waves and Spallation in Homogeneous Brittle Materials," in *Shock Compression of Condensed Matter—1991*, edited by S.C. Schmidt, R.D. Dick, J.W. Forbes, and D.G. Tasker, Elsevier Science Publishers B.V., Amsterdam, 451–454 (1992b).

Kanel, G.I., Razorenov, S.V., Utkin, A.V., Baumung, K., Karow, H.U., and Licht, V., "Spallation Near the Ultimate Strength of Solids," in *High-Pressure Science and Technology—1993*, edited by S.C. Schmidt, J.W. Shaner, G.A. Samara, and M. Ross, AIP Press, Woodbury, NY, 1043–1046 (1994a).

Kanel, G.I., Razorenov, S.V., Utkin, A.V., Fortov, V.E., Baumung, K., Karow, H.U., Rush, D., and Licht, V., "Spall Strength of Molybdenum Single Crystals," *J. Appl. Phys.*, **74**(12), 7162–7165 (1993).

Kanel, G.I., Razorenov, S.V., Utkin, A.V., and Grady, D.E., "The Spall Strength of Metals at Elevated Temperatures," in *Shock Compression of Condensed Matter—1995*, edited by S.C. Schmidt, and W.C. Tao, AIP press, Woodbury, NY, 503–506 (1996b).

Kanel, G.I., and Shcherban, V.V., "Plastic Deformation and Cleavage Rupture of Armco Iron in a Shock Wave," *Combustion, Explosion, Shock Waves* **16**(4), 439–446 (1980).

Kanel, G.I., Tolstikova, Z.G., and Utkin, A.V., "Effect of Filler-Particle Size on the Cleavage Strength of Elastomers," *Appl. Mech. Tech. Phys.*, **35**(3), 399–403 (1993b).

Kanel, G.I., and Utkin, A.V., "Cavitation Dynamic in Reflection of a Compression Pulse from the Interface of Two Media," *J. Appl. Mech. Tech. Phys. (USSR)*, **32**(4), 23–26 (1991).

Kanel, G.I., Utkin, A.V., and Tolstikova, Z.G., "Response of the High-Filled Elastomers to Shock-Wave Loading," in *High-Pressure Science and Technology—1993*, edited by S.C. Schmidt, J.W. Shaner, G.A. Samara, and M. Ross, AIP Press, Woodbury, NY, 1123–1126 (1994b).

Kanel, G.I., Vakhitova, G.G., and Dremin, A.N., "Metrological Characteristics of Manganin Pressure Gauges Under Conditions of Shock Compression and Relief," *Fiz. Goreniya i Vzryva*, **14**(2), 130–135 (1978).

Keough, D.D., DeCarli, P.S., and Cooper, T., "Distant Mountain Test Series (Test Bed Designs and Flatpack Stress Gage Measurements)," Defense Nuclear Agency Technical Report DNA-TR-93-103, Alexandria, VA (1993).

Kipp, M.E., and Grady, D.E., "Dynamic Fracture Growth and Interaction in One Dimension," *J. Mech. Phys. Solids*, **33**(4), 399–415 (1985).

Kipp, M.E., and Grady, D.E., "Elastic Wave Dispersion in High-Strength Ceramics," in *Shock Compression of Condensed Matter—1991*, edited by S.C. Schmidt, R.D. Dick, J.W. Forbes, and D.G. Tasker, Elsevier Science Publishers B.V., Amsterdam, 459–462 (1992).

Kipp, M.E., and Grady, D.E., "Experimental and Numerical Studies of High-Velocity Impact Fragmentation," in *High Pressure Shock Compression of Solids II: Dynamic Fracture and Fragmentation*, edited by L. Davison, D.E. Grady, and M. Shahinpoor, Springer, New York (1996).

Kipp, M.E., Grady, D.E., and Chen, E.P., "Strain-Rate Dependent Fracture Initiation," *Int. J. Fracture*, **16**(5), 471–478 (1980).

Kobayashi T., and Shockey, D.A., "FRASTA: A New Way to Analyze Fracture Surfaces—Part 1: Reconstructing Crack Histories," *Advanced Materials Processes*, **140**(5), (1991a).

Kobayashi T., and Shockey, D.A., "FRASTA: A New Way to Analyze Fracture Surfaces – Part 2: Determining Fracture Mechanisms and Parameters," *Advanced Materials Processes*, **140**(6), (1991b).

Kobayashi, T., Shockey, D.A., and Jones, R.L., "Deriving SCC Initiation Times and Growth Rates from Posttest Fractographic Analysis," *Corrosion*, **47**(7), 528–535 (1991).

Kobayashi, T., Shockey, D.A., Ogundele, G., McNabb, D.D., and Sidey, D., "Deducing Crack History in an Aged Boiler Tube from Fracture Surface Topography," *J. Testing and Evaluation, JTEVA*, **22**(4), 309–317 (1994).

Kobayashi, T., Shockey, D.A., Schmidt, C.G., and Klopp, R.W., "Assessment of Fatigue Load Spectrum from Fracture Surface Topography," *Int. J. Fatigue*, **19**(Supp. No. 1), S237–S244 (1997).

Kohn, B., "Compilation of Hugoniot Equations of State," Technical Report AFWL-TR-69-38, Air Force Weapons Laboratory, Kirtland Air Force Base, NM (1969).

Kolski, H., *Stress Waves in Solids*, Dover, New York (1963).

Kornfeld, M., *Elasticity and Strength of Liquids*, State Publisher of Scientific and Technical Literature, Moscow–Leningrad, USSR (1951).

Krajcinovic, D., and Fonseka, G.U., "The Continuous Damage Theory of Brittle Materials, Part 1: General Theory," *J. Appl. Mechanics,* **48**, 809–815 (1981).

Kumano, A., and Goldsmith, W., "Analytical and Experimental Investigation of the Effect of Impact on Coarse Granular Rocks," *Rock Mechanics,* **15**(2), (1982).

Lade, P.V., "Effects of Voids and Volume Changes on the Behavior of Frictional Materials," *Int. J. Num. Analyt. Methods Geomech.*, **12**, 351–370 (1988).

Landau, L.D., and Lifshitz, E.M., *Fluid Mechanics*, Pergamon Press, Oxford, UK (1959).

Lassila, D.H., Leblanc, M., and Gray, G.T. III, "High-Strain-Rate Deformation Behavior of Shocked Copper," in *Shock-Wave and High-Strain-Rate Phenomena in Materials*, edited by M.A. Meyers, L.E. Murr, and K.P. Staudham, Marcel Dekker, New York, 587-596 (1992).

Lee, L.M., "Nonlinearity in the Piezoresistance Coefficient of Impact-Loaded Manganin," *J. Appl. Phys.,* **44**, 4017–4022 (1973).

Liebowitz, H., editor, *Fracture,* Vols. I–VII, Academic Press, New York (1968–1972).

Lubliner, J., Oliver, J., Oller, S., and Onate, E., "A Plastic-Damage Model for Concrete," *Int. J. Solids Struct.*, **25**(3), 299–326 (1989).

Malvern, L.E., "The Propagation of Longitudinal Waves of Plastic Deformation in a Bar of Materials Exhibiting a Strain-Rate Effect," *J. Appl. Mech.,* **18**, 203–208 (1951).

Malvern, L.E., *Introduction to the Mechanics of a Continuous Medium*, Prentice-Hall, Englewood Cliffs, NJ (1969).

Margolin, L.G. "Calculations of Cratering Experiments with the Bedded Crack Model," in *Proceedings of the American Physical Society Conference on Shock Waves in Condensed Matter—1981*, AIP Press, Woodbury, NY, 465-479 (1982).

Margolin, L.G., "Elastic Moduli of a Cracked Body," *Int. J. Fracture,* **22**, 65–79 (1983).

Marsh, S.P., *LASL Shock Hugoniot Data*, University of California Press, Berkeley, CA (1980).

Marston, P.L., and Unger, B.T., "Rapid Cavitation Induced by the Reflection of Shock Waves," in *Shock Waves in Condensed Matter—1985*, edited by Y. Gupta, Plenum Press, New York, 401–405 (1986).

Mathews, A.R., Boat, R.M., Hemsing, W.F., Warnes, R.H., and Whittemore, G.R., "Full-Field Fabry-Perot Interferometer," in *Shock Compression of Condensed Matter—1991*, edited by S.C. Schmidt, R.D. Dick, J.W. Forbes, and D.G. Tasker, Elsevier Science Publishers B.V., Amsterdam, 759–762 (1992).

McGlaun, J.M., Thompson, S.L., and Elrick, M.G., "CTH: a Three-Dimension Shock Wave Physics Code," *Int. J. Impact Eng.*, **10**, 351–360 (1990).

McQueen, R.G., and Marsh, S.P., "Equation of State for Nineteen Metallic Elements from Shock-Wave Measurements to Two Megabars," *J. Appl. Phys.,* **31**(7), 1253–1269 (1960).

McQueen, R.G., and Marsh, S.P., "Ultimate Yield Strength of Copper," *J. Appl. Phys.*, **33**(2), 654–665 (1962).

McQueen, R.G., Marsh, S.P., Taylor, J.W., Fritz, J.N., and Carter, W.J., in *High-Velocity Impact Phenomena*, edited by R. Kinslow, Academic Press, New York, NY (1970).

Me-Bar, Y., Boas, M., and Rosenberg, Z., "Spall Studies on Ti-6Al-4V," *Mat. Sci. Eng.*, **85**, 77–84 (1987).

Merzhievsky, L.A., and Titov, V.M., "The Criterion of Metal Lasting in Microsecond Range," *Acad. of Sciences USSR—Doklady (USSR)* **286**(1), 109–112 (1986).

Meyers, M.A., *Dynamic Behavior of Materials*, John Wiley and Sons, New York (1994).

Meyers, M.A., and Aimone, C.T., "Dynamic Fracture (Spalling) of Metals," *Progress in Materials Science*, **28**, 1–96 (1983).

Meyers, M.A., Murr, L.E., and Staudhammer, K.P. (editors), *Shock wave and High-Strain-Rate Phenomena in Materials*, Marcel Deckker, New York (1992).

Murri, W.J., Curran, D.R., and Seaman, L., "Fracture Model for High Energy Propellant," in *Proceedings of the American Physical Society Conference on Shock Waves in Condensed Matter—1981*, AIP Press, Woodbury, NY, 460–464 (1982).

Murri, W.J., Grady, D.E., and Mahrer, K.D., "Equation of State of Rock," Stanford Research Institute Technical Report (SRI Project no. PYU-1883) (July 1975).

Murri, W.J., Young, C., Shockey, D.A., Tokheim, R.E., and Curran, D.R., "Determination of Dynamic Fracture Parameters for Oil Shale," SRI International, Final Report to Sandia Laboratories, Albuquerque, NM (1977).

Nahme, H., Hohler, V., and Stilp, A., "Determination of the Dynamic Material Properties of Shock Loaded Silicon-Nitride," in *High-Pressure Science and Technology—1993*, edited by S.C. Schmidt, J.W. Shaner, G.A. Samara, and M. Ross, AIP Press, Woodbury, NY, 765–768 (1994).

Needleman, A., and Tvergaard, V., "An Analysis of Dynamic, Ductile Crack Growth in a Double Edge Cracked Specimen," *Intl. J. of Fracture*, **49**, 41–67 (1991).

Nemat-Nasser, S., and Hori, M., "Void Collapse and Void Growth in Crystalline Solids," *J. Appl. Phys.*, **62**(7), 2746–2757 (1987).

Nemat-Nasser, S., and Hori, M., *Micromechanics: Overall Properties of Heterogeneous Materials*, North-Holland, Amsterdam (1993).

Nemes, J.A., and Eftis, J., "Rate-Dependent Modeling of Multidimensional Impact and Post-Spall Behavior," in *Shock-Wave and High-Strain-Rate Phenomena in Materials*, edited by M.A. Meyers, L.E. Murr, and K.P. Staudham, Marcel Dekker, New York, 723–731 (1992a).

Nemes, J.A., and Eftis, J., "Pressure-Shear Waves and Spall Fracture Described by a Viscoplastic-Damage Constitutive Model," *Intl. J. Plasticity*, **8**, 185–207 (1992b).

Nigmatulin, R.I., and Friedly, J.C. (editors), *Dynamics of Multiphase Media*, Hemisphere Publishing, Washington (1991).

Novikov S.A., Divnov, I.I., and Ivanov, A.G., "The Study of Fracture of Steel, Aluminum, and Copper under Explosive Loading," *Phys. Metals Metal Science (USSR)*, **21**(4), 608–615 (1966).

Osher, J., Barnes, G., Chau, H., Lee, R.S., Lee, C., Speer, R., Weingart, R.C., "Operating Characteristics And Modeling of the LLNL 100-kV Electric Gun," *Plasma Science*, **17**(3), 392–402 (1989).

Paisley, D.L., Warnes, R.H., and Kopp, R.A., "Laser-Driven Flat Plate Impacts to 100 GPa with Sub-Nanosecond Pulse Radiation and Resolution for Material Property Studies," in *Shock Compression of Condensed Matter—1991*, edited by S.C.

Schmidt, R.D. Dick, J.W. Forbes, and D.G. Tasker, Elsevier Science Publishers B.V., Amsterdam, 825–828 (1992).

Parhomenko, I.P., and Utkin, A.V., In *Matter Under Extreme Conditions*, IVTAN, Moscow, Russia, 126–130 (1990).

Perzyna, P., "Internal State Variable Description of Dynamic Fracture of Ductile Solids," *Int. J. Solids Structures*, **22**(7), 797–818 (1986).

Poritsky, H., "The Collapse or Growth of a Spherical Bubble or Cavity in a Viscous Fluid," in *Proceedings of the First U.S. National Congress on Applied Mechanics*, edited by E. Sternberg, *ASME*, New York, 813 (1952).

Postnov V.I., "The Use of Manganin Gauges for Measurement of Pressure on the Level of 1 Mbar," in *Unsteady Problems of Hydrodynamics*, Novosobirsk, **48**, 116–120 (1980) (in Russian).

Prevost, J.H., and Loret, B., "Dynamic Strain Localization in Elasto-(Visco-)Plastic Solids, Part 1. General Formulation and One-Dimensional Examples," *Methods Appl. Mech. Eng.*, **83**, 247–273 (1990).

Raiser, G., and Clifton, R.J., "Failure Waves in Uniaxial Compression of an Aluminosilicate Glass," in *High-Pressure Science and Technology—1993*, edited by S.C. Schmidt, J.W. Shaner, G.A. Samara, and M. Ross, AIP Press, Woodbury, NY, 1039–1042 (1994).

Rajendran, A.M., Dietenberger, M.A., and Grove, D.J., "A Void Growth-Based Failure Model to Describe Spallation," *J. Appl. Phys.*, **65**, 1521–1527 (1989).

Rajendran, A.M., Grove, D.J., and Bless, S.J., "A New Yield Function Based Dynamic Failure Model," in *Shock Waves in Condensed Matter—1987*, edited by N.H. Holmes, and S.C. Schmidt, North Holland, Amsterdam, 359–362 (1988).

Razorenov, S.V., Bogatch, A.A., Kanel, G.I., Utkin, A.V., Fortov, V.E., and Grady, D.E., "Elastic-Plastic Deformation and Spall Fracture of Metals at High Temperatures," in *Shock Compression of Condensed Matter—1997*, edited by S.C. Schmidt, D.P. Dandekar, and J.W. Forbes, AIP Press, Woodbury, NY, 447–450 (1998).

Razorenov, S.V., and Kanel, G.I., "Spall Strength of Metals over a Wide Range of Magnitudes and Durations of Shock Load," in *Chemical Physics of Combustion and Explosion: Detonation*, Institute of Chemical Physics, Chernogolovka, Russia, 46–49 (1986).

Razorenov, S.V., and Kanel, G.I., "The Study of Edge Effects Under Spallation," *Problems of Strength (USSR)*, (9), 52–55 (1991).

Razorenov, S.V., and Kanel, G.I., "The Strength of Copper Single Crystals and the Factors Governing Metal Fracture in Uniaxial Dynamic Stretching," *Physics Metals Metallography*, **74**(5), 526–530 (1992).

Razorenov, S.V., Kanel, G.I., Anufriev, V.G., and Loskutov, V.F., "Deformation and Failure of Structural Steels in Pulsed Loading," Translated from *Problemy Prochnosti,* (3), 42–48 (1992).

Razorenov, S.V., Kanel, G.I., Fortov, V.E., and Abasehov, M.M., "The Fracture of Glass Under High-Pressure Impulsive Loading," *High Pressure Research,* **6**, 225–232 (1991).

Razorenov, S.V., Kanel, G.I., and Yalovets, T.N., "Dynamic Strength of Ruby," *J. Chem. Phys. (USSR),* **12**(2) 175–177 (1993).

Razorenov, S.V., Utkin, A.V., Kanel, G.I., Fortov, V.E., Yarunichev, A.S., Baumung, K., and Karow, H.U., "Dynamic Deformation and Fracture of High-Purity Titanium," in *Metallurgical and Materials Applications of Shock-Wave and High-Strain-Rate*

Phenomena, edited by Murr, L.E., Staudhammer, K.P., and M.A. Meyers, Elsevier Science Publishers B.V., Amsterdam, 235–242 (1995a).

Razorenov, S.V., Utkin, A.V., Kanel, G.I., Yarunochev, A.S., Fortov, V.E., Baumung, K., and Karow, H.U., "Response of High-Purity Titanium to High-Pressure Impulsive Loading," *High Pressure Research,* **13**, 367-376 (1995b).

Read, H.E., and Cecil, R.A., "Rate-Dependent Constitutive Model for Shock-Loaded S-200 Beryllium," 3SR-949, Systems, Science, and Software, La Jolla, California, DNA 2845F (1972).

Reed-Hill, R.E., *Physical Metallurgy Principles*, D. Van Nostrand, Princeton, NJ (1964).

Rice, M.H., McQueen, R.G., and Walsh, J.M., "Compression of Solids by Strong Shock Waves", in *Solid State Physics* **6**, edited by F. Seitz and D. Turnbull, Academic Press, New York, 1–63 (1958).

Rice, J.R., and Tracey, D.M., "On the Ductile Enlargement of Voids in Triaxial Stress Fields," *J. Mech. Phys. of Solids*, **17**, 201–217 (1969).

Rinehart J.S., and Pearson, J., *Behavior of Metals under Impulsive Loads*, American Society for Metals, Dover (1964).

Rogers, C., "Adiabatic Shearing—General Natural and Material Aspects," in *Material Behavior Under High Stress and Ultra High Loading Rates*, edited by J. Mascall, and V. Weiss, Plenum Press, New York, 101–118 (1983).

Romanchenko, V.I., and Stepanov, G.V., "The Dependence of Critical Stresses Upon the Time Parameters of Load at Spalling in Copper, Aluminum, and Steel," *J. Appl. Mech. Tech. Phys.,* **21**(4), 141–147 (1980).

Rozenberg, Z., "Determination of the Dynamic Response of W-2 Tungsten by Commercial Manganin Stress Gauges," *J. Appl. Phys.,* **62**(5), 1745–1749 (1987).

Rozenberg, Z., "The Response of Ceramic Materials to Shock Loading," in *Shock Compression of Condensed Matter—1991*, edited by S.C. Schmidt, R.D. Dick, J.W. Forbes, and D.G. Tasker, Elsevier Science Publishers B.V., Amsterdam, 439–444 (1992).

Runesson, K., Larsson, R., and Sture, S., "Characteristics and Computational Procedure in Softening Plasticity," *J. Eng. Mech, ASCE,* **115**(8) 1628–1646 (1989).

Saltykov, S.A., in *Stereometric Metallography*, 2nd Ed., Metallurgizdat, Moscow, Russia, 446 (1958).

Sandler, I.S., and Wright, J.P., "Summary of Strain-Softening, Theoretical Foundations for Large-Scale Computations of Nonlinear Material Behavior," in *DARPA-NSF Workshop*, edited by S. Nemat-Nasser, Northwestern University, Evanston, IL, 225–241 (1983).

Sandler, I.S., and Wright, J.P., "Stain-Softening," in *Proceedings of the Workshop on the Theoretical Foundation for Large-Scale Computations of Nonlinear Material Behavior*, Evanston, IL, October 24–26, 1983, edited by S. Nemat-Nasser, R.J. Asaro, and G.A. Hegemier, Martinus Nijhoff Publishers, Dordrecht/Boston/London, 285–300 (1984).

Scheil, E., "Die Berechnung der Anzahl und Grössenverteilung kugelförmiger Kristalle in undurchsichtigen Körpern mit Hilfe der durch einen ebenen Schnitt erhaltenen Schnittkreise," *Z. Anorg. Allgem. Chem.* **201**, 259 (1931).

Scheil, E., "Statistische Gefügeuntersuchungen I," *Z. Metallk.* **27**, 199 (1935).

Schmidt, C.G., Kobayashi, T., Shockey, D.A., and Flournoy, T.H., "Fatigue in New and Ageing Aircraft: Correlation of Fracture Surface Topography with Fatigue Load History," in *Int. Comm. on Aeronautical Fatigue '97 Proceedings (ICAF '97)*, Edinburgh, Scotland (June 16–20, 1997).

Seaman, L., SRI International Final Report to the U.S. Army Ballistic Research Laboratory, Report No. ARBRL-CR-00420 (March 1980).

Seaman, L., "Development of Computational Models for Micro Structural Features," in *Proceedings of the American Physical Society Conference on Shock Waves in Condensed Matter—1981*, AIP Press, Woodbury, NY, 118–129 (1982).

Seaman, L., Barbee, T.W., Jr., and Curran, D.R., "Dynamic Fracture Criteria of Homogeneous Materials," Technical Report No. AFWL-TR-71-156, Air Force Weapons Laboratory, Kirtland Air Force Base, NM (1971).

Seaman, L., and Curran, D.R., "SRI PUFF 8 Computer Program for One-Dimensional Stress Wave Propagation," SRI International Final Report (Volume II) Prepared for U.S. Army Ballistics Research Laboratory, Contract No. DAAK11-77-C-0083 (1978a).

Seaman, L., and Curran, D.R., "Mesh-Independent Ductile Fracture Calculations," in *International Conference on Shock Waves in Condensed Matter*, Saint-Petersburg, Russia, 12–17 (1998).

Seaman, L., and Curran, D.R., "Inertia and Temperature Effects in Void Growth," presented at the American Physical Society Conference on Shock Compression of Condensed Matter, held in Atlanta, GA (June 25–29, 2001).

Seaman, L., Curran, D.R., and Crewdson, R.C., "Transformation of Observed Crack Traces on a Section to True Crack Density for Fracture Calculations," *J. Appl. Phys.*, **49**, 5221 (1978b).

Seaman, L., Curran, D.R., and Erlich, D.C., "An Anisotropic Plasticity Model for Shear Band Damage," *18th Annual Meeting, Society of Engineering Science, Inc.,* Brown University, Providence, RI (September 2–4, 1981).

Seaman, L., Curran, D.R., Erlich, D.C., Cooper, T., and Dullum, O., "Shear Band Observations and Derivations of Requirements for a Shear Band Model," *DYMAT International Conference on Mechanical and Physical Behavior of Materials Under Dynamic Loading*, Paris (September 2–5, 1985).

Seaman, L., Curran, D.R., Erlich, D.C., and Shockey, D.A., "Shear Band Processes in High Rate Deformation," *SES meeting*, San Diego, CA (August 1992).

Seaman, L., Curran, D.R., and Murri, W.J., "A Continuum Model for Dynamic Tensile Microfracture and Fragmentation," *J. Appl. Mech.,* **107**, 593–600 (1985).

Seaman, L., Curran, D.R., and Shockey, D.A., "Computational Models for Ductile and Brittle Fracture," *J. Appl. Phys.,* **47**(11), 4814 (1976).

Seaman, L., Curran, D.R., and Shockey, D.A., "Calculation of Failure by Shear Bands," in *Proceedings of the 29th Sagamore Conference,* Lake Placid, NY, edited by J. Mescall, and V. Weiss, Plenum Press, New York (1983).

Seaman, L., and Dein, J.L., "Representing Shear Band Damage at High Strain Rates," in *Proceedings of the IUTAM Symposium on Nonlinear Deformation Waves*, Tallinn, Estonia, held August 1982, Springer, Berlin (1983).

Seaman, L., Simons, J.W., Erlich, D.C., Olsen, E.M., Rosenberg, J.T., and Matheson, E.R., "Development of a Viscous Internal Damage Model for Energetic Materials Based on the BFRACT Microfracture Model," in *11th International Detonation Symposium*, Snowmass Village, CO (August 31–September 4, 1998).

Seaman, L., Tokheim, R., and Curran, D.R., "Computational Representation of Constitutive Relations for Porous Materials," Report DNA 3412F, Defense Nuclear Agency, Washington, DC (May 1974).

Senf, H., Straßburger, E., and Rothenhäusler, H., "Visualization of Fracture Nucleation During Impact in Glass, in *Metallurgical and Materials Applications of Shock-Wave and High-Strain-Rate Phenomena*, edited by Murr, L.E., Staudhammer, K.P., and M.A. Meyers, Elsevier Science Publishers B.V., Amsterdam, 163–170 (1995).

Sheffield, S.A., and Fisk, G.A., "Particle Velocity Measurements in Laser Irradiated Foils Using ORVIS," in *Shock Waves in Condensed Matter—1983*, edited by J.R. Asay, R.A. Graham, and G.K. Straub, Elsevier Science Publishers B.V., Amsterdam, 243–246 (1984).

Shockey, D.A., Curran, D.R., Seaman, L., Caligiuri, R.D., Erlich, D.C., Giovanola, J.H., and Kobayashi, T., "Development of Improved Dynamic Failure Models," Final Technical Report prepared by SRI International for U.S. Army Research Office, Research Triangle Park, NC (1985).

Shockey, D.A., Curran, D.R., Seaman, L., Rosenberg, J.T., and Petersen, C.F., "Fracture of Rock Under Dynamic Loads," *Int. J. Rock. Mech. Sci. Geomech. Abstr.,* **11**(1), 303–317 (1974).

Shockey, D.A., Erlich, D.C., and Dao, K.C., "Particle Impact Damage in Silicon Nitride at 1400°C," *J. Mater. Science,* **16**, 477–482 (1981).

Shockey, D.A., Seaman, L., and Curran, D.R., "Dynamic Fracture of Beryllium Under Plate Impact and Correlation with Electron Beam and Underground Test Results," Final Report No. AFWL-TR-73-12, Kirtland Air Force Base, NM (1973).

Shockey, D.A., Seaman, L., Curran, D.R., DeCarli, P.S., Austin, M., and Wilhelm, J.P., "A Computational Model for Fragmentation of Armco Iron Under Ballistic Impact," Final Report, prepared by SRI International for the U.S. Ballistic Research Laboratories, Aberdeen Proving Ground, MD (1973b).

Shockey, D.A., Seaman, L., and Curran, D.R., "Microfracture Models and Their Applications to Nonlinear Dynamic Fracture Problems," *Nonlinear and Dynamic Fracture Mechanics, ASME, AMD*, **35**, 79 (1979).

Shockey, D.A., Seaman, L., and Curran, D.R., "A Computational Shear Band Model for Armor Design," in *Proceedings of the 1984 Army Symposium on Advances in Solid Mechanics for Design and Analysis*, Newport, RI, (October 1984).

Skidmore, I.C., "An Introduction to Shock Waves in Solids," *Appl. Mater. Res.,* **4**, 131 (1965).

Smith, J.H., "Three Low Pressure Spall Thresholds in Copper," *ASTM Spec. Tech. Publ. No. 336*, 264–281 (1962).

Smith, J.H., "Three Low Pressure Spall Thresholds in Copper," in *Symposium on Dynamic Behavior of Materials*, ASTM, Philadelphia, PA, 264 (1963).

Sneddon, I.N., and Lowengrub, M., *Crack Problems in the Classical Theory of Elasticity*, John Wiley and Sons, New York (1969).

Speight, C.S., and Taylor, P.F., "Dynamic Fracture Criteria from Free Surface Velocity Measurements," in *Metallurgical Applications of Shock-Wave and High-Strain-Rate Phenomena*, edited by L.E. Murr, K.P. Staudhammer, and M.A. Meyers, Marcel Dekker, New York, 805–820 (1986).

Speight, C.S., Taylor, P.F., and Wallace, A.A., "Observation of Spallation and Attenuation Effects in Aluminum and Beryllium from Free-Surface Velocity Measurements," in *Metallurgical Effects at High Strain Rates*, Edited by R. W. Rohde et. al., Plenum Press, New York, 429–442 (1973).

Staehler, J.M., Predeborn, W.W., and Pletka, B.J. "The Response of High Purity Alumina to Plate-Impact Testing," in *High-Pressure Science and Technology—1993*, edited by S.C. Schmidt, J.W. Shaner, G.A. Samara, and M. Ross, AIP Press, Woodbury, NY, 745–748 (1994).

Steinberg, D.J., and Lund, C.M., "A Constitutive Model for Strain Rates from 10^{-4} to 10^6 s^{-1}," *J. Appl Phys.*, **65**(4), 1528–1533 (1989).

Stellingwerf, R.F., and Wingate, C.A., "Impact Modeling with Smooth Particle Hydrodynamics," *Int. J. Impact Engng.*, **14**, 707–718 (1993).

Stepanov, G.V., "Spall Fracture of Metals by Elastic-Plastic Loading Waves," *Problemy Prochnosti*, (8), 66–70, (1976), (in Russian).

Sugak, S.G., Kanel, G.I., Fortov, V.E., Ni, A., and V.G. Stel'makh, "Numerical Modeling of the Effect of an Explosion on an Iron Plate," *Soviet J. Fizika Goreniya i Vzriva*, (2), 121 (1983).

Sugiure, S.H., Kondo, K., and Sawaoka, A., "Dynamic Response of Fused Quartz in the Permanent Densification Region," *J. Appl. Phys.*, **52**(5), 3375–3382 (1981).

Swegle, J.W., and Grady, D.E., "Shock Viscosity and the Prediction of Shock Wave Rise Times," *J. Appl. Phys.*, **58**(2), 692–701 (1985).

Tarver C.M., and Maiden, D.E., "Experimental Measurements and Numerical Simulations of Metal Spallation by Detonating Solid Explosives," in *Shock Waves in Condensed Matter—1987*, edited by N.H. Holmes, and S.C. Schmidt, North Holland, Amsterdam, 363–366 (1988).

Taylor, G.I., *Proc. Royal Soc.* **A200**, 235–247 (1950).

Taylor, J.W., in *Dislocation Dynamics*, edited by A.R. Rosenberg, New York (1968).

Thissell, W.R., Zurek, A.K., Rivas, J.M., Tonks, D.L., and Hixson, R.S., "Damage Evolution and Clustering in Shock Loaded Tantalum," in *International Conference on Shock Waves in Condensed Matter*, Saint-Petersburg, Russia (1998).

Tobolsky, A., and Eyring, H., "Mechanical Properties of Polymeric Materials," *J. Chem. Phys.*, **11**, 125–134 (1943).

Tollier, L., Bartnicki, E., Fabbro, R., "Experimental and Numerical Study of Laser-Driven Spallation with Visar Diagnostic," in *Shock Compression of Condensed Matter—1995*, edited by S.C. Schmidt, and W.C. Tao, AIP press, Woodbury, NY, 1265–1268 (1996).

Tollier, L., Cottet, F., Boustie, M., and Ansart, J.P., "Experimental and Numerical Studies of Ti-6Al-4V Spallation by Laser and Impact Loadings," in *High-Pressure Science and Technology—1993*, edited by S.C. Schmidt, J.W. Shaner, G.A. Samara, and M. Ross, AIP Press, Woodbury, NY, 1047–1050 (1994).

Tong, W., and Ravichandran, G., "Inertial Effects on Void Growth in Porous Viscoplastic Materials," *J. Appl Mech.*, **62**, 633–639 (1995).

Tonks, D.L., Hixson, R., Gustavsen, R.L., Vorthman, J.E., Kelly, A., Zurek, A.K., and Thissell, W.R., "Spallation Studies on Shock Loaded Uranium," in *Shock Compression of Condensed Matter—1997*, edited by S.C. Schmidt, D.P. Dandekar, and J.W. Forbes, AIP Press, Woodbury, NY, 239–242 (1998).

Tonks, D.L., Zurek, A.K., and Thissell, W.R., "Ductile Damage Modeling Based on Void Coalescence and Percolation Theories," in *Metallurgical and Materials Applications of Shock-Wave and High-Strain-Rate Phenomena*, edited by Murr, L.E., Staudhammer, K.P., and M.A. Meyers, Elsevier Science Publishers B.V., Amsterdam, 171–178 (1995).

Trevena, H.D., "The Behavior of Liquids Under Tension," *Contemp. Phys.*, **8**, 185–191 (1967).

Tsai, Y.M., "Exact Stress Distribution, Crack Shape and Energy for a Running Penny-Shaped Crack in an Infinite Elastic Solid," *Int. J. Fracture*, **9**(2), 157–169 (1973).

Tsukrov, I., and Kachanov, M., "Solids with Holes of Irregular Shapes—Effective Moduli and Anisotropy," *Int. J. Fracture*, **64**(1), R9–R12 (1993).

Tuler, F.R., and Butcher, B.M., "A Criterion for the Time Dependence of Dynamic Fracture," *Int. J. Fract. Mech.* **4**(4), 431–437 (1968).

Utkin, A.V., "Influence of the Damage Rate on the Dynamics of Interaction of Shock Load Pulse with a Body Surface," *J. Appl. Mech. Tech. Phys. (USSR),* **33**(6) , 82 (1992).

Utkin, A.V., "Influence of the Initial Damage Rate on the Forming Spall Pulse," *J. Appl. Mech. Tech. Phys. (USSR),* **34** (4), 578–584 (1993).

Venugopalan, V., "The Thermodynamic Response of Polymers and Biological Tissues to Pulsed Laser Irradiation," Ph.D. Dissertation, Massachusetts Institute of Technology, Cambridge, MA (May 1994).

Wang, Y., and Mikkola, D.E., "Response of Alpha-Aluminum Oxide to Shock Impact," in *Shock-Wave and High-Strain-Rate Phenomena in Materials*, edited by M.A. Meyers, L.E. Murr, and K.P. Staudham, Marcel Dekker, New York, 1031–1040 (1992).

Wang, Z.-P., "Void-Containing Nonlinear Materials Subject to High-Rate Loading," *J. Appl. Phys.*, **81**(11), 7213–7227 (1997).

Wang, Z.-P., and Lam, K.Y., "Evolution of Microcracks in Brittle Solids Under Intense Dynamic Loading," *J. Appl. Phys.*, **77**(7), 3479–3483 (1995).

Wilkins, M.L., "Calculations of Elastic-Plastic Flow," in *Methods of Computational Physics, Vol. 3: Fundamental Methods of Hydrodynamics*, edited by B. Adler, Academic Press, New York (1964).

Wilkins, M.L., Streit, R.D., and Reaugh, J.E., "Cumulative-Strain-Damage Model of Ductile Fracture: Simulation and Prediction of Engineering Fracture Tests," LLNL Report UCRL-53058 (October 3, 1980).

Winkler, W.-D., and Stilp, A.J., "Spallation Behavior of TiB_2, SiC, and B_4C Under Planar Impact Tensile Stresses, in *Shock Compression of Condensed Matter—1991*, edited by S.C. Schmidt, R.D. Dick, J.W. Forbes, and D.G. Tasker, Elsevier Science Publishers B.V., Amsterdam, 475–478 (1992).

Wright, T.W., "Adabatic Shear Bands: Modeling and Scaling Laws," *Explomet 2000 International Conference on Fundamental Issues and Applications of Shock-Wave and High-Strain-Rate Phenomena*, Albuquerque, NM (June 18–22, 2000).

Wu, F.H., and Freund, L.B., "Deformation Trapping Due to Thermoplastic Instability in One-Dimensional Wave Propagation," *J. Mech. Phys. Solids*, **32**(2), 119–132 (1984).

Yaziv, D., Bless, S.J., and Dandekar, D.P., "Shock Compression and Spall in Porous Beryllium Oxide," in *Metallurgical Applications of Shock-Wave and High-Strain-Rate Phenomena*, edited by L.E. Murr, K.P. Staudhammer, and M.A. Meyers, Marcel Dekker, New York, 793–803 (1986).

Zaretsky, E., and Kaluzhny, M., "Fracture Threshold and Shock Martensite Formation in Stainless Steel," in *Metallurgical and Materials Applications of Shock-Wave and High-Strain-Rate Phenomena*, edited by Murr, L.E., Staudhammer, K.P., and M.A. Meyers, Elsevier Science Publishers B.V., Amsterdam, 203–210 (1995).

Zel'dovich, Ya.B., and Raizer, Yu.P. *Physics of Shock Waves and High-Temperature Hydrodynamic Phenomena*, Vol. II, edited by W.D. Hayes, and R.F. Probstein, Academic Press, New York (1967).

Zhang, L., and Jin, X., "Predicting of Fragment Number and Size Distribution," in *Shock Compression of Condensed Matter—1997*, edited by S.C. Schmidt, D.P. Dandekar, and J.W. Forbes, AIP Press, Woodbury, NY, 227–230 (1998).

Zhao, S.D., Ling, Z., Shen, L.T., Chen, S.X., and Bai, Y.L., "An Experimental Approach to Evolution of Spallation," in *Shock Compression of Condensed Matter—1991*, edited by S.C. Schmidt, R.D. Dick, J.W. Forbes, and D.G. Tasker, Elsevier Science Publishers B.V., Amsterdam, 523–526 (1992).

Zhurkov, S.N., "Kinetic Concept of the Strength of Solids," *Int. J. Fract. Mech.*, **1**(4), 311 (1965).

Zurek, A.K., Embury, J.D., Kelly, A., Thissell, W.R., Gustavsen, R.L., Vorthman, J.E., and Hixson, R.S., "Microstructure of Depleted Uranium under Uniaxial Strain Conditions," in *Shock Compression of Condensed Matter—1997*, edited by S.C. Schmidt, D.P. Dandekar, and J.W. Forbes, AIP Press, Woodbury, NY, 423–426 (1998).

Zurek, A.K., and Follansbee, P.S., "Microstructural Effects on Spall Fracture in 1008 Steel," in *Shock Compression of Condensed Matter—1989*, edited by S.C. Schmidt, J.N. Johnson, and L.W. Davison, Elsevier Science Publishers B.V., Amsterdam, 433–436 (1990).

Zurek, A.K., Follansbee, P.S., and Hack, J., "High Strain-Rate-Induced Cleavage Fracture in Mild Carbon Steel," *Metallurgical Transactions A,* **21A**, 431–439 (1990).

Zurek, A.K., Frantz, Ch.E., and Gray, G.T., III, "The Influence of Shock Pre-Strain and Peak Pressure on Spall Behavior of 4340 Steel," in *Shock-Wave and High-Strain-Rate Phenomena in Materials*, edited by M.A. Meyers, L.E. Murr, and K.P. Staudham, Marcel Dekker, New York, 759–767 (1992).

Index

A

adiabatic shear, 8

B

BFRACT, 28, 236, 272
  coalescence and fragmentation, 251, 256
  crack growth, 251
    constant stress threshold, 254
    fracture mechanics threshold, 255
  crack nucleation, 251
  crack size distribution, 252
  damage processes, 249
  determination of the crack size distribution, 239
  fracture initiation, 251
  nucleation and growth processes, 253
  solution procedure, 262
  spallation, 262
  stress calculations coupled with damage, 260
  stress–strain relations, 260
bipolar stress pulse, 55
brittle damage growth, 18
brittle fracture, 236
  in Armco iron, 239
  experimental aspects, 237

C

centered wave, 41
characteristic equations, 39
classical fracture mechanics, link to microstatistical fracture mechanics, 304
coalescence
  of cracks, 273
  of cracks in XAR30 steel, 289
  of voids, 270
compression waves, 41
computer codes, 34
  advection requirements, 34–35
computer simulation, 175
  calibration of the simulation tool, 213
conservation equations, 37, 44
conservation laws, 34
constitutive equations, 37–38
constitutive modeling, 175
  general approaches, 175
  Hooke's law, 179
  nonlocal behavior, 193
  plasticity, 179
  response of intact material, 175
  stability, 194
  stress and strain
    basic concepts, 175
    deviatoric strain, 179
    deviatoric stress, 178
    maximum shear strain, 178
    maximum shear stress, 178
    notations and conventions, 175
    pressure, 178
    principal strain, 178
    principal stress, 178
    volumetric strain, 178

constitutive modeling (*continued*)
  temperature computations, 185
  thermodynamic requirements, 190
  yield criteria, 179
constitutive models
  empirical, 127, 302
  nucleation and growth, 302
constitutive relations, 15, 17
continuity condition, 53
cracks, 13
crack size distribution, 245, 252
  areal to volumetric transformation, 244
creep test, 14

D

damage, 3
  accumulation, 11
  by crack propagation, 273
  criterion, 2
  kinetics, 12–14
  measurement scatter, 269
  microscopic, 13
  nucleation mechanisms, 18
  by void growth, 268
damage evolution, 15
  kinetics, 301
damage growth, 302
damage kinetics, 120, 132
damage nucleation, 145, 302
damage nucleation sites, 10
damage rate, 130, 142
data analysis, 11
deviatoric stress, 38
DFRACT, 219, 268
  coalescence of voids, 234
  damage characterization parameters, 226
  damage processes, 227
  determination of the void size distribution, 221
  nucleation void size, 226
  solution procedure, 229
  stress–strain relations, 228
  thermal effects, 234
  variable growth threshold, 231
  void growth rate, 225
  void nucleation rate, 225
diagnostic measurements, 13, 28
distance-time diagram, 49, 101
ductile damage growth, 18
ductile fracture, 218
  experimental aspects, 219
dynamic failure process, 18
dynamic fracture, 1, 34
dynamic strength, *see* spallation
dynamic tensile strength, *see* spall strength

E

element-size dependence, 192
energy deposition, 10
equation of state, 38
equations of motion, 38
Eulerian codes, 34–35
Eulerian coordinates, 38
Eulerian–Lagrangian codes, 34–35
experimental methods, 11
experimental techniques, 59
explosives, 41

F

failure, 2
failure process, 3
failure waves in glass, 167
finite element, 34
flow field, 39
fracture damage, 43
fracture kinetics, 132, 143
fracture mechanics, 32
fracture model implementation, 203
  recommended features, 205
    advection, 210
    data transfer, 208
    data types, 208
    locality, 210
    modularity, 205
    units, 206
fracture modeling, 197
  active approaches, 200

Grady's energy criteria, 198
Gurson model, 201
microstatistical approach, 201
passive approaches, 197
fracture properties, 301
fracture stress, 33
fracture surface, 143
fracture time, 140–141, 144
fracturing stress, 97, 137
fragmentation, 3
fragmentation model, 279
FRASTA, 9
free-surface velocity, 137

G
gas gun, 11
graphical analysis, 49
Griffith criterion, 4
gun, electric, 63

H
Hopkinson bar, 14
Hugoniot, 45
Hugoniot elastic limit, 97

I
impactor, 51
impedance, 41
in-contact explosive, 30
isentrope, 40, 45
isotherm, 45

L
Lagrangian
coordinates, 38
sound velocity, 39, 47
laser interferometer, 13
liquid cavitation, 107
load duration, 141
localization, 193

M
macrocrack, 3
manganin stress gauge, 73
effect of strain, 74
hysteresis effects, 76
material characterization, 5
material element, 16
material strength, 1
mesomechanics, 3
constitutive equations, 34
continuum theory, 17
model, 4
metallographic examination of voids, 221
method of characteristics, 37
microdamage evolution, 13
microfractures, 2
microscopic
damage, 14
examination, 11
failure kinetics, 17
failure process, 16
nucleation sites, 18
microstructural characterization, 9
microstructural damage, 14
Mie–Grüneisen equation of state, 228
model instabilities, 192
molecular dynamics, 4
multidimensional dynamic experiments, 90
Murnaghan equation, 182

N
nucleation and growth, 2, 217
brittle fracture, *see* brittle fracture
ductile fracture, *see* ductile fracture
inherent assumptions, 218
model for brittle materials, *see* BFRACT
model for ductile materials, *see* DFRACT
nucleation and growth model, 35
nucleation and growth modeling
applications, 267
armco iron, 293
beryllium, 286
commercially pure aluminum, 268
iron, 289

nucleation and growth modeling, applications (*continued*)
  polycarbonate, 272
  quartzite, 276
  solid rocket propellant, 281
  steel, 289
 guidelines for use of, 267
numerical instabilities, 192

O

overstress, 143

P

phase transition, 7
plastic flow, 18
plate impact, 43, 50
plate impact test, 11
 experimental configuration, 27
porous compaction, 6
post-test examination, 14
post-test microscopic evaluation, 13
post-test microscopic examinations, 301
process zone, 3
properties
 chemical, 10
 mechanical, 5
 thermal, 7

Q

quartzite fragmentation, 279
quasi-static experiments, 90

R

radiation sources, 53
Rankine–Hugoniot jump conditions, 45
rarefaction
 fan, 42–43
 waves, 26, 41
relevant volume element, 3, 15
Riemann integral, 40
Riemann invariants, 109, 113–114, 122
Riemann trajectories, 95, 99
rise time, 107

S

scabbing, 1
sensor, 12
shear banding, 13–14, 18
shear strength, 6
shock adiabat, 45
shock front, 44
shock impedance, 13
shock loading, 34
shock viscosity, 48
shock wave diagnostics, 66
 particle velocity measurements, 67
  capacitor gauge, 67
  electromagnetic gauge, 69
  laser velocimeter, 69
 stress history measurements, 73
shock wave generation, 59
 using explosive devices, 59
 using gas and powder guns, 62
 using radiation, 64
shock waves, 37
simple wave, 40
simulation codes, 203
 ALE, 204
 Lagrangian, 203
 meshless, 204
 smooth particle hydrodynamics, 204
size distribution
 of cracks in beryllium, 286
 of cracks in MIL-S-12560B Armor steel, 290
 of cracks in polycarbonate, 273
 of cracks in XAR30 steel, 289
 of preexisting flaws in rock (quartzite), 276
 of voids in aluminum, 268
size distribution of voids, 218, 222
 areal to volumetric transformation, 221
softening, 3
soft recovery, 63
spall, 1
 effect of temperature, 145, 147
 in porous materials, 6

spall analyses uncertainties, 134
effect of constitutive properties, 135
effect of material properties, 135
limitations of continuum mechanics, 135
spall damage, 26, 28
spall data (Russian), 305
aluminum AD1, 310
aluminum AMg6M, 313–316
aluminum D16, 311–312
Armco iron, 357–359
copper M2, 333–335
copper single crystal, 336–338
epoxy EDT-10, 362
lead, 359–360
magnesium Ma1, 355–356
molybdenum polycrystal, 340–341
molybdenum single crystal, 341–343, 345–351
deformed, 344–345
nickel NP-2, 339
niobium single crystal, 351–353
deformed, 353–355
PMMA, 362–365
propellant simulant, 368–370
quartz (x-cut), 374–375
rubber, 366–368
ruby, 371–373
sapphire (z-cut), 373–374
stainless steel (Kh18N10T), 318–320
steel 3, 316–317
steel 45, 317–318
steel–KhVG, 321–323
tin, 361
titanium, 324–327
titanium carbide, 375–377
titanium VT5-1, 327–328
titanium VT6, 329–331
titanium VT8, 332
spall fracture, 1, 32, 34, 137
compared to quasi-static fracture, 302
experimental procedures, 76
active measurements, 77, 85
metallographic observations, 78
passive measurements, 78
kinetics, 127
spall initiation, 144
spall pulse, 95, 97, 134
spall strength, 4, 33, 95, 301
compared to ultimate theoretical strength, 159
edge effects, 161
effect of elastic-plastic properties, 97, 99–100
effect of load history, 160
effect of sample thickness, 98
effect of shock heating, 159
effect of strain rate, 119, 159
effect of velocity gradients, 99
estimation errors, 104, 107
estimation methods
discrete measurement of surface velocity, 111
free surface motion, 95
inspection, 110
motion of the interface with a soft buffer, 105
thickness of spalled layer, 111
of metal alloys, 142
of metals, 142
spall stress, 93
spall studies, new applications, 303
spall test, 34
spall threshold, *see* spall strength
spallation, 1
spallation in brittle materials, 162
alumina, 162
boron carbide, 162
glass, 165
silicon carbide, 162
titanium carbide, 163
titanium diboride, 162
spallation in liquids, 173
glycerol, 173
water, 173
spallation in metallic alloys, 139
spallation in metals, 139
aluminum, 145, 149

spallation in metals (*continued*)
  Armco iron, 147
  copper, 145, 147–148
  lead, 147
  magnesium, 145, 150
  molybdenum, 147
  nickel, 149
  stainless steel, 140
  steel, 139, 141
  tin, 147
  titanium, 145
spallation in polymers and elastomers, 169
  PMMA, 169
  propellant simulant, 170
  rubber, 170
spallation in single crystals
  copper, 152
  molybdenum, 152
  niobium, 155
  ruby, 165
  x-cut quartz, 164
  z-cut sapphire, 165
spallation models, 34
SRI PUFF, 28
stability, 3
strain
  definition, 176
  engineering, 177
  infinitesimal, 177
  longitudinal, 176
  shearing, 176
strain rate, 11, 49
strength, 1
stress histories, 28
stress relaxation, 112, 119–120, 137, 143, 301
stress-particle velocity diagram, 49
surface imperfections, 32

T

Taylor wave, 93, 106
temperature, 11, 34
temperature in shock and rarefaction waves, 57
tensile failure, 4
tensile strength, 33
tensile test, 14
theoretical strength, 4
theory of characteristics, 34; *see also* characteristic equations
thermal softening, 8
thermoelastic stress, 8

U

ultimate strength, 11
uniaxial strain, 11
units, 206

V

velocity pullback, 95, 97
VISAR, 13, 70
void damage
  effect of load duration, 224
  effect of peak stress, 224
  quantitative analysis, 220
void growth, 218
  effect of void shape, 235
  effect on void volume, 227
  inertial effects, 232
  microvoid interaction, 236
  nucleation void size, 226
  transition from voids to cracks, 235
void growth rate, 225
void nucleation, 218
  effect on void volume, 227
void nucleation rate, 225
void recompaction, 110
voids, 13

W

wave interaction, 28, 93
  distance-time diagram, 93, 107, 118
  influence of damage kinetics, 112
  particle velocity diagram, 107
  stress-particle velocity diagram, 95, 118
  triangular pulse, 93, 107, 120
wave propagation, 28, 37
wave reverberations, 31